Water Resources Management

Other McGraw-Hill Books of Interest

AWWA • *Water Quality and Treatment*

BRATER, KING, LINDELL & WEI • *Handbook of Hydraulics*

DODSON • *Storm Water Pollution Control: Industry and Construction NPDES Compliance*

MAIDMENT • *Handbook of Hydrology*

MAYS • *Water Resources Handbook*

PARMLEY • *Hydraulics Field Manual*

ZIPPARO & HASEN • *Davis' Handbook of Applied Hydraulics*

Water Resources Management

Principles, Regulations, and Cases

Neil S. Grigg

Department of Civil Engineering
Colorado State University
Fort Collins, Colorado

McGraw-Hill

New York San Francisco Washington, D.C. Auckland Bogotá
Caracas Lisbon London Madrid Mexico City Milan
Montreal New Delhi San Juan Singapore
Sydney Tokyo Toronto

Library of Congress Cataloging-in-Publication Data

Grigg, Neil S.
    Water resources management : principles, regulations, and cases /
Neil S. Grigg.
       p.   cm.
    Includes index.
    ISBN 0-07-024782-X
    1. Water resources development—Economic aspects. 2. Water-
supply—Management.   I. Title.
HD1691.G75   1996
333.91—dc20   95-51376
                                             . CIP

McGraw-Hill

*A Division of The **McGraw·Hill** Companies*

1 2 3 4 5 6 7 8 9 0 DOC/DOC 9 0 1 0 9 8 7 6

ISBN 0-07-024782-X

*The sponsoring editor for this book was Larry S. Hager, the editing supervisor
was Stephen M. Smith, and the production supervisor was Suzanne W. B.
Rapcavage. It was set in Century Schoolbook by Victoria Khavkina of
McGraw-Hill's Professional Book Group composition unit.*

Printed and bound by R. R. Donnelley & Sons Company.

McGraw-Hill books are available at special quantity discounts to use as premiums and sales promotions, or for use in corporate training programs. For
more information, please write to the Director of Special Sales, McGraw-Hill,
11 West 19th Street, New York, NY 10011. Or contact your local bookstore.

This book is printed on acid-free paper.

BUSCA

Contents

Preface

This book presents a comprehensive management framework for the water industry. The "principles" that form the comprehensive framework include technical topics such as hydrology and systems analysis, and management topics such as law, finance, and political science. Case studies show how these principles apply to problems faced by the water industry. The important aspect of the comprehensive framework is to recognize that it has many interdependent parts. Recognizing this helps to explain why no single academic discipline is able to fully prepare a water resources manager.

The material is designed for the professional development of engineers, scientists, planners, policy analysts, lawyers, and managers who work in the water industry. It can be used for university courses, for continuing education and training, or for self-study by managers. None of these disciplines has all the answers—they must work together to find feasible strategies.

The principles and case studies have been tested on undergraduate and graduate engineering students at Colorado State University. They have produced lively discussion and stimulated interest in careers in the water industry. However, when a student says: "I want to study water resources management because I believe that the solutions to environmental problems are in management, not technology," I reply: "You are right, but you must work through a technical route to become a player." The route can be in a field other than engineering, but one normally does not gain entry to the playing field of coordinating, mediating, or making decisions about water without a technical background of some kind.

I do not suggest that this material should replace a basic undergraduate course such as hydrology or sanitary engineering, but I do claim that the engineer or manager can gain an appreciation from it for the world of water management. The material goes well beyond the scope of a single graduate course. The principles provide enough material for a course at the senior or master's degree level, and the case studies provide enough additional material for a graduate course or two.

Of necessity, the case studies are mostly situations where I have personal experience, but I hope that the approach taken will stimulate others to write cases. The case study method has been proved to be effective, but not many cases have been prepared about water resources management.

In the twenty-first century, water resources managers must pursue sustainable development with measures that manage water for human systems, but at the same time protect and nuture natural systems. This laudable goal masks deep conflicts, and to deal with them water resources managers will require skills well beyond technical training. At a conference in 1992, I said: "Engineers working primarily on technical problems will take back seats in water conflicts unless they combine technical expertise with in-depth knowledge of policy, planning, communications, finance and public involvement." That holds true even more today.

Case studies describe problems that bind players in the water industry together. I have compiled about 50 of them, and they form the major part of the second part of the book. The cases are not all local situations; rather, they include general situations such as water conservation, drought, estuary management, and a number of others.

Managers must understand the industry in which they work, whether they come from engineering, law, or another field. To aid in understanding, I present a description of water industry structure with a four-sector model consisting of service providers, regulators, planners, and support organizations.

Tasks, roles, and principles form the intellectual core of the water management field. Management thinker Peter Drucker (1973) wrote: "The many, many books on management are skill-focused, discipline-focused, or function-focused," but the "tasks of management are the reason for its existence, the determinants of its work, and the grounds of its authority and legitimacy." I believe that if the water industry's players undertake their proper roles, solutions will be a lot easier.

The text presents basic quantitative techniques, but it does not focus on mathematics. Many texts present quantitative work, and the niche sought by this book is the framework for coordinated management, focusing on principles and applications.

The material is intended for both private-sector and public-agency water management workers. Perspectives of all three levels of government (local, state or provincial, and national) and all three branches of government (executive, legislative, and judicial) are included. Also, roles of the press and interest groups are included.

In the 11 years since I wrote *Water Resources Planning* (Grigg, 1985), the field of water resources management has changed. New experiences that I have incorporated in the book include service on the Fort Collins, Colorado, Water Board; involvement in the politics of Colorado water; work as the River Master of the Pecos River, a Supreme Court appointment; studies of U.S. droughts; consulting with the Alabama state government on water agency organization and interstate water protection; and consulting on relicensing disputes on the Platte River in Nebraska. Although many cases are from the United States, the principles apply worldwide. International experiences reflected in the book include technical tours to China, Japan, and Vietnam, and collaborative visits to England, Germany, and France. Participation in several UNESCO water projects is also included, and the book reflects my association with students from Egypt, Greece, Indonesia, Russia, Jordan, Iraq, Saudia Arabia, China, Pakistan, India, South Africa, Brazil, the Netherlands, Chad, Ethiopia, Venezuela, Argentina, Peru, Spain, Tunisia, Turkey, Syria, Iran, Nigeria, Korea, the Philippines, Thailand, and Sri Lanka.

I want to acknowledge association with the Water Resources Planning and Management Division of the American Society of Civil Engineers (ASCE) and with a founder of the division, Victor A. Koelzer. I co-taught a course with Vic for several years, and was pleased to see Vic elected as an Honorary Member of the ASCE just before he passed away in October 1994.

Finally, I owe a lot to my colleagues at Colorado State University. The tremendous water resources research and education programs built up at that university continue to inspire me to look deeper into the field of water resources management, and I hope that this book motivates others in the same way.

Neil S. Grigg

References

Drucker, Peter F., *Management: Task, Responsibilities, Practices*, Harper & Row, New York, 1973, pp. x, 36.
Grigg, Neil S., *Water Resources Planning*, McGraw-Hill, New York, 1985.

Water Resources Management

Water Management Principles

1

Management in the
Water Industry

Introduction

During the past few decades a dramatic change has occurred in the water resources industry. Whereas the last generation of engineers and managers focused on building projects, tomorrow's engineers and managers will be confronted with a more complex arena. Executives from business and government told engineering educators in 1995 that "tomorrow's engineers must be people who can go beyond the numbers to understand the impact their projects are likely to make on society. They must be able to communicate, to solve problems in teams, to speak other languages and work with other cultures, to understand environmental impacts, and to resolve conflicts" (*Prism,* 1995). The message is clear: The times are different, and the practices of water resources managers will be different if they are to succeed.

This shift in practice was described by Gilbert White (1994) in looking back on 60 years of practice. He wrote that when he began work in 1934, the mood was the same as in John Wesley Powell's nineteenth-century West, where there was "the conviction that the earth was awaiting further development for human good, and that the challenge was in providing prudent stewardship." By 1994, however, he saw that the mood had changed and that the quest is now for a "lasting, healthy, balanced place for existence. It was to be a home for all in sustenance and in spirit."

Managing water resources requires skills and approaches that go beyond pure engineering, science, management, or law. In the twenty-first century, water managers will deal with complexity and conflict. They will have to confront this complexity by analysis that enables them to unravel interdependency of systems, and they will have to confront conflict with cooperation, coordination, and communication, especially with the public.

The stakes are high. The U.S. water industry, depending on how it is defined, accounts for around 2 percent of gross domestic product, and has more control over the natural environment than other industries. If the United States and other nations are to succeed in solving the problems of their water industries, they will need effective water industry managers who straddle the technical, administrative, and political worlds.

This is definitely a sea change under way in the requirements of managers who work to achieve public purposes, and it has occurred in other arenas as well. However, water resources is unique because of the growth-versus-environment conflict, the relationship of water quality to health, conflicts over property values and business objectives, the need for water to sustain agriculture and fisheries, and various emotional issues related to water resources.

Some years ago, it was thought that water resources management was primarily an engineering task—build dams, lay pipelines, install pumps, and operate systems. This era lasted into the 1950s and 1960s. No more. Although in that era lip service was given to the need for comprehensive approaches, it was with the birth of environmentalism that the complexity and conflict of water management started to grow.

By 25 years after Earth Day we could see that no "comprehensive and coordinated" paradigm for water management was in place. This is not satisfactory, because water management requires a comprehensive framework of tasks that involve science, law, finance, public administration, and systems analysis—along with engineering.

In contrast to the view of some, water management is not purely a "command and control" activity, nor is it something that can be entirely "privatized." It requires a comprehensive framework with elements of both systems.

The aim of this book is to outline and describe the tasks of the manager who works in the water industry. Peter Drucker (1973), a well-known management thinker and writer, wrote: "The many, many books on management are skill-focused, discipline-focused, or function-focused," but the "tasks of management are the reason for its existence, the determinants of its work, and the grounds of its authority and legitimacy." Tasks, roles, and principles form the intellectual core for the professional who seeks to manage in the water industry and form the subject for this book.

Who will be the professionals who seek to manage in the water industry? In the past they came from engineering and from the ranks of workers in the industry, but in the future more may come from other disciplines such as law, biology, or finance. Regardless of their educational discipline, they will need the skills and knowledge outlined in this book.

Throughout the book, I refer to "water resources managers" and "water industry managers" interchangeably. The terms are meant to describe in a single phrase the manager who is prepared to face the special challenges of the water industry. As the terms "water resources" and "management" have several meanings, the book seeks to explain how they combine to make up "water resources management" as a field of practice.

Although I use these terms interchangeably, I would like to acknowledge that the term "water resources management" seems to be somewhat ambiguous and not recognized by the public. The term "water industry management" is not in widespread use, so we have a dilemma in explaining the tasks of water management, and more effort is needed to explain them better.

Water resources managers in the twenty-first century will face a confusing playing field. Whereas engineers and managers seek an orderly world where systematic approaches can be applied to solve problems, this does not describe the complexity and conflict in today's world of water resources management. It is not just a behind-the-scenes world of work where technicians design operational systems and implement them in controlled environments. It involves technical systems, but it is also a world of governmental meetings, public hearings, appeals to regulators, and lawsuits. To succeed in this world, participants need technical skills, communication and political skills, and a lot of patience.

This message, the need for skills that go beyond the technical realm, comes to us over and over in universities. Graduates say: "We appreciate the technical skills you taught us, but we need more political and communications skills, and we need to know about the law and finance." With technical subjects becoming more complex, there simply is not room in the curriculum to teach everything. Thus the approach must be through integrated instruction. That is the approach taken in this book—principles to illustrate areas of knowledge, and case studies to illustrate how they come together in the real world.

For water managers, the saying, "If you don't like the heat, get out of the kitchen," translates into, "If you don't like complexity or conflict, find another field to work in." Hydrology, ecology, water networks, and interlocking water laws add technical complexity to decision making. Political complexity happens because diverse water organizations serve diverse groups in diverse regions. Complexity requires competence, through education, training, and professional development. Conflict arises because water travels freely across political and property boundaries, leading to interjurisdictional disputes over uses and pollution.

When the water resources manager enters the political and legal arenas, media attention follows. Problems such as drought, flood, groundwater pollution, and unsafe drinking water create opportunities to fashion victim–culprit stories, and the water resources manager must avoid becoming the culprit. "Water crisis" articles lump urgent problems with management issues such as bureaucratic delays and financial problems. The real crisis in water is a "creeping crisis"—it comes on slowly but demands a response now. Unless we deal with it, we destroy natural systems and fall behind on goals such as providing safe drinking water to citizens of the world.

Complexity and conflict characterize most public-sector problems today. The difference between water resources management and other fields—such as transportation or education, for example—is that water involves environmental interdependencies.

The playing field for water resources was aptly described by California's Water Education Foundation (1993):

> Water has become one of the hottest political and scientific issues in California. If our students, who are our future citizens and voters, are to make intelligent decisions about the fate of this precious resource, they must be taught not only the scientific facts about water, but the skills necessary for gathering and evaluating information. They must also be given the opportunity to practice problem-solving strategies on real-life environmental issues.

The complexity and conflict of water were summed up in a quote that is attributed to John F. Kennedy: "Anyone who solves the problem of water deserves not one Nobel Prize but two—one for science and the other for peace."

Figure 1.1 illustrates the playing field of water resources today: a river basin as a playing field, computers to illustrate how complexity is worked out, and a courthouse to illustrate conflict resolution.

What Is Water Resources Management?

As the field of water resources management has evolved, it has taken on a structure. Compare it to building a house. You would need policies, plans, and specifications, you would have to follow rules and codes, you would need materials, you would require a team to build and operate the house, you would need certain skills, there would be customers to buy and live in the house, and the house would provide for different functions of living and working.

Water resources management is somewhat like building and oper-

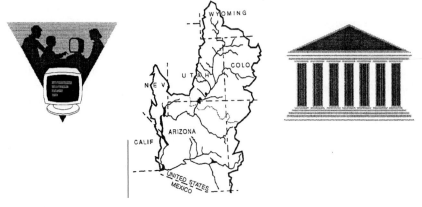

Figure 1.1 Complexity and conflict in water resources management.

ating a house, but much more complex. As a house has a set of bound-aries, a water resources system has boundaries—usually contained within a river basin or urban area. There are policies and plans for guidance, rules and codes, materials for construction and operation, teamwork, skills, customers and water users, and functions of the water resources system.

The major difference between the simple case of house building and management and the complex case of the water system is the interde-pendence inherent in water resources systems. Water users are de-pendent on each other; even the natural ecology is closely interwoven with the water systems. You might extend the house analogy by com-paring a neighborhood or a city to a water system, but regardless of the scale, the features that render the water system unique are com-plexity and interdependence.

The famous environmentalist John Muir summed up interdepen-dency by saying: "When we try to pick out anything by itself, we find it hitched to everything else in the universe" (Chesapeake Bay Program, 1994).

Philosophical Foundations for Water Resources Management

Leading thinkers have concluded that sustainable development must be the organizing principle for the balance between economic and en-vironmental goals. Sustainable development maintains and preserves resources for future generations, and does not degrade the natural en-vironment. Most agree on this general principle, and it provides strong guidance for professional practice.

However, while sustainable development is a widely shared value, actions to achieve it generate conflict because they involve property rights, jobs, taxation, and land uses, and because decisions are often made without understanding interactions between and among complex natural and social systems. Regardless of our conceptual views or general values, the only realistic way to achieve sustainable development is to deal with these realities which frame the external environment for the practice of water resources management.

Explaining complex systems, shaping feasible solutions, and managing conflict will be the key tasks for water resources managers who seek to balance economic development and environmental protection in over 150 nations around the globe. If water managers are to succeed in these tasks, they need more than technical understanding: They need the ability to blend knowledge from diverse fields with judgment and experience. I believe that to apply these skills requires a new type of professional—such as the engineer who can apply judgment along with technical knowledge, or the public works manager who is able to blend engineering with law, policy, finance, and management skills.

How can this activity be coordinated to assure the best results without excessive conflict and costs? What new skills will be required to succeed? How can the public be involved meaningfully during a communications revolution that flattens organizations, changes management tasks, and creates instant access to information for everyone? This book seeks to answer these questions and to provide the core knowledge needed by the water manager.

Sustainable Development

The quest for sustainable development is a key concept in water resources management, and it may be the strongest driving force of the water industry. The consensus definition of sustainable development seems to be the one promulgated by the World Commission on Environment and Development in 1986, that sustainable development is a process that "meets the needs of the present without compromising our ability to meet those of the future" (Environmental and Energy Study Institute Task Force, 1991).

An important aspect of sustainable development is accounting for environmental assets (Repetto, 1992). Standard economic measures such as GNP do not account for environmental capital. Originally, economic accounting was to deal with natural resources, labor, and capital, but natural resources have dropped out of the accounts. Repetto showed that three accounts in Costa Rica (forestry, fisheries, and soils) accounted for $4.1 billion in accumulated depreciation from

1970 to 1989, a tremendous loss to the small country. Causes were deforestation, overfishing, and soil degradation.

In the past few years, momentum has been building for greater attention to the environment. There have been a number of global working groups, such as the World Commission on Environment and Development (1983–1986). Policies recommended had to do with population and human resources, food security, species and ecosystems, energy, industry, and the urban challenge. The International Conference on Water and the Environment (1986) set the stage for the United Nations Conference on Environment and Development (UNCED) by formulating a "Dublin Statement" on water.

Agenda 21 from UNCED captures much of the sustainable development issue (UN, 1992). It provided action programs for six categories in freshwater management: integrated water resources development and management; water resources assessment; protection of water resources, water quality and aquatic ecosystems; drinking-water supply and sanitation; water and sustainable urban development; water for sustainable food production and rural development; and impacts of climate change on water resources.

In 1994 the President's Council on Sustainable Development had "committed itself to move the United States toward sustainability by the year 2050." It defined sustainable development as "to meet the needs of the present without compromising the ability of future generations to meet their own needs." This was explained further by a group organizing a seminar on sustainable development in the Colorado River Basin (see Chap. 19): "A sustainable United States will have an economy that equitably provides opportunities for satisfying lifestyles in a safe, healthy, high-quality environment for current and future generations (President's Council on Sustainable Development, 1994).

Sustainability was expressed in another way by the Water Quality 2000 team (see Chap. 14): *a society living in harmony with healthy natural systems.*

Closer to home, after the 1992 election, a group met in Colorado to prepare a suggested water agenda for the Clinton administration. Their report was entitled: "America's Waters: A New Era of Sustainability (Long's Peak Working Group, 1992). Their report is aimed directly at reforming water management to achieve sustainability.

The group began with this statement:

> Sound water policy must address the contemporary and long-term needs of humans as well as the ecological community. Nationally, we have not been using water in a manner that meets these needs on a sustainable basis. Examples include the endangered Columbia River salmon, the overtaxed San Francisco Bay Delta, the poisoned Kesterson National

Wildlife Refuge, the salt-choked Colorado River, the vanishing Ogalalla Aquifer, Louisiana's eroding Delta, New York's precarious Delaware River water supply, and the dying Florida Everglades. The environmental costs of current water policy are extraordinary, both to this and future generations.

This is quite an indictment of the nonsustainability of current practices. The group attributes the problems this way:

Intensive economic uses—agriculture, hydropower, flood control, navigation, and urban development—became the dominant forces in managing water. All too often, other concerns—including sound fiscal policy and the needs of Indian Tribes, other ethnic communities, and ecosystems—were ignored.

They recognized some positive developments, including:

[S]tate and federal programs for instream flow protection, pollution prevention, recognition of the public interest, development of watershed and regional water management approaches, and comprehensive settlements of tribal reserved water rights.

In advising the new administration, they called for a new water policy based on sustainability, which would feature "re-examination of federal policy affecting water quality and aquatic ecosystems consistent with social equity, economic efficiency, ecological integrity, and continued commitment to federal trust responsibilities to tribes."

Interestingly, they recognized the need for more coordination, and called for:

Implementation of a truly national, not "federal," water policy requires the federal government to facilitate, support, and help coordinate efforts to optimize the effectiveness of all levels of government—federal, state, tribal, and local.

To implement this new policy the group identified four objectives: water use efficiency and conservation, ecological integrity and restoration, clean water, and equity and participation in decision making; and they stated that institutional reform to advance these objectives "must be sensitive to human economic needs and to the government's financial constraints."

Recommended principles to implement these objectives were generally as follows.

Water use efficiency and conservation: Efficient use of water as a central aspect of water policy; federal leadership to make conservation an explicit part of each program and policy; facilitation of water transfers; efficient use of water through cooperation and open participatory processes.

Ecological integrity and restoration: Watersheds as basic management units to protect and sustain aquatic biological diversity, including in-stream, wetland, riparian, and upland resources, and to integrate with large-scale watershed systems; replacement of crisis management with preventive and integrative responses; adaptive action for positive environmental restoration while knowledge about systems improves; restoration actions at local levels to reestablish links between communities and ecology.

Clean water: Watershed basis for water quality management; aggressive action to deal with polluted runoff; link of water quality with protection and restoration of aquatic ecosystems; prevention of pollution at its source; link of water quality with water quantity management; integrated resource planning and funding tied to water quality goals.

Equity and participation in decision making: Federal government to fulfill special trust relationship with Indian tribes; decision making to include all affected interest groups; decision-making bodies to provide the public with clear information; respect for existing equities when reallocation of water is necessary.

Institutional reform: Make most effective use of government and strengthen incentives for private action; integrate decisions and actions at lowest levels where problems are posed and impacts felt; integrated resource management to consider demand reduction, supply enhancement, full consideration of economic and environmental costs, with full public participation; federal agency organization to promote efficiency in decision making, consistency in administration, and public understanding of federal actions.

Of course, the term "sustainable development" has attracted a number of different definitions. Here are a few ideas about what it means from candidates for a prize for sustainable development, as published by the Woodlands Forum (Center for Global Studies, 1993):

> "A form of smart growth that employs the high-tech revolution and economic restructuring to manage all this growth in a more sophisticated manner that is ecologically benign."

> "Produce nothing until we have a way of fully integrating the product, its by-product, and any by-product from the production process into the system in a positive manner."

> "The basic principle is that we live from flows not from the stocks."

> "The proper integration of environmental concerns into the development process."

> "In a perfect world, an end service like social security would be sustainable. In other words, there is a circular pattern of continual renewal."

"Pass on to each generation a population level, a set of technologies, and a stock of fertile land and fossil fuels which would enable them to do at least what we have done."

The Water Resources Manager

In the future, some water resources managers will be civil engineers, but water management will no longer be their exclusive province; they will be joined by other professionals and officials who have learned the lessons about technology, ecology, law, and management sciences applied to water resources.

Is the practice of water resources management a profession? Some would say yes, and some would say no. A profession is an occupation requiring study in an advanced field, and water resources management fulfills this requirement, but it is not as focused as engineering, law, or medicine, and it lacks its own academic departments in universities. It does have its own industry and specialized regulatory structure.

Regardless of whether it is a profession, water resources management is an occupation involving professionals who manage within the water industry. They may begin as engineers, lawyers, biologists, or from another specialty, but they are united by a focus on the interdisciplinary problem of managing water resources systems. They must put together the whole picture and make it work, and this requires skill in coordination, in public involvement, in law, in finance, and in technical areas.

Here are a few examples of water resources managers drawn from my own experience: an engineer who has left behind technical work to become a financial specialist in an urban water utility; an engineer who has become a generalist and manages an urban water supply utility; a lawyer who demonstrated skill in negotiation and policy who became director of an urban water supply utility; an engineer with an MBA who directs a large irrigation district; a history major who worked up from a policy position in state government to become director of a state water agency; a business major who worked in state government and formed and directed a state government water agency; a biologist who became a water resources policy analyst at a major multilateral lending bank; a career army officer who became director of a major irrigation district; a technology graduate who worked in state government and became director of a major water agency; an economist in the federal government, who became director of a major water agency and then a professor of water politics; an attorney who became the president of a major water board; and a legislator who learned about water during a legislative career, and became a manager.

In the United States, a nation of over 250 million people, I estimate that there about 100,000 practicing water managers in the sense of this book. In addition to these, another 100,000 to 300,000 are preparing for work in water management. They work in, regulate, or support state, local, and federal government organizations, and consist of managers of the water, wastewater, and stormwater utilities; regulators, planners, and support personnel in the federal and state governments; and consultants, policy analysts, attorneys, and scientists who work in support of the industry. Also included are the elected and appointed officials who, once they begin to work in the water industry, discover their own needs to learn more about the planning and decision-making process.

What skills will these water managers need? To begin, their scope of work will be expanded to include both providing water and protecting the environment. Their educational background should be broad. The American Society of Civil Engineers (ASCE) stated (1994) that future engineers must be strong in their technical and scientific knowledge base, and supplemented by exposure to a global vision and approach to problem solving; a basic management knowledge base (business, resource, cost, and time management); and a solid foundation in personal and interpersonal attributes, ethics, and social science/humanities. The ASCE believes that the BS degree should reflect the real world and a marketplace which is interdisciplinary and involves teamwork, social, and communication skills. Some skills needed by graduates, according to the ASCE, are social skills (communication, leadership, appreciation for the needs of society, and ability to lead or participate in a multidisciplinary team); ability to frame problems in terms of legal, social, political, environmental, sustainable, and life-cycle systems; personal skills such as critical thinking (synthesis and analysis), mastery of computer skills, appreciation of professional ethics, time management, and a basic appreciation of life-long learning; and flexibility skills that include teamwork, presentation styles, community involvement, articulation of ideas, and creative display of results to sell ideas.

A strong theme that is implicit in these skills is the creation of citizenship skills, the ability to work for the public good on teams and build trusting relations with others.

Definitions

Defining terms will help to explain and focus the work of water resources managers and overcome potential confusion over the use of general terms such as "water resources" and "management." I offer the following definitions, meant to be concise but comprehensive.

Water resources management is the application of structural and nonstructural measures to control natural and man-made water resources systems for beneficial human and environmental purposes.

Water industry management is the practice of water resources management within the water industry.

The water industry consists of water service organizations (water supply, wastewater, flood control, hydropower, recreation, navigation, environment), regulators, and support organizations.

Structural measures for water management are constructed facilities used to control water flow and quality.

Nonstructural measures for water management are programs or activities that do not require constructed facilities.

A water resources system is a combination of water control facilities and environmental elements that work together to achieve water management purposes.

A natural water resources system is a set of hydrologic elements in the natural environment that includes the atmosphere, watersheds, stream channels, wetlands, floodplains, aquifers and groundwater systems, lakes, estuaries, seas, and the ocean.

A man-made water resources system is a set of constructed facilities put into place to control water flow and quality.

Man-made water resources systems, consisting of structural water control facilities, provide for control of water flow and quality, and involve constructed facilities for water supply and wastewater management, for drainage of land and control of floods, and for water control in rivers, reservoirs, and aquifers. Examples include conveyance systems (channels, canals, and pipes); diversion structures; dams and storage facilities; treatment plants; pumping stations and hydroelectric plants; wells; and all appurtenances. These are described in Chap. 3.

Examples of nonstructural measures include pricing schemes, zoning, incentives, public relations, regulatory programs, or insurance.

The Water Industry

One of the central features of this book is its introduction of a concept for the structure of a comprehensive water industry. This structure can be observed in different water industry models, as outlined in some detail in Chap. 8. It has four parts: the *service providers,* the *regulators,* the *planners,* and the *support* organizations. Figure 1.2 shows the linkages between different parts of the water industry.

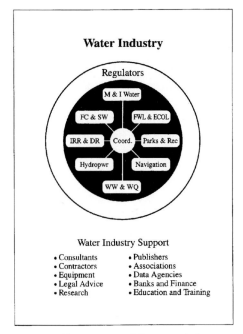

Figure 1.2 Components of the water industry.

Service providers are those having direct responsibility for water management or services. *Regulators* may be responsible for regulating rates, water quality, health issues, or service levels. The *planners and coordinators* have responsibility for planning and coordination functions other than in the course of providing services or regulating. *Support organizations* provide a diverse array of services and goods and include those providing support such as research, technical services, supplies, information and data, and others.

These organizations may be classified according to the sector in which they work or by their economic and political form. The industry sectors discussed in the book are water supply utilities; wastewater management utilities; stormwater and flood-control organizations; irrigation and drainage agencies; in-stream flow-control groups (hydropower, navigation, recreation, and environmental); regulators; support organizations (mainly data management, research, and education); and planners/coordinators. The economic and political forms are local, state, federal government and international organizations, and private-sector entities.

Purposes of Water Resources Management

The purpose of water resources management is to provide adequate water for humans and the natural environment. Humans may perish due to scarce or contaminated water, and society has complex economic, health, and amenity needs that add to water requirements and demands. Plant life and the food chain require water supplies, both on farms and in nature; wild and domestic animals rely on water for drinking and to nourish their food supplies; and natural water systems require water to sustain their functions in humid or arid areas.

To translate the water needs of humans and nature into management terms, we refer to "purposes" of water resources management: water supply, wastewater and water quality management, storm- and floodwater control, hydropower, transportation, recreation, and water for the environment, fish, and wildlife. These purposes line up pretty well with the sectors of the water industry, as it was just defined.

The purposes serve four categories of water users: people, industries, farms, and the general environment. Thus, we speak of water supply for people (domestic water supply); for cities (urban water supply); for farms (irrigation); for industries (industrial water supply); for cities and industries (municipal and industrial or M&I water); or we speak of water for the environment (water for natural systems).

We can also speak of wastewater management for the same categories, as urban wastewater, industrial wastewater, and drainage for farms.

Storm and floodwater control is a different type of activity that involves handling excessive water. As a "protective" service it does not provide water, but it removes or stores the excess water.

Meeting these purposes requires trade-offs, and although we always hope for an "optimum" system, we strive for the best-balanced outcome among the constraints of the real world. Achieving this balance is the main challenge to water resources managers.

Roles of the Players in Water Resources Management

The water industry evolved with subsectors that line up with the interest groups associated with the purposes of water resources management. Because of this strong role of interest groups, we must clarify the roles of the players—the professionals, businesses, elected officials, and activists who influence water management.

For the most part, the service providers are local government agencies or special districts that furnish water supply, wastewater and water quality management, storm- and floodwater control, hydropower, transportation, and water for the environment, fish/wildlife, and recreation. Although for the most part the players work in government, there is increasing interest in privatization, and private provision of services may be increasing.

Players representing the service providers are found in organized groups such as the American Water Works Association, the Water Environment Federation, the National Association of Water Companies, and many others. More details are provided in App. C.

Regulators deal with health, environment, finance, and service quality, but in the water industry the main regulators administer health and environment statutes. Although environmental organizations are considered in the support category, the regulatory system is their access to power. The Environmental Protection Agency and the "state EPAs" are main regulators, as are water quantity regulators such as state engineers.

Planning and coordination organizations provide needed coordination in the water industry. They are few in number, and their roles are explained in detail later.

Support organizations are numerous and diverse. They provide the goods and services that are not provided by the service organizations, items such as research, data, technical assistance and training, public information through the press, financial assistance, equipment, professional services, legal support, construction services, and yes, environmental opposition.

Environmental organizations are influential in the water industry because of their dedication, focus, and zeal. They, along with the press, provide a balance to actions of service providers and regulators, and in some ways represent a "fourth branch of government" operating in the water industry.

Integrating the Viewpoints of Water Resources Management

One of the central ideas in the book is the need for *integrated water resources management,* a term that captures the key issue of balance in decisions and allocation of resources. One of the difficult problems facing managers is to integrate viewpoints into workable strategies, and the starting place must be to understand the viewpoints.

The term is very comprehensive, and means different things to different people, so I will attempt a definition. First, the verb "to integrate" means to bring together the parts of something, and as there are several parts of water management to be brought together, the term covers a broad range of issues.

Integrated water resources management balances the views and goals of affected political groups, geographical regions, and purposes of water management; and protects the water supplies for natural and ecological systems.

This rather awkward definition strives to show the balancing of four different viewpoints of water management, with a fifth behind the scenes.

The *political viewpoint* has horizontal and vertical dimensions. Horizontal issues arise between government agencies at the same level, usually local government, and vertical issues deal with relationships between layers of government agencies, as in state–federal issues. Political integration is important because much of water resources management is by government agencies.

The *geographic viewpoint* refers to scale and accounting units: global, river basin, country, water body, locale, region, etc. Later in the book I will show that coordination, or integration, within river basins and metropolitan areas is one of the most important unresolved policy issues in the water industry.

Each *purpose or function* of water management has its own viewpoint, including the service categories of water supply for cities, wastewater management, irrigation, and others. As the water industry evolved to serve the separate purposes, we might look at functional integration as a cutting across the sectors of the water industry. This is not easy because of the subdivision into interest groups.

Protecting the water supply for natural and ecological systems refers to the *hydroecological viewpoint.* There are many examples, such as conjunctive use for joint use of surface and groundwaters; or integrated consideration of water, plant, biological, and wildlife issues in a river basin.

The fifth dimension of integration is the *disciplinary viewpoint,* which blends different branches of knowledge: technology, law, finance, economics, politics, sociology, life science, mathematics, and others. Rather than being a necessary condition for integration, I consider it to be a supporting condition, one that stimulates the other four dimensions of integration: political, geographical, functional, and hydroecological.

In practical situations, integrated solutions must be assembled piece by piece. This disjointed task can be frustrating, but to understand it better, some further explanation may help.

We could say, for example, that integration in water management brings together its parts, such as the water supply and wastewater functions; or brings together cooperating parties in a region, creating area-wide integration.

A key element in integration is *cooperation,* or working together with others. Thus, cooperation in water management is any form of working together to manage water, as in a region (a distinct district of a state, country, or city), resulting in *regionalization* in water management (the act or process or regionalizing water management—meaning to manage water with regional cooperation or integration).

Regionalization is an important policy issue in water management. In general, it means integration or cooperation on a regional basis. Examples include a regional management authority, consolidation of systems, a central system acting as raw water wholesaler, joint financing of facilities, coordination of service areas, interconnections for emergencies, and sharing of managers, operators, training, purchasing, data collection, emergency equipment, or water conservation advertising.

As I will show in this book, seeking ways to integrate water management is an excellent management strategy, and although it is not easy, partial successes are often quite worthwhile.

Need for Collaborative Leadership

I believe that the water industry requires a new approach to collaborative leadership to deal with its complexity and conflicts. The prescription for dealing with complexity is competence, to be built through education, training, and professional development. Conflict is managed by promoting cooperation, coordination, and communication. These concepts make up a "Six-C" model for water resources management: "competence to overcome complexity; cooperation, coordination and communication to overcome conflict."

The difficulty of solutions should not be underestimated. Some political systems are overwhelmed with water-related problems, and others struggle against odds to develop water policies that work. Taking professional approaches based on the Six-C model is the best way to design rational and workable management systems that also provide for sustainable natural systems.

According to Chrislip and Larson (1994): "Leaders and citizens in this country's cities and regions face unprecedented challenges in addressing public problems of shared concern. Despite differences in culture, place, and circumstance, these challenges are strikingly similar in terms of the political dynamics of the issues." This is certainly a valid description of the conflicts in the water arena. The authors describe how citizens are angry and leaders are powerless to respond to the rapid change that is upon us. Their book provides insight into three issues: dealing with complex issues, engaging frustrating and angry citizens, and generating civic will to break gridlocks. According to one researcher in their field, it boils down to a simple statement: civics matters. How well a community pulls together to solve its problems is a measure of its "social capital" or "civic infrastructure." These principles are as valid for water communities, working within watersheds, as they are for cities.

Chrislip and Larson ask:

> Just what is collaboration? That concept, as we use it, goes beyond communication, cooperation, and coordination. As its Latin roots—com and laborare—indicate, it means "to work together." It is a mutually beneficial relationship between two or more parties who work toward common goals by sharing responsibility, authority, and accountability for achieving results. Collaboration is more than simply sharing knowledge and information (communication) and more than a relationship that helps each party achieve it own goals (cooperation and coordination). The purpose of collaboration is to create a shared vision and joint strategies to address concerns that go beyond the purview of any particular party.

According to Chrislip and Larson, what we need is a better "civic community" to deal with complex, public-sector problems. They proposed a "collaborative premise"—"if you bring the appropriate people together in constructive ways with good information, they will create authentic visions and strategies for addressing the shared concerns of the organization or community."

These concepts are directly applicable to the water industry, where complexity is a big problem. As an example, Chrislip and Larson cite the Clark Fork River in Montana. There, 11 federal agencies are involved, along with a host of local and state agencies. The question is: "Who's in charge here? Everybody is, and nobody is." Finding ways to work successfully in watersheds is a key issue in this book. This issue—no one in charge—is a critical problem in water resources. Solving it is a key theme of this book.

Chrislip and Larson sum up the keys to successful collaboration, and they turn out to be important principles for water resources planning and decision making. They include good timing and clear need; strong stakeholder groups; broad-based involvement, credibility, and openness of process; commitment and/or involvement of high-level, visible leaders; support or acquiescence of "established" authorities or powers; overcoming mistrust and skepticism; strong leadership of the process; interim successes; and a shift to broader concerns.

As I began to work on this book, I had already developed the 6-C model of water resources management. After working through the idea, I now think that the problem goes deeper. As Chrislip and Larson state: "collaboration. . . goes beyond communication, cooperation, and coordination. . . to create a shared vision and joint strategies to address concerns that go beyond the purview of any particular party." I am not optimistic that the water industry will go that far in the near future, but it is a worthwhile goal.

Water Resources Management Scenarios

At detailed levels, we design water structures, analyze cash flow statements, or interpret statutes, but how do these management details fit into a strategy for a water resources system?

An understanding of how the details fit into the system requires scenario building—that is, illustrating how the complete system will function in its natural and political environments.

One of the assumptions of this book is that water management problems cannot be fully understood from their parts: They must be examined holistically. It is the same with business problems, and that is one of the reasons that the case study method is used in some business schools. The mechanism used to portray water management cases is the "scenario," following the assumption that these can be studied to see the main issues and options facing water managers, and the skills needed to solve the problems.

A scenario is a concept for a hypothetical or projected chain of events. It must include the functioning of the system itself and the response of the system's environment. It is a framework in which to show how a water resources system will influence and be influenced by the external environment and the players in the management game. Later in the book, I show how planning and decision making using the systems approach enable us to analyze the water resources system within its external environment.

1. *Planning and coordination.* Water management might be undertaken by single entities with control of planning, but this is becoming less common. More often, water management is a multijurisdictional activity, requiring careful attention to political integration. Scenarios described in the book include river basin planning and coordination, drought water management, regional integration of investment, water use, environment, and water and sanitation in developing countries.

2. *Organization.* As in any management enterprise, organizational issues arise constantly. A chapter is devoted to scenarios about the organization of agencies in the water industry.

3. *Water operations management.* A chapter is devoted to reservoir operations, one of the most important water operational problems encountered. Other operating decisions involve selecting system pressures to maintain, allocating water to users, moving water from place to place, generating power, and selecting modes of treatment.

4. *Regulation.* Regulatory programs have evolved ("exploded" might be a better word) to control water development and operations.

Water and environmental laws are the basis for regulation, and a number of case studies are included, such as water quality management and nonpoint source control, and estuary management. Water supply regulation is discussed in a chapter on water supply and environment and another on water conservation and efficiency. Water quantity regulation is covered in a chapter on water allocation, control, transfers, and compacts. Environmental issues are discussed in a chapter on watersheds, wetlands, and riparian zones. Groundwater regulation and floodplain regulation are special problems addressed in separate chapters.

5. *Capital investments in facilities.* Capital facilities may be classified as conveyance systems (channels, canals, pipes, bridges); dams; reservoirs; treatment plants for water supply and wastewater management; pumping stations; hydroelectric plants; spillways, valves, and gates; wells; river training systems (diversion structures, boat chutes, levees, and locks); and appurtenances. These are discussed in a chapter on infrastructure planning and management.

6. *Policy development scenarios.* Unique policy problems require special attention, such as developing laws and policies to solve political issues dealing with resource use. Two of these, Western water management and water and sanitation in developing countries, are discussed in separate chapters.

Knowledge Needed by Water Managers

Like other professionals, water managers need a great deal of knowledge to apply to problems and decisions. The categories that seem most critical are the following.

Hydroecology. The engineering and natural sciences that underlie water management actions are hydrology, hydraulics, water quality, and ecology. A basic understanding of them will help the water manager to assess options and impacts of decisions. Without this understanding, water management may be based purely on politics.

Infrastructure of water management: structures and systems. A water resources system is, by definition, a combination of water control facilities and/or environmental elements. Management strategies must consider how the system's components work together. This requires system-wide decision making and control from an integrated viewpoint.

Planning and decision making. While technical plans may follow logical steps, it is often difficult to gain the approval and resources needed to implement them. As a result, planning and decision making are difficult tasks for water managers, as in the scenarios described earlier. The planning and decision-making process deals with the questions: Given a goal for water management, what is the best way to accomplish it; and can approval and support be gained? Finding the best plan is technical in the sense that financial, institutional, economic, legal, and engineering aspects are "technical." Gaining approval requires dealing with the public, politicians, and regulatory processes. The best way to view the steps is as a technical process functioning inside of a political environment.

Organizational theories. Along with planning and decision-making techniques, the water manager needs insight into organizational issues, both as they affect internal aspects of organizations and for interorganizational dynamics.

Systems analysis and decision support systems. To accomplish technical planning, it is essential to use quantitative techniques for analysis of water resources systems. The set of techniques used fit into the framework of "decision support systems" (DSS), a term which generally means the databases, models, and communication systems necessary to provide the manager with good advice for decision making.

Water and environmental law. Water managers need to know a lot about water and environmental law. With conflicts over values and rights on the increase, the venue for solving them is the legal system.

Financial management. Finance is one of the key steps in implementing any water project or action, and one of the most important subjects in the field of water resources planning and management. Financial tasks are planning, programming, budgeting, accounting, cost control, and revenue management.

Principles for water resources management. After considering the tasks and scenarios of water resources management, a group of principles for effective management emerges. Some of these, such as the enterprise principle for managing a water utility, were learned only after mistakes were made.

Special Themes of the Book

As the reader proceeds through the book, a number of special themes will emerge. I consider these to provide a skeleton for the many theories and principles that make up the field of water resources management. They are the following.

A management system that can be applied in many states, regions, and countries. The concepts of water industry structure, role definition, finance, law, decision making, and coordination make up a management system that can be broadly applied for policy development or review. It is applicable to all countries, regardless of political system or status of development.

An explanation of water industry structure. The water industry model explains how service provision, regulation, coordination, and water industry support go together to meet the multiple purpose needs of water users and the environment.

Sustainable development in water resources management—infrastructure and environment. Sustainable development is explained, and applying water resources management to meet the dual but often conflicting needs of people and natural systems is a principal theme of the book.

Illumination of sources of conflict in water industry. The book explains why the conflicts over water resources management are so intense, and relates the conflicts to the application of sustainable development, to processes of democratic decision making, and to the public–private arena of water management.

Paradigm for process of water resources decision making. Whereas engineers learn about a rational model of problem solving, the book explains why such a model is necessary, but not sufficient, in the conflict-filled water industry, and it advances a process model that deals with the political and scientific nature of water resources management.

Identification of key role of DSS to deal with complexity. The decision support system, along with its components of data, models, analysis, and communications/dialog, is shown to be the answer to the complex of water decision making.

Explanation of integrated water resources management. Jargon of water resources management, such as "integrated management," is explained and simplified by breaking it into its component parts.

Key but underappreciated task of coordination. Coordination, an underappreciated and misunderstood management concept, is shown to be the key to making management in the water industry work well.

Review of policy themes and explanation of past policy failures. The book describes the major policy failures of the past 50 years, and reviews current findings and recommendations of study panels.

Overview of water resources finance, including pricing.   An introduction to public finance, as it applies to the water industry, is provided.

Integrated view of law and water administration.   Environmental law has become quite complex and multifaceted. It is reviewed and related to the major themes of water administration such as the role of state engineers, water quality regulation, and the water master function.

Exposition of public–private issues in the water industry.   During the past 10 years, major lessons about the effectiveness of the public and private sectors in the water industry have been demonstrated. These are reviewed to form a basis for understanding the range of allocation of roles in the water industry.

Case studies to integrate principles.   The book includes a large collection of actual case studies to present major, cross-cutting topics such as water quality and river basin management, and to provide interesting material for the study of the principles of water management.

Unique view of civics and science in water management.   The book's presentation of complexity and conflict in water management illustrates how civics and science are the keys to improved water management. This presentation can be used at varying levels in schools and universities to illustrate citizenship and the need for scientific understanding.

A Comprehensive Framework for Water Resources Management

As I prepared the book, I was struck over and over by the need for a comprehensive framework for water resources management. Chapter 9 discusses the reasons for the framework in some detail, and Fig. 1.3 illustrates key features of it that are explained in the book. These include:

- Coordinated actions between water agencies within a regulated water industry structure (see Wagner, 1995)

- A watershed focus for problem solving

- Local responsibility to the maximum extent, including the "enterprise principle" (see Chap. 7)

- Voluntary and cooperative actions rather than a "command-and-control" approach (see Chap. 8)

- Maximum practical use of market mechanisms to allocate and price water services and resources

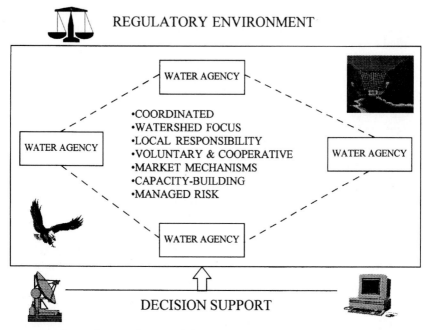

Figure 1.3   Comprehensive framework for water resources management.

- Emphasis on capacity building to encourage local responsibility
- Managed risk for water development and operating actions

A decision support system is illustrated to show that the comprehensive framework should be supported by data, analysis, and communication systems (see Chap. 5).

Questions

1. Explain the concept of "integration" in water management and give examples of the following types of integration: political, functional, hydrologic. Describe how "comprehensive, coordinated joint planning" relates to integration.

2. Water scientist Gilbert White used the following phrase relating to water management in his book, *Strategies of American Water Management:* "multiple purposes and multiple means." "Purposes" of water resources management are objectives or categories of service; "means" refers to ways of achieving goals or methods, either structural or nonstructural. List the main purposes of water resources management. Which is most important to society and why? List some of the "means" of water resources management. Which are of most interest in the United States today? Why?

3. It has been said that in the United States we have a water "crisis." Of all the water problems you can think of, which are of most concern to the public and why? Does this, in your opinion, make a "crisis"?

4. Water resources management was defined as *the application of structural and nonstructural measures to control natural and man-made water resources systems*. Can you propose an alternative definition?

5. What is your assessment of the U.S. water industry's status, and what policy changes would you propose to improve it?

6. What is your definition of "sustainable development," and how does the water industry affect it? Does any other industry, by itself, have as much impact on sustainable develop as water? Which?

7. Are those who call for "collaborative leadership" just dreamers, or can it be applied in the water industry? How?

8. From your own experience, can you think of water management scenarios that fit the following classifications: planning and coordination, organization, water operations management, regulation, capital investments, or policy development.

9. Is it better for water managers to be engineers? Why or why not?

10. Chapter 9 discusses a "comprehensive framework" for water management, but some of the features, as listed below, are introduced in this chapter. Do you agree with them? Why or why not?

   a. Coordinated actions between water agencies within a regulated water industry structure.
   b. A watershed focus for problem solving.
   c. Local responsibility to the maximum extent, including the "enterprise principle."
   d. Voluntary and cooperative actions rather than a "command-and-control" approach.
   e. Maximum practical use of market mechanisms to allocate and price water services and resources.
   f. Emphasis on capacity building to encourage local responsibility.
   g. Managed risk for water development and operating actions.

References

American Society of Civil Engineers, 1994 Civil Engineering Workshop Committee, 1994 Civil Engineering Workshop Report, Re-Engineering Civil Engineering Education: Goals for the 21st Century, New York, September 22–25, 1994.

Center for Global Studies, Towards Understanding Sustainability, *Woodlands Forum,* Vol. 10, No. 1, 1993.

Chesapeake Bay Program, *A Work in Progress: A Retrospective on the First Decade of the Chesapeake Bay Restoration,* Annapolis, MD, 1994.

Chrislip, David D., and Carl E. Larson, *Collaborative Leadership: How Citizens and Civic Leaders Can Make a Difference,* Jossey-Bass, San Francisco, 1994.

Drucker, Peter F., *Management: Task, Responsibilities, Practices,* Harper & Row, New York, 1973, pp. x, 36.

Environmental and Energy Study Institute Task Force, *Partnership for Sustainable Development, A New U.S. Agenda for International Development and Environmental Security,* Washington, DC, May 1991.

International Conference on Water and the Environment, Development Issues for the 21st Century, Dublin, Ireland, January 26–31, 1986.

Long's Peak Working Group, *America's Waters: A New Era of Sustainability,* Natural Resources Law Center, University of Colorado, Boulder, December 1992.

President's Council on Sustainable Development, brochure for Workshop on Challenges to Natural Resource Management and Protection of the Colorado River Basin, University of Nevada, Las Vegas, December 12, 1994.

Prism (American Society for Engineering Education, Washington, DC), Educating Tomorrow's Engineers, May/June 1995.

Repetto, Robert, Accounting for Environmental Assets, *Scientific American,* June 1992.

United Nations, Agenda 21 of U.N. Conference on Environment and Development, New York, 1992.

Wagner, Edward O., Integrated Water Resources Planning Approaches the 21st Century, presented at the 22nd Annual Conference of the Water Resources Planning and Management Division, American Society of Civil Engineers, Cambridge, MA, May 8, 1995.

Water Education Foundation, Materials and Publications (brochure), Sacramento, CA, 1993.

White, Gilbert F., Reflections on Changing Perceptions of the Earth, *Annual Review Energy and the Environment,* Vol. 19, 1994, pp. 1–13.

2

Hydrology and the Water Environment

Introduction

As water conflicts increase, water accounting becomes more important as the scientific basis for decisions. Water managers need a broad understanding of the principles of the water cycle and ecology, and they must understand how to make and interpret basic water computations and practical aspects of water science and engineering. This basic knowledge will help to assess options, deal with stakeholders, evaluate consultant reports, and design public and school education programs.

The purpose of this chapter is to present basic concepts of water science and ecology to aid in management tasks. Without a valid concept of the actual hydrology and ecology of a water system, the decision process will be confusing. This is often the case because understanding and data of hydrology and the ecology are inadequate and will always be so to some extent. Accurate concepts of sustainability of natural systems are especially important. Most "natural" systems have been modified, and the present ecosystems have adapted to current conditions. Unless managers understand systems, there may be political and legal gridlock rather than rational solutions.

As an interdisciplinary field, hydrologic science belongs both to the earth sciences and to engineering. Although the emphasis is on physical hydrology and aspects such as rate and volumes of water occurrence, hydrology also covers chemistry and statistics, and it relates to biology in the context of supporting ecological systems. In an attempt to delineate it as a separate field, the National Research Council published a volume outlining issues and opportunities in the hydrologic sciences (Committee on Opportunities in Hydrologic Sciences, 1991).

Going beyond hydrology itself, we can see the emergence of a new, integrated science called hydroecology or biological hydrology, which seeks to explain the impact of hydrologic modifications on ecosystems.

Important subjects to be covered include the hydrologic cycle, statistical methods, watershed characteristics, precipitation, groundwater, flood analysis, channel and reservoir routing, water yield, evaporation, erosion and sedimentation, and modeling (see, for example, McCuen, 1989.)

An essential "skill requirement" for water managers is the ability to compute basic parameters and to assess order-of-magnitude issues such as water savings and allocation. In all cases, knowledge is needed of the limits of quantitative analysis in solving problems to enable proactive programs to go ahead even in the face of scientific uncertainty. Much of the activity in court cases is to determine who is right in their hydrologic estimates.

The most useful aspect of hydrology is often its water accounting feature—the computation of water budget quantities. This is the basis for Western water rights engineering, for example, and is necessary to administer water law and make decisions on water transfers, as described in Chap. 15. Closely coupled with that is the need for water supply studies (Chap. 10) and the ability to handle water supply yield calculations (Chap. 20). Flood control management requires a different kind of hydrology; the hydrology of extreme values and storm rainfall. Chapter 11 includes several case studies. Reservoirs are the most important storage features of water systems, and managing them requires various skills in hydrology, hydraulics, and water scheduling. Chapter 13 discusses these tasks in some detail. In fact, the analysis of water facilities, rivers, estuaries, and groundwater systems requires skill in hydraulic computations. Groundwater systems require additional skills relating to hydrogeology. Chapter 18 introduces the subject.

Water quality analysis is another topic requiring knowledge of the chemical, physical, and biological aspects of water as well as hydrology and hydraulics. This chapter and Chap. 14 discuss the fundamental requirements. Sediment is a closely related subject, affecting both river morphology and water quality.

Finally, one might say that ecology is the coordinator of the water sciences: If natural systems are in balance, the ecological systems will be healthy. This chapter presents some aspects of ecology, and Chaps. 16 and 22 carry the water management aspects of the subject a bit further.

The Hydrologic Cycle

The starting point for understanding hydrology is the hydrologic cycle, shown in Fig. 2.1. Note that the cycle has three basic parts: the atmosphere, the surface water network, and the groundwater net-

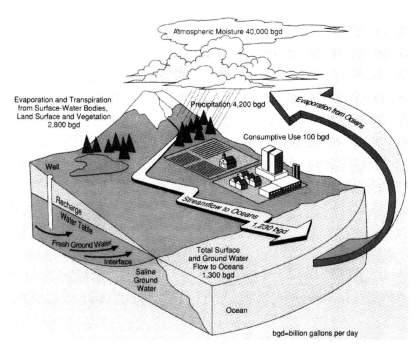

Atmospheric Moisture 40,000 bgd

Evaporation and Transpiration
from Surface-Water Bodies,
Land Surface and Vegetation
2,800 bgd

Precipitation 4,200 bgd

Evaporation from Oceans

Consumptive Use 100 bgd

Well

Recharge
Water Table

Fresh Ground Water

Interface

Saline
Ground
Water

Streamflow to Oceans
1,230 bgd

Total Surface
and Ground Water
Flow to Oceans
1,300 bgd

Ocean

bgd=billion gallons per day

Figure 2.1   Hydrologic cycle. (*Source: U.S. Geological Survey, from Frederick, 1995.*)

work. Quantity of water is measured by rates of flow and stocks of water in storage. Also, the quality of the water varies and must be measured.

From a systems standpoint, the cycle can be shown as in the flow diagram of Fig. 2.2. Block diagrams like this are the starting point for mathematical simulation models (see Chap. 5). This one is very general; specific diagrams are necessary for each water resources system.

The nutrient cycles and the mineral cycles occur in parallel with the hydrologic cycle, as shown in Fig. 2.3. The three major cyclic nutrients are carbon, phosphorus, and nitrogen. In the mineral cycle, sediment travels from mountains to ocean over long periods of time.

Atmospheric water

In the hydrologic cycle, atmospheric water is part of the stocks and flows of water. Nace (1964) estimated that of the global water budget, some 12,900 km^3 is in the atmosphere, or 5.6 percent of the total surface water of 230,700 km^3. Atmospheric water is only 0.0009 percent of the sum of global water, but that sum includes all oceanic and ice-cap water.

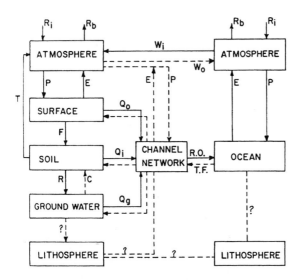

Figure 2.2 Hydrologic cycle: block diagram view. (*Source: Dooge, 1973.*)

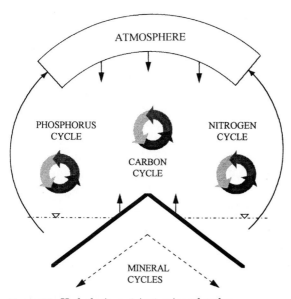

Figure 2.3 Hydrologic–nutrient–mineral cycles.

Different regional hydrologic regimes have developed because of atmospheric water patterns. For example, Georgia is a state with abundant rainfall, whereas Nevada is poor in rainfall. Saudia Arabia has no significant rivers and streams because of moisture patterns, whereas Bangladesh is plagued with floods.

Common terms in atmospheric water analysis are meteorology, weather, and climate. *Meteorology* is the science of atmospheric weather and weather conditions. *Weather* is the state of the atmosphere in terms of measurable variables such as temperature, precipitation, wind velocity, barometric pressure, and humidity. *Climate* refers to the prevailing meteorological conditions in a region.

Think of the atmosphere as having three components: dry air, water vapor, and impurities. Dry air is made up of nitrogen, argon, carbon dioxide, and trace gases. Nitrogen and oxygen make up over 99 percent of the total. Other than slight variations in CO_2, the composition of the atmosphere is constant, indicating complete mixing. CO_2 results from the carbon cycle, an important ecological issue that will be discussed later. Water vapor depends on temperature; thus there will be more atmospheric water in hot climates than in cold. The mass of the atmosphere is about 5.6×10^{15} metric tons. The water vapor accounts for 1.5×10^{14} metric tons, or 2.7 percent. The weight of the atmosphere is equivalent to a water depth of 33 ft all around the globe (1 atm). Precipitation occurs as a result of different weather systems. Average rainfall around the world varies from over 60 in at the equator to below 5 in at the South Pole, which is a desert in terms of precipitation. There are many variations, of course, and at both 40–60° south and north latitudes, there are increases in average precipitation (Petterssen, 1962).

Climate varies over time. Just because average conditions in a place have been relatively fixed for, say, the last 50 years, this does not mean that they will stay at that level. Climate change, in the language of the press, has received much attention. In 1991, for example, the Environmental Protection Agency (EPA) sponsored the first national conference on climate change and water resources management. The purpose statement explained the concept: "Margins of safety adequate for past climate may be insufficient for a changed climate. . . . Only two things are certain about the world's climate: it is unpredictable and it is in perpetual flux." The topics of the conference were framed with this language:

> Presently water management deals with droughts, floods and random patterns, but climate change introduces a new and additional degree of uncertainty. Some call it a coming ecological and economic scare tactic: As responsible water planners and managers we need to know: what would climate change mean to water management; how do we determine

if the threat is real; what are the risk factors; when and where might these factors occur; what problems and opportunities does this new uncertainty offer? (U.S. EPA, 1991)

Of course, global climate change is controversial. Some believe in it; some don't. Basic issues include an increase in carbon dioxide and increased trace gases that will result in effects of greenhouse/global warming, sea-level rise, altered weather patterns, and loss of the ozone layer. These effects remain controversial, and will be studied in research programs of the 1990s. One school of thought is alarmist: a projected global warming of 4°C; another thinks the warming will be much less, and will even be beneficial in terms of increased crop yields.

Two aspects of atmospheric water will be illustrated briefly, one on water balance and one on storm rainfall.

Atmospheric water balance: monthly precipitation and evaporation

Climatological data is published in the United States by the National Weather Service. Table 2.1 shows a record of five years of monthly precipitation and evaporation depths at Lake Sumner in east-central

TABLE 2.1 Precipitation and Pan Evaporation Depths at Sumner Lake, New Mexico

	Jan.	Feb.	Mar.	Apr.	May	June	July	Aug.	Sept.	Oct.	Nov.	Dec.	Total
					Precipitation (in)								
1988	0.16	0.07	0.08	1.01	1.65	1.95	5.70	2.00	2.88	0.18	0.02	0.24	15.94
1989	0.32	0.59	0.14	0.26	0.38	0.95	0.72	4.14	0.60	0.24	0.00	0.36	8.70
1990	0.53	0.94	1.18	1.41	1.16	0.05	2.42	5.29	1.42	0.64	0.43	0.29	15.76
1991	0.36	0.00	0.00	0.00	2.01	0.27	6.12	4.65	4.16	0.21	1.42	1.86	21.06
1992	0.31	0.27	0.29	0.75	2.98	2.68	0.35	3.32	0.50	0.08	0.07	.079	12.39
Avg.	0.34	0.37	0.34	0.69	1.64	1.18	3.06	3.88	1.91	0.27	0.39	0.71	14.77
					Pan evaporation (in)								
1988	2.44	3.74	8.36	10.79	12.29	12.51	12.02	9.62	8.78	6.97	7.08	4.41	99.01
1989	4.78	5.50	8.24	11.23	14.25	13.19	14.80	10.77	9.11	8.28	7.71	3.58	111.44
1990	4.29	5.06	6.13	9.26	15.03	18.68	13.15	10.61	8.02	8.08	5.18	3.68	107.17
1991	2.33	5.26	9.39	13.73	16.39	15.51	11.38	10.14	8.22	7.81	3.50	2.18	105.84
1992	2.12	4.22	8.49	8.95	9.76	11.88	13.91	11.85	10.76	8.52	4.62	2.52	97.60
Avg.	3.19	4.76	8.12	10.79	13.54	14.35	13.05	10.60	8.98	7.93	5.62	3.27	104.21

SOURCE: National Weather Service climatological records.

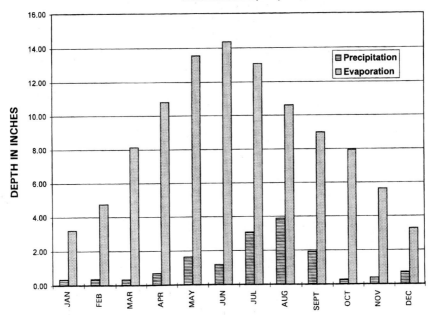

Figure 2.4 Precipitation and evaporation, Sumner Lake, New Mexico. (*Source: National Weather Service, Washington, DC.*)

New Mexico. Note that both precipitation and evaporation are highly variable, but seasonal. The fact that evaporation exceeds precipitation in every month of the year is typical for dry regions. By plotting the data in the table (Fig. 2.4), one can examine monthly fluctuations as well as see the randomness in climate variation.

Storm rainfall intensity

Data to illustrate rainfall intensity include rainfall depths and rainfall intensities for different time periods. A graph of rainfall depth for a storm duration is called a *hyetograph*. The example in Table 2.2 shows a remarkable rainfall event at the Birmingham, Alabama, Weather Forecasting Office (part of the National Weather Service) gauge on December 2–3, 1983. Note that the total storm reached 9.28 in after 26 hr from 8 p.m. on December 2 to 9 p.m. on December 3. In this case, river flooding was the result (see Holt Reservoir case study, Chap. 11), so long-duration, heavy rainfall was of most concern. If urban flooding was also being analyzed, short-duration, intense rain would be the focus. Table 2.2 shows that in terms of hourly intensi-

TABLE 2.2   Rainfall Depths, Birmingham WFSO Gauge, December
2–3, 1983

Date	Hour ending	Depth (in)	Daily total (in)
12/2/83	8:00 p.m.	0.02	
	9:00 p.m.	0.29	
	10:00 p.m.	0.89	
	11:00 p.m.	1.22	
	12:00 p.m.	0.91	3.33
12/3/83	1:00 a.m.	0.95	
	2:00 a.m.	0.54	
	3:00 a.m.	1.25	
	4:00 a.m.	1.02	
	5:00 a.m.	0.51	
	6:00 a.m.	0.23	
	7:00 a.m.	0.24	
	8:00 a.m.	0.26	
	9:00 a.m.	0.05	
	10:00 a.m.	0.01	
	11:00 a.m.	0.01	
	12:00 a.m.		
	1:00 p.m.		
	2:00 p.m.	0.01	
	3:00 p.m.		
	4:00 p.m.		
	5:00 p.m.		
	6:00 p.m.	0.39	
	7:00 p.m.	0.28	
	8:00 p.m.	0.16	
	9:00 p.m.	0.04	5.95

SOURCE: National Weather Service climatological records.

ties, the maximum intensity occurred between 2 and 3 a.m. Figures
2.5 and 2.6 show the storm in graphical and map form. Note that the
maximum depth of 11.5 in occurred just to the west of Birmingham's
gauge.

Surface water including pipe flow

The surface water network, as opposed to groundwater, involves watersheds, tributaries and streams, and related parts of the riverine-riparian system. Much of hydrology deals with predicting rates and
volumes of surface water flow that results from rainfall or snowmelt.

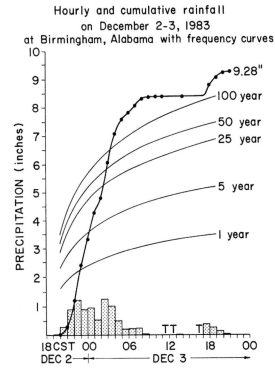

Figure 2.5 Hourly and cumulative rainfall, Birmingham, Alabama, December 2–3, 1983. (*Source: National Weather Service, Washington, DC.*)

In mountain regions, snowpack forms a big part of the total water reserve, and is a gift of water storage from nature. Figure 2.7, showing snowpack near Bradley Lake, Alaska, illustrates the water yield potential of snowpack.

Hydraulics is the main water science that applies in the surface water network. It uses basic fluid mechanics to analyze flows through natural and man-made systems. The reader seeking a full treatment should consult an appropriate text on engineering hydraulics.

Sometimes surface water is diverted temporarily to underground pipes. Pipes, conduits, and tunnels are used to convey water. The hydraulics of pipe flow are rather simple compared to flow in open channels. Still, various complexities arise, such as measurement of flow, determining the resistance to flow in a pipe, flow in pipe networks, and transient flows.

Measurement of flow in a pipe is done by various types of meters. In a water treatment or industrial plant, flow may be metered accu-

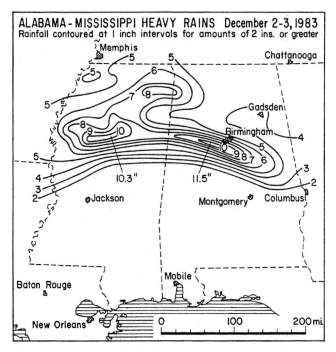

ALABAMA-MISSISSIPPI HEAVY RAINS December 2-3, 1983
Rainfall contoured at I inch intervals for amounts of 2 ins. or greater

Figure 2.6 Rainfall contours, Birmingham, Alabama, December 2–3, 1983. (*Source: National Weather Service, Washington, DC.*)

rately with an orifice or venturi meter, a device with a constriction that causes a pressure change and enables the flow to be measured as a function of the change in pressure. Meters on a residential water supply connection measure cumulative volume of flow and must be read from time to time to determine water use.

The energy required to overcome resistance to flow and push water through a pipe is supplied by a pump, by a head of water from a tank or reservoir, or by water flowing from a higher elevation. The resistance to flow in a pipe is caused by resistance due to roughness on the inner pipe walls and by constrictions caused by chemical or mineral deposits or obstructions. When a pipe is new, its roughness can be predicted from standard tables based on laboratory measurements. However, as the pipe ages, roughness becomes unpredictable without in-place measurements. Taking these measurements is often not practical, as for example in a section of buried pipe. This problem makes the determination of a pipe's condition difficult. Pipe networks pose particular challenges to the analyst which are overcome through the use of computer simulation models. Several such models have been improved to the point that they are run on personal computers.

Figure 2.7   Snowpack near Bradley Lake, Alaska. (*Courtesy Bechtel Corporation.*)

Measurement of flow in open channels is done by introducing a constriction or change in channel cross section that enables a correlation to be made between depth and discharge. A well-known device for open channel measurement is the Parshall flume, developed at Colorado State University (see Chap. 3).

The basic problem to be solved in open channel analysis is to find the depth and velocity of water, given the discharge. For most cases, an estimate of depth and velocity is made using a simple formula called the *Manning equation,* which relates velocity to depth of water and to size, roughness, and slope of the channel. This formula gives only an approximation, however, and for long channels with changes in shape, size, and slope, it is necessary to do a "backwater analysis" or to analyze flow that is "gradually varied." The tool for doing a backwater analysis is a computer program that uses the energy principle to relate depth and velocity at different points to the conditions downstream or upstream. A number of such programs exist, but in the United States one developed by the Corps of Engineers' Hydrologic Engineering Center (HEC) has become the most popular. This model, called HEC-2, has been improved over a number of years and is in wide use by firms and agencies.

A complex situation in an open channel is to analyze the case of "unsteady flow," where a flood wave travels down a channel or depth and velocity change due to other factors. There are a number of computer programs for making such an analysis, but using them is not considered routine and requires a specialist.

Rivers and estuaries pose special problems to the analyst because they have uneven banks and channels and conditions change rapidly. One perplexing problem in rivers is to compute the depth of water in an alluvial channel when the sand bed changes form from time to time and the bed roughness is not constant. Along with this problem is the challenge of computing the rate of sediment transport or the depth of scour. Specialists are able to estimate these quantities, but most estimates are not very reliable.

Another complex hydraulics problem is the variation of velocity in an estuary. The velocity varies across the channel and vertically in the water profile. For this reason, water quality modeling, which depends on hydraulic modeling, is only approximate. Models of the hydraulics of estuaries must be applied by specialists, and their results must be used with caution (see Chap. 22).

Groundwater hydrology

Groundwater systems are key components of hydrologic systems. The management of groundwater systems is discussed in Chap. 18.

As shown in Fig. 2.7, groundwater is a dynamic part of the hydrologic cycle, just as surface water is. The main differences are that groundwater moves much more slowly and is exposed to different chemical and biological environments. Some groundwater, called "fossil water," may have been in storage for thousands and even millions of years. Other groundwater, in tributary aquifers, may flow almost as quickly as surface waters.

Figure 2.8 shows types of aquifers with flowing and pumping wells, and Fig. 2.9 shows typical water well construction. In addition to wells for withdrawing water, there are also wells for recharge, that is, to return water to aquifers. Recharge is an emerging water manage-

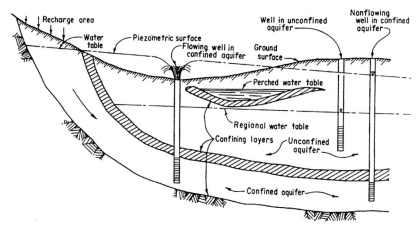

Figure 2.8 Aquifer types, showing wells. (*Source: Ground Water Manual, U.S. Bureau of Reclamation, U.S. Dept. of the Interior, Washington, DC, 1977, p. 8.*)

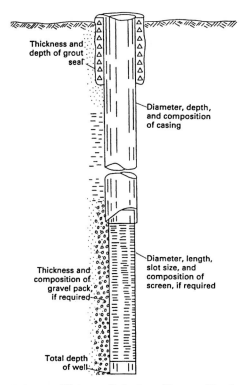

Figure 2.9 Water well design. (*Source: Heath, 1989, p. 56.*)

ment technique that merits careful study and consideration (see Chap. 18).

Hydrology texts provide explanations of groundwater flow, and a few special academic programs concentrate on it. For example, Colorado State University offers graduate degrees with specialization in groundwater engineering. In these programs there is close cooperation between groundwater engineers and the hydrogeologists.

In a text on groundwater hydrology, Heath (1989) explains that groundwater has been utilized from springs or tapped through wells beginning hundreds of years before Christ. They define the field of ground-water hydrology (there is a continuing controversy as to whether groundwater is one word or two), and state that it "deals with the unseen" because the only place you can see groundwater in its natural form is in limestone caverns or other large, subterranean openings.

Key concepts presented by Heath and Trainer are the hydraulic characteristics of rocks and aquifers, transmissibility and storage coefficients, and the quality of groundwater. The variability and characteristics of the rocks and aquifers make up the complexity of groundwater. Transmissibility and storage coefficients enable us to apply theories of flow and storage to manage groundwater withdrawals. Quality of groundwater has become an important issue in recent years with the discovery of many contaminants in long-term storage in aquifers.

Computations to predict the flow of groundwater start with simple equations and proceed to complex mathematical models. More complex models utilize digital and analog techniques to account for water flows and volumes in aquifers that can take on irregular characteristics and dimensions. Collecting data to determine these characteristics is difficult.

Water Accounting and Water Budgets

Much of practical hydrology deals with water accounting, or how much, when, and where water appears. We account for stocks of water with a *balance sheet,* and for flows of water over periods of time with a *water budget.*

Water budgets can involve geographic scales such as global water budgets, national budgets, or budgets for regions, states, river basins, urban areas, catchments, reservoirs, and stream reaches. The geographic scale depends on the use of the water budget, and normally corresponds to the area for which a decision is to be made—for example, a river basin. As we will see in Chap. 19, one of the main management problems in water is caused by lack of congruence between river basins (water accounting units) and political accounting units (cities, counties, states).

Quantities that require computation for water budgets are those that measure the flows and stocks of water systems, and include surface, ground, and atmospheric inputs and outputs, diversions, returns, storage inputs, releases and levels, and watershed yields.

In the water budget, all credits and debits must be accounted for, both in a time period (as in an income statement) and at any point in time (as in a balance sheet). Like a financial budget, water budgeting provides the planning information used to allocate the available supply over competing demands during an accounting period, such as a month, season, or year.

The water budget for a period of time for a catchment would be

$$\text{Inflow} - \text{outflow} = \text{change in storage}$$

Inflows could include precipitation, imported water, and imports from adjacent groundwater basins. Outflows could include stream discharge, consumptive uses, evaporation and evapotranspiration losses, and any seepage losses that left the basin via groundwater channels. Water diverted, or withdrawn, can either be consumed or returned. If it is returned, it might be returned to the same stream, at some other point, to the groundwater system, or to another basin.

The water budget equation becomes the familiar storage equation of hydrology when the time increment is included:

$$I - O = \frac{dS}{dT}$$

where I and O are rates of flow and dS/dT is the rate of change of storage with time. This equation might be used in flood routing—that is, to study the time rate of change of water flows and levels during a flood.

National and global water budgets

For a national water budget, an example can be taken from the United States, where the first national assessment of the nation's water resources showed that the United States had an annual resource of 4200 billion gallons per day, provided by an average of 30 in of precipitation (Fig. 2.10). About 70 percent was said to be consumed by evaporation and transpiration, with the remainder, about 9 in, being runoff of about 1200 billion gallons per day (U.S. Water Resources Council, 1968, 1978).

The global water budget has been the subject of much study by hydrologic scientists. One source classified the global water resources as

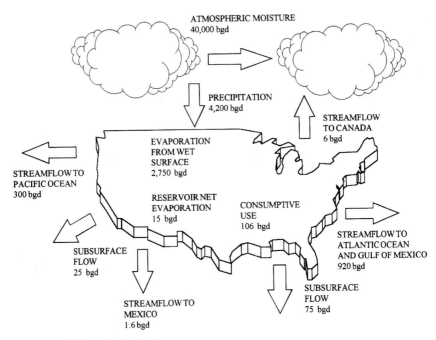

ATMOSPHERIC MOISTURE
40,000 bgd

PRECIPITATION
4,200 bgd

STREAMFLOW
TO CANADA
6 bgd

EVAPORATION
FROM WET
SURFACE
2,750 bgd

STREAMFLOW TO
PACIFIC OCEAN
300 bgd

RESERVOIR NET
EVAPORATION
15 bgd

CONSUMPTIVE
USE
106 bgd

STREAMFLOW TO
ATLANTIC OCEAN
AND GULF OF MEXICO
920 bgd

SUBSURFACE
FLOW
25 bgd

SUBSURFACE
FLOW
75 bgd

STREAMFLOW TO
MEXICO
1.6 bgd

Figure 2.10 U.S. water budget.

shown in Table 2.3. As you can see, almost all of the water is contained in the oceans, icecaps, and glaciers. The quantity that we usually manage, annual flows through the hydrologic cycle, is a small percentage of the total. Computing the world's water balance is a problematical endeavor, and one can find different estimates by different groups. Gleick (1993) gives six different estimates, but the totals do not vary much.

Rates of flow

Rates of flow are used for many different purposes including streamflow, flow in pipes, discharge of a pumped well, rate of consumption, and flow through a treatment plant. Rates of flow can be expressed in a number of different units. Consider, for example, the flow of a small stream at, say, 100 ft^3/sec. That might also expressed as, for example (see App. C), 2.83 m^3/sec, 64.63 mgd, or 44,884 gal/min.

Example of flow rates. A city of 150,000 population has an average annual withdrawal of 167 gallons per capita per day (gpcd). Convert this rate of withdrawal in gpcd to other appropriate units. Note that 1 million gallons per day (1 mgd) is a convenient unit, 25.05 mgd for

TABLE 2.3   World Water Quantities (from Nace, 1964)

Water volumes	(km³)	% of total
Surface water:		
Freshwater lakes	125,100	0.009
Saline lakes	104,300	0.008
Stream channels	1,300	0
Subsurface water:		
Shallow groundwater*	4,171,400	0.307
Deep groundwater	4,171,400	0.307
Soil moisture, etc.	66,700	0.005
Icecaps and glaciers	29,199,700	2.147
Atmosphere (at sea level)	12,900	0
Oceans	1,322,330,600	97.217
Totals	1,360,183,400	100.000

*Depth less than ½ mi.

the city of 150,000. The unit cubic feet per second (cfs) is also convenient, but mgd seems to reflect usage better, even though it is actually a flow rate. Cubic meters per second is a small number, but liters per second (lps) and gallons per minute (gpm) are too large for convenience in this case. Each has applications elsewhere, however. In the metric system, million cubic meters (MCM) per year is a convenient unit of measurement.

$$167 \text{ gpcd} = 167 \times 150,000 \div 10^6 = 25.05 \text{ mgd}$$

Unit	Flow rate
mgd	25.05
ft³/sec (cfs, cusec)	38.76
m³/sec (cumec)	1.10
liters/sec (lps)	1,097.49
gal/min (gpm)	17,395.7
MCM/day	94,823.8

Storage volumes

Storage volumes are a key management quantity in water budgets. Water can be stored in surface water reservoirs or groundwater systems. When the water is stored and released is important to the ecosystem balance. Where in the reservoir the released water comes from is important to water quality computations. How much storage water is lost to seepage and evaporation is important in water management computations.

The following example illustrates the computation of volumes using several different systems of units.

Example of water storage. According to the city's bathymetric chart, Cross Lake in Shreveport, Louisiana, holds 5.6 billion gal at water level 162.0 ft and 24.6 billion gal at elevation 170.0 ft. Convert these storage quantities to other appropriate units. Along with storage volume, bathymetric charts present surface area. Acres and square miles are in general use in the United States, and hectares (ha) in most other places around the world. Compute the lake's surface area for the two elevations.

Elevation (ft)	162.0	170.0
Volume (BG)	5.60	24.60
Volume (ft^3)	7.486E + 8	3.289E + 9
Volume (m^3)	2.120E + 7	9.312E + 7
Volume (Mm^3)	21.20	93.12
Volume (AF)	17186	75495
Volume (TAF)	17.19	75.49
Volume (MAF)	0.0172	0.0755
Volume (km^3)	0.02120	0.09312
Volume (mi^3)	0.00509	0.02234
Surface area (mi^2)	7.7	12.6
Surface area (acres)	4928	8064
Surface area (ha)	1994	3263

See the conversion factors given in App. C. The unit billion gallons is sometimes used for municipal water supply. The unit cubic feet requires numbers that are too large to be useful for this case, but cubic feet (and thousand gallons) are units used to express the volume of usage by an individual water user.

Million cubic meters and acre-feet are common units for water storage, use, and streamflow yields. The units cubic kilometers and cubic miles are used only at continental scales, or to express the volumes of huge water bodies. For example, the annual flow of the Nile River, expressed in an agreement between Egypt and Sudan, is about 90 km^3. This is also referred to in Egypt as "milliards," or 10^9 m^3. A "milliard of water" means 1 billion m^3 or 1 km^3. Americans should note that England and some other countries do not use the term "billion" but state instead a "thousand million" or milliard. Also, Spanish speakers refer to "thousand million" (*mil millones*), and billion (*billón*) in the Spanish language means 1 million millions, or a trillion as we say in American English.

Watershed, Aquifer, or System Yield

Predicting yields from a watershed, a water resources system, or a groundwater source is an important element in all phases of water resources management: planning for supplies, operation of facilities, and rationing during drought. See Chap. 20 for a discussion of safe yield.

Planning for supply requires that the combined yield and reliability of all water supply sources be known. Water supply organizations must be conservative and not take much risk that supplies will run out. The necessity for risk aversion, and the lack of incentives to cooperate to share supplies or draw from a common water supply pool, encourages overdevelopment of streams and aquifers and increases conflict in water management. This institutional issue is discussed in Chaps. 20 and 21.

Yield, a measure of water supply, can be defined as the expected quantity of water to be provided by a surface water or groundwater source on a periodic basis, usually annually. For example, it could be said that the average annual yield of a certain watershed is 50,000 acre-ft of water. When used properly, yield includes a statement of water supply reliability in the form of a statistic which is based on some period of observed record.

Risk is measured by assigning statistical values to yield. The term "safe yield" is being used increasingly for this purpose. The concept of safe yield is different in surface and groundwater systems. In surface water systems, *safe yield* means the minimum expected available water for a particular period of time. A watershed might have a yearly safe yield, for example, of 30,000 acre-ft, whereas the annual average yield would be 50,000 acre-ft. If the safe yield was determined on the basis of a 40-year record, then we would say that the watershed has a 1-in-40 year safe yield of 30,000 acre-ft. This means that the probability that in a given year the yield will be less than 30,000 acre-ft is 1/40 or 0.025. If we then are looking to manage risk, we can say: If the demand is 30,000 acre-ft per year, there is a 1-in-40 chance we will be short in any given year. The annual average yield is the quantity that is expected, on the average, every year. Of course, the actual quantity in a given year will be more or less, but over a period of years the annual average will be experienced.

In the case of groundwater, yield is keyed to aquifer recharge. If we say that the annual yield of a given groundwater system is, say,

10,000 acre-ft, we mean that if 10,000 acre-ft are withdrawn per year, there will be no decline in the aquifer water level. Groundwater yield is therefore a different concept than surface water yield, because it depends on withdrawing the water to produce the yield. If the groundwater is not withdrawn, the aquifer may not store the water; the water may flow to springs or to other aquifers.

Safe yield for a complex water supply system consisting of multiple sources and storage reservoirs is more complex to analyze. To estimate the yield, it is necessary to simulate the system's operation under the assumption of changing inputs and operational strategies to determine the statistics of yield.

Example of water supply yield. Assume that a city of 100,000 has an average daily consumption of 150 gpcd. The average daily demand is then 15 mgd (46.1 acre-ft/day) or 16.1 thousand acre-feet (TAF) (20.7 MCM) per year. If the surface supply sources yield an annual average of 20 TAF, there will be a reserve. If, however, once in 50 years the yield drops to 12 TAF, there will be a shortage. To state the reliability of the water supply, it is necessary to analyze the statistics of the yield using a probability distribution, as shown in Fig. 2.11. The once-in-50-year event of the water supply dropping to 12 TAF is shown on the curve at the point where 2 percent of the area (1/50) is to the left of the value 12 TAF.

Example of Colorado River flow. Chapter 19 outlines some aspects of the dispute and resulting compact that is used to allocate water on the Colorado River. In a U.S. Geological Survey circular, Leopold (1959) presented 61 years of records of the reconstructed annual flows of the Colorado River at Lees Ferry, Arizona. (See Table 2.4.) The pur-

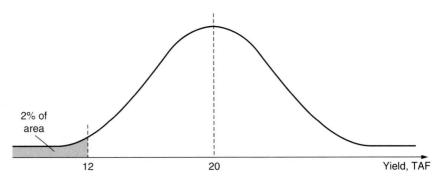

Figure 2.11 Statistical distribution of watershed yield.

TABLE 2.4   Reconstructed Annual Flows, Colorado River at Lee's
Ferry, Arizona

Year	Discharge (1000 acre-ft)	Year	Discharge (1000 acre-ft)
1896	10,089	1927	18,616
1897	18,009	1928	17,279
1898	13,815	1929	21,428
1899	15,874	1930	14,885
1900	13,228	1931	7,769
1901	13,582	1932	17,243
1902	9,393	1933	11,356
1903	14,807	1934	5,640
1904	15,645	1935	11,549
1905	16,027	1936	13,800
1906	19,121	1937	13,740
1907	23,402	1938	17,545
1908	12,856	1939	11,075
1909	23,275	1940	8,601
1910	14,248	1941	18,148
1911	16,028	1942	19,125
1912	20,520	1943	13,103
1913	14,473	1944	15,154
1914	21,222	1945	13,410
1915	14,027	1946	10,426
1916	19,201	1947	15,473
1917	24,037	1948	15,613
1918	15,364	1949	16,376
1919	12,462	1950	12,894
1920	21,951	1951	11,647
1921	23,015	1952	20,290
1922	18,305	1953	10,670
1923	18,269	1954	7,900
1924	14,201	1955	9,150
1925	13,033	1956	10,720
1926	15,853		
		Average	15,180
		Std. dev.	4,217

SOURCE: Leopold, 1959.

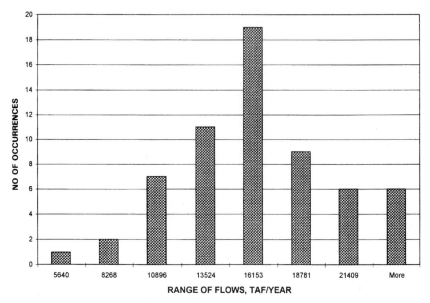

Figure 2.12 Statistical distribution of Colorado River flows. (*Source: Leopold, 1959.*)

pose was to illustrate a statistical analysis of runoff records that enable us to make statements about the yield of the river system. Leopold showed how simple statistical calculations can be used to make inferences about the mean value, variance, and persistence of hydrologic records. His presentation was used in the *Arizona v. California et al.* case, which involved the Colorado River compact.

Leopold presented a 61-year series of flows which had been reconstructed to produce total river water availability or "virgin flow." The reconstruction amounts to accounting for the diversions and abstractions that take away from the virgin flow. The flows are arrayed into the statistical distribution that fits them best, in this case a normal or "bell curve" distribution, as shown in Figs. 2.12 and 2.13.

Flood Flow

Floods (Chap. 11) and droughts (Chap. 20), as extreme hydrologic events, affect both human and ecologic systems. For the water manager, the main issue with floods is risk. The sequence of flood analysis is: (1) What is the expected flood discharge? (2) What will be the depth and velocity of the water? (3) What will the damage be? Analysis thus requires first a hydrologic analysis, then a hydraulic analysis, and, finally, an analysis of the depth–damage relationships for the flooded area.

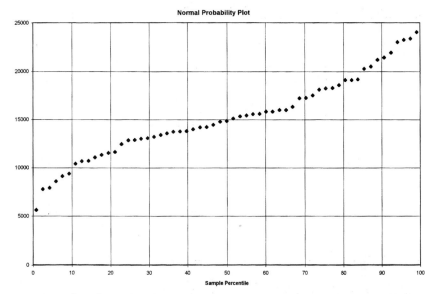

Figure 2.13 Cumulative distribution of Colorado River flows. (*Source: Leopold, 1959.*)

Expected flood discharge is basically a statistical concept. There are many techniques for the analysis of runoff from different kinds of watershed land uses, such as rural, urban, agricultural, forest, and wetland; depending on the tributary watershed, the magnitude and timing of flood discharge will vary.

Risk of flooding is reported by the related concepts of *exceedance probability* and *flood frequency*. Let P be the probability that a flood of a particular magnitude Q will be equaled or exceeded in a given year. For example, if the probability is .01, then there is a 1 percent chance in any year that Q will be equaled or exceeded at least once. On the basis of frequency, this is said to be a 100-year flood, or a flood with a return period T of 100 years. To compute this we use the formula $T = 1/P$.

Computing the probability for a multiyear period requires different thinking. If a 100-year flood has a nonexceedance probability of $P = .01$, then intuition might tell you that the probability that the flood will occur at least once in a two-year period would be $P + P = .02$. That wouldn't work, however, because if it was true, then the probability that the flood will occur at least once in 100 years would be 1.0—in other words, a certain event—which it is not, according to the laws of probability. We get around this problem by considering the complement of the problem. The probability that the flood will not occur in two successive years is simpler to compute: It is P(not 1 and not 2) = P(not 1) $\times$ P(not 2); in other words, the product of the probabilities of two independent events. This is simple to compute for any series of years: P(not in a hundred years) = $(1 - P)^{100}$. Then P(at least once in 100 years) = $1 - (1 - P)^{100}$. This generalizes to the formula

$$J = 1 - (1 - P)^n$$

where J = the risk that a flood with exceedance probability P will occur at least once in a period of n years.

Sediment Transport

Sediment transport is, on the one hand, a water quality issue and, on the other, a matter of stream and surface water network dynamics. The mineral cycle, a concept for showing the movement of sediments, shows how sediments are produced and moved, go into storage, and are discharged, much the same as their carrier, the water resource. However, sediment transport is much more complex than water flow.

A detailed treatment of sediment transport is beyond the scope of this discussion, but a few concepts will be described briefly. See texts such as Simons and Senturk (1992) for more detail.

Erosion of sediment is a major problem in watershed management and generates sediments that clog streams and waterways and cause much trouble. Watershed treatment and best management practices are ways to reduce erosion; see Chaps. 14 and 16.

As sediment is transported in streams, it affects the flow in natural systems. A certain amount is normal and helps to nourish the banks and ecosystems, but too much can destroy fisheries, cause flooding, clog intakes, and cause other serious problems. Stream sediment includes that moved along the bed (bed material), that suspended by turbulence (suspended load), and the fine sediments which are permanently suspended and which make the stream appear muddy (wash load).

Reservoirs act as traps to capture sediment. If the rate of capture is too great, the reservoir is quickly filled and valuable storage space is lost. Many of the world's largest reservoirs were completed during the twentieth century, and will become filled in the decades ahead. This will be a serious problem when it occurs. See Chap. 13 for a discussion of reservoirs.

Sediment has water quality impacts. Certain contaminants and nutrients can become lodged in the sediment and not be released for long periods of time. Chemicals can adhere to sediment particles and be transported with them. Water quality is discussed in more detail in the next sections.

Sediment has implications for ecological systems because habitat can be destroyed or created as a result of erosion and sedimentation. In the Platte River, for example (Chap. 19), too much sediment (and too little flood flow to flush it out) can destroy fish habitat and lead to channel narrowing, and a loss of habitat for sandhill cranes. Too little sediment can lead to channel erosion and a loss of habitat. The balance is delicate.

The overall impact of sediment on ecosystems and on geologic features illustrates that to have sustainable water management requires responsible watershed management and sediment management measures in riverine systems.

Water Quality

Many parameters make up "water quality," and no index exists to provide a single, all-encompassing indicator. Rather, water quality is a comprehensive variable with chemical, biological, and physical parameters, and it requires both descriptive and quantitative analysis.

Water quality management issues are discussed in Chap. 14. This section presents key elements of water quality as they relate to hydrology and ecology. The quality of water as it moves through the hydrologic cycle is determined by natural and man-made influences. Natural influences result mostly from the cycles of chemicals and nutrients and physical forces. Man-made influences include the discharge of substances into the water and changes in natural flow regimes due to intervention. Although water quality can be described in many ways, the language of the Clean Water Act calls for maintaining "the chemical, physical, and biological integrity of the Nation's waters," so it is useful to present water quality in terms of its chemistry, biology, and physical characteristics.

Chemistry of water

Chemical parameters of water that are controlled are oxygen content, inorganic chemicals, organic chemicals, and radionuclides.

The oxygen content of water reflects the ability of natural waters to sustain aquatic life and the extent of oxygen-demanding wastes. The oxygen content of natural waters is related to the temperature and the quantity of oxygen-demanding waste in the water. Water will reoxygenate itself through natural action as it flows downstream. Thus, in the absence of wastes, it will seek a level of oxygen content that depends on the temperature and flow conditions. Dissolved oxygen contents above about 5 mg/liter can sustain a warm-water fishery, while values above about 7 mg/liter are necessary for a cold-water fishery.

The saturated value of dissolved oxygen depends on temperature and is about 9.0 mg/liter at atmospheric pressure and 70°F. It rises to over 14.0 mg/liter at 0°C, and falls to below 8.0 mg/liter at 30°C.

One use of dissolved oxygen is as an indicator variable for determining how much waste should be allowed to discharge into a stream. If a stream is rated as a cold-water fishery, for example, then only that quantity of wastewater which can be assimilated without driving the oxygen too low is allowed. This level is set as a stream standard, and water quality simulations tell how much waste can be allowed.

Inorganic chemicals that are regulated in drinking water include arsenic, cadmium, chromium, lead, mercury, selenium, and silver, among others. These inorganic chemicals, mostly metals, can be harmful to humans and toxic to fish. The toxic effects of the heavy metals lead and mercury on humans have been publicized much in recent years, and research has expanded to determine the lethal concentrations of various chemicals on different species of fish.

Salt compounds, expressed as part of total dissolved solids, are regulated both in drinking water and in streams. Salt and salinity are also very important aspects of the quality of irrigation waters.

Organic chemicals in water can be quite toxic. They include chlorinated hydrocarbons such as Endrin, Toxaphene, and other agricultural chemicals. Also, industrial discharges can contain a variety of organic chemicals that are produced in manufacturing or other operations.

Radionuclides are regulated by the Safe Drinking Water Act. They include familiar compounds such as tritium, as well as any number of isotopes that might get into the water from nuclear power plants, research, medical, or other activities.

Physical characteristics of water

Qualities of water such as taste, odor, color, temperature, floating objects, and sediment load can be considered physical in nature. Some are regulated under the Safe Drinking Water Act. For example, color, odor, and pH are included among the secondary standards for drinking water, while taste and suspended solids are identified as qualities of water that are important for preserving water for domestic use (Hammer and MacKichon, 1981).

The mechanism of sediment transport takes sediments from mineral sources and passes them through the hydrologic cycle to their ultimate resting place in the sea. Sediment transport is an important issue in water quality as well as in stream management for natural uses such as fisheries. One person told me that sediment was the "ultimate pollutant."

Sediment is produced in nature by weathering processes and is transported to streams by water and wind actions. Then sediment begins its trip to the ocean. Along the way, it forms sand waves, causes stream resistance and water depths and velocities to rise and fall, is associated with various contaminants, goes into storage, and is subjected to further abrasive forces. The science of sediment is called *sedimentology,* and is a subject that belongs to geology. Ancient stream deposits give clues to the location of oil and gas, and are of interest to petroleum geologists as well as for the search for clues to ancient geological history.

To study sediment mechanics, it is well to study the properties of sediment, transport mechanisms, river bed forms, and channel stability. Sediment begins with large sizes in mountain regions and is reduced in size progressively to beach sand, found in coastal areas, having mean diameters of around 0.2 mm or less. Normally, sediment is quartz with a specific gravity of about 2.65, but it can contain other minerals. Also, various clays, silts, and other soil particles are part of the sediment load.

Biology of water

In water quality the biological constituents that gain most attention are those that cause or indicate the possibility of disease, coliform bacteria, or organisms such as Giardia, Legionella, or Shistosomiasis snails. These various microbiological agents are in the categories of bacteria, viruses, protozoa, and helmiths (Krenkel and Novotney, 1980).

The primary natural biochemical influences on water are reflected in the carbon, phosphorous, and nitrogen cycles. The carbon cycle starts with the incorporation of carbon molecules from atmospheric carbon dioxide, CO_2, into plant life through the mechanism of photosynthesis, resulting in the creation of glucose and the movement of the carbon molecules into plant tissues.

The phosphorous cycle involves the soil–plant–water part of ecology, but not the atmosphere. Plants need inorganic phosphate, PO_4^{-3}, which in turn is bonded into organic phosphate and passes through the food chain. Inorganic phosphate is available from rocks and minerals. When an organism uses the phosphate, it may be recycled back to the soil or water as waste. Unless phosphate is deposited right at the point of consumption, it may get into the waters and move down the hydrologic cycle. The nitrogen cycle involves both a gas and a mineral phase. The atmosphere is 78 percent nitrogen, but plants cannot take nitrogen directly from the atmosphere; they must get it from ammonium, NH_4^+, or nitrate, NO_3^-. Fertilizers provide artificial sources of nitrogen to the soil–water system, and some bacteria associated with legumes will "fix" nitrogen, that is, convert atmospheric nitrogen to ammonium. The plants then convert the ammonium form to organic nitrogen, and it begins its trip through the food chain. Organisms release nitrogen as waste generally in the ammonium form, making it available for uptake by plants or release into the water, where it may be taken up by aquatic weeds. Some nitrogen may also be fixed in the atmosphere and fall as rain. This may increase the nitrogen available to a water body, and exacerbate an overnutrient problem.

Sources of contaminants

Sources of contaminants and changes in water quality can be characterized as point sources, nonpoint sources, and natural sources which may be point or nonpoint. Point sources are those where the discharge is from a single pipe or channel which can be identified. Nonpoint sources, sometimes called diffuse sources, are those where the contaminants are dispersed on the land and carried to the stream by the action of rainwater. Examples include any runoff resulting from a use of land, such as urban areas, rural lands, including cropland and forests, seepage from adjacent groundwater sources, highways, airports, and other land uses. Natural sources can include runoff of sediments, volcanic ash, salt discharge from springs, and any other source that arises from a natural process.

Monitoring and modeling of water quality

Monitoring programs are necessary to assess water quality and make sure that management programs are working. Mathematical models are necessary to understand systems and to provide information for management decision making. Monitoring and modeling are discussed in Chap. 14.

Ecological Systems

With today's emphasis on environmental issues, no water manager can afford to ignore ecological concepts. Ecosystems add a biological component to hydrologic systems, but until recently they were neglected in engineering education.

Ecology is the use of biological and other sciences to explain the relationships between living organisms and their environment. An ecosystem is a group of varied species of plants, animals, and microbial populations that interact in a common environment. "Ecosystems" is a broad term, and ecosystems come in different geographic scales.

Water managers deal with both types of ecosystems, terrestrial and aquatic. Although water management deals mostly with the aquatic environment, much decision making also involves the terrestrial features of transition zones between land and water, such as in watersheds, wetlands, or riparian zones (see Chap. 16). Aquatic ecosystems include streams, lakes, aquifers, lakes, estuaries, and oceans.

Key ecosystem concepts are the *food chain,* or the dependence of living things for food on other living things, and *ecological communities,* where there is natural competition for food and other habitat resources. *Habitat* is where the species lives, and its *niche* is its occupation, that is, where, when, and on what it feeds, where it builds its nest, reproduces, etc. *Competition* is a key factor in ecosystems, where one species competes with others for survival. Conditions within ecosystems can be favorable or adverse, depending on the stress of environmental and habitat conditions. Levels of stress can be used to identify zones of stress and tolerance limits.

All of these principles can be observed in today's water management issues. Bovee (1992) explained the environment of an organism as the habitat, the food supply, and the competitive environment of other species. An ecological system extends the concept to include the food chain and the habitat and competitors of organisms making up the food base. Ecological systems range from small to global: from population ecology, dealing with an individual species in a local place; to community ecology, dealing with associated species, such as fish; to ecosystem ecology, dealing with multiple species, such as fish, birds, and associated vegetation; to higher-level systems and global ecology.

Ecosystem sustainability is the key to achieving sustainable development. Woodmansee (1992), in defining sustainable development, started with the simple definition of the World Commission on Environment and Development: "to ensure that development meets the needs of the present without compromising the ability of future generations to meet their own needs." Woodmansee then attempted to be more specific about sustainability and suggested that sustainability of ecosystems is a function of six factors: physical/biological properties, climate/water, energy, economic viability, cultural viability, and organizational and political viability.

In the case studies, I will show how the principles of ecology suggest sustainable water management practices, including in-stream flow management (Chap. 16) and water quality (Chap. 14).

People in harmony with healthy natural systems

Healthy aquatic ecosystems require that people be "in harmony with healthy natural systems," a motto of Water Quality 2000 (see Chap. 14) and another statement of sustainable development.

Adaptation of ecosystems begins with an existing situation, as before human arrival in large numbers at some particular place. For example, where I live, along the Front Range of the Rockies, this situation is reflected in the settlement patterns prior to about 1840. At that time, mountain and foothill ecosystems were based on water management without human intervention. Plains ecosystems were quite arid, with only 10–15 in of precipitation per year. Native American populations had low impacts on the ecosystems and apparently lived in harmony with them. After dense settlement by European-Americans, the ecosystems were changed dramatically. Mountain streams were dammed up, hydrographs were altered, irrigation systems were developed, creating wetlands where there had been none before, and plains streams began to run all year instead of only during the spring runoff.

This adaption of ecosystems has been repeated in many locations. At any point in the development of water utilization patterns, the natural ecosystem will adapt to the new regime and communities of organisms will develop. Intervention then changes the balance point and new communities develop (U.S. EPA, 1991).

Aquatic ecology in streams, watersheds, wetlands, estuaries, and oceans

Two ecological accounting units come into play for aquatic and terrestrial ecology: the watershed and the stream reach (see Chap. 16).

The stream is a hydroecological environment that integrates or cumulates all aspects of land use and water resources management. In the stream environment the food chain begins with microorganisms. These feed on organic matter that is provided by runoff and deposition. The balance must be right for the microorganisms to be the right types and healthy for the fish and macroinvertebrates to prosper. The larger fish feed on the smaller fish, and riparian birds and animals feed on all of the fish.

Watersheds are divided into upland and bottomland zones. In upland zones the environment may be different due to changes in elevation, slope, soil type, and living organisms. For example, in the Rocky Mountains it is not uncommon for watersheds to extend up to 12,000 and even 14,000 ft, with steep slopes down to the 5000- to 6000-ft elevation. In flatter country, such as Florida, the maximum elevation is not more than 300 ft and the lower elevations reach sea level. In upland zones, the birds and animals will be dependent on the hydrologic cycle, on grazing, and on each other, but they may be of different types than in bottomland zones, where they will be dependent on the riparian aquatic ecosystem. Wildlife of all kinds are dependent on the stream and aquatic ecosystem.

Wetlands serve as protective and nursery habitat for diverse species of fish, birds, and other wildlife, protect groundwater supplies, purify surface water by filtration and natural processes, control erosion, provide storage and buffering for flood control, and provide sites for recreation, education, scientific studies, and scenic viewing. They are key natural elements for ecological systems.

Estuarine and marine environments also provide important ecological systems, and are discussed in Chap. 22.

Questions

1. Make your own sketch of the hydrologic cycle. Can you explain it to others?

2. Sketch the three principal nutrient cycles. Can you explain how they work?

3. How is sediment produced, and what happens to it as it travels along the rivers? Can you explain the linkages between sediment and water pollution?

4. For a stream reach, formulate an equation for a water balance and explain the terms you use.

5. Sketch a bell curve for a water supply source. Mark on the graph where the mean and standard deviation should be placed.

6. If a stream has an annual average of 950 cfs available for irrigation water, and the crop needs a total annual application of 750 mm, how many hectares can be irrigated? Assume that you can store and release water so that all of the annual runoff can be applied to irrigation. (Some conversion factors: 1 m³/sec = 35.3 cfs; 1 acre = 0.405 ha; 1 ha = 10^4 × m²; 1 in = 25.3 mm.)

7. Explain the physical, chemical, and biological aspects of water pollution.

8. What are the linkages between aquatic and terrestrial ecology?

9. Use the following questions to practice some of the units and conversions in water resources planning.

    a. City population = 250,000; average consumption = 175 gpcd; what is the annual consumption in: AF, mgd, MCM?

    b. A 5-mg storage tanks serves a city. What is the volume in AF, MCM?

    c. You irrigate 10,000 acres. What is the area in hectares? If you apply 700 mm of water in a year, what is the consumption in MCM, AF?

    d. Annual average Q = 2000 cfs. Head = 200 ft. Use metric units and the formula $P = QH$ to find the kilowatt-hours (kWh) per year.

    e. Find the AF/month and annual total for the following monthly flows in cfs: 112, 187, 375, 500, 650, 575, 387, 305, 261, 185, 150, 112.

    f. Annual average well pumping is 500 gpm. How many cfs, AF/year?

    g. Q = 1000 cfs. What is this in CMS?

    h. A watershed is 100 SM in tributary area. Effective storm runoff is 20 mm. What is the flood runoff in MCM, AF?

    i. Lake evaporation is 70 percent of the annual pan evaporation of 750 mm. What is annual lake evaporation in AF?

    j. The following are average annual discharges of rivers in cfs; convert to km³/year. Colorado, 23,000; Alabama, 31,600; Mississippi, 620,000; Nile, 420,000; Amazon, 7,200,000.

References

Bovee, Ken, Problems in River Management, Concepts in Ecology, unpublished, March 11, 1992.

Committee on Opportunities in Hydrologic Sciences, Water Science and Technology Board, *Opportunities in the Hydrologic Sciences,* National Academy Press, Washington, DC, 1991.

Dooge, James C. I., *Linear Theory of Hydrologic Systems,* Agricultural Research Service Technical Bulletin 1468, U.S. Department of Agriculture, Washington, D.C., 1973.

Frederick, Kenneth D., America's Water Supply: Status and Prospects for the Future, *Consequences,* Saginaw Valley State University, Vol. 1, No. 1, 1995.

Gleick, Peter H., ed., *Water in Crisis, A Guide to the World's Fresh Water Resources,* Oxford University Press, New York, 1993.

Hammer, Mark J., and Kenneth A. MacKichon, *Hydrology and Quality of Water Resources,* John Wiley, New York, 1981.

Heath, Ralph C., *Basic Ground-Water Hydrology,* U.S. Geological Survey Water Supply Paper 2220, U.S. GPO, Washington, DC, 1989.

Krenkel, Peter A., and Vladmir Novotney, *Water Quality Management,* Academic Press, New York, 1980.

Leopold, Luna B., Probability Analysis Applied to a Water Supply Problem, U.S. Geol. Survey Circular, USGS, Washington, DC, 1959.

McCuen, Richard H., *Hydrologic Analysis and Design,* Prentice-Hall, Englewood Cliffs, NJ, 1989.

Nace, R. L., Water of the World, *Natural History,* Vol. 73, No. 1, January 1964.

Petterssen, Sverre, Meteorology, in Ven T. Chow, ed., *Applied Hydrology,* McGraw-Hill, New York, 1962.

Simons, Daryl B., and Fuat Senturk, *Sediment Transport Technology: Water and Sediment Dynamics,* Water Resources Publications, Littleton, CO, 1992.

U.S. EPA, First International Conference on Climate Change and Water Resources Management, Albuquerque, NM, November 1991.

U.S. Water Resources Council, *National Water Assessment,* Washington, DC, 1968, 1978.

Woodmansee, Robert G., Ecosystem Sustainability, unpublished working paper, Colorado State University, 1992.

3

Water Infrastructure
and Systems

As is made clear throughout this book, nonstructural water management systems have less impact on natural systems than structural systems. Nevertheless, water resources management requires a heavy investment in man-made structures and infrastructure systems that capture, process, transport, and store water. These structures and systems provide the basis for economic and social uses of water, and many of them enhance water quality.

This chapter describes the structural components and systems, explains how they work, and identifies issues related to them. The goal of the chapter is to describe the types and functions of the components so that their applications in water systems can be analyzed. The chapter does not present detailed information for analysis and design of the structures and components. That information is readily available in engineering texts and handbooks. In a companion chapter (Chap. 12), aspects of planning, design, and project management are presented.

In this chapter, water resources structures and components are presented in the context of the overall system, then the individual structures and system components are described, including: conveyance systems, including channels and canals, pipes, and bridges; diversion structures; dams; reservoirs; locks; treatment plants for water supply and wastewater management; pumping stations; hydroelectric plants; spillways, valves, and gates; and aquifers and wells. Table 3.1 places these structures and components into an organizing system that classifies them by purpose and by function.

TABLE 3.1   Typical Structures for Water Resources Management

Purpose	Conveyance	Storage	Treatment*	Pump or generate	Flow control
Water supply	Supply pipes	Tanks	WTPs	System pumps	Valves
Wastewater	Sewer pipes	Equalization	WWTPs	Sludge pumps	Gates
Stormwater	Drainage pipes	Detention	SWTPs	Flood pumps	Inlets
Hydropower	Penstock	Reservoir	—	Turbine	Gates
Navigation	Locks	Lakes	—	—	Valves
Environment	Rivers	Natural storage	Wetlands	—	—

*WTP, water treatment plant; WWTP, wastewater treatment plant; SWTP, stormwater treatment plant.

Systems of Water Resources Structures and Facilities

Water resources management must be considered from the systems viewpoint because of the many interdependencies between water uses. A water resources system is, by definition, a combination of water control facilities and/or environmental elements, and it requires system-wide decision making and control that considers the integrated viewpoint. This is easier said than done, and a comprehensive or "holistic" view of managing water resources systems is still beyond our grasp, but with modern information technologies, we are getting closer. Chapter 5 describes systems analysis techniques that can be applied for management purposes.

Figure 3.1 presents a view of a complex, multipurpose water resources system. It includes a watershed as a source of water and numerous features for water utilization. Let us consider the parts in terms of function from the top of the basin to the bottom.

First, the watershed is a prominent feature of the system, and you can see two catchments, the larger one on the left with snowmelt feeding the reservoir, and the smaller one on the right with a small tributary stream. Protecting these watersheds through *watershed management* is very important; see Chap. 16.

Several types of dams and reservoirs are shown. Most obvious is the large, multiple-purpose reservoir, with an arch dam and a hydroelectric plant connected to electric transmission lines. Just below is a diversion dam, which enables the high line canal to take irrigation water from the stream. At the upper right is a beaver dam and lower down is a regulating basin. On the main stem of the river is a reregulating reservoir with a lockage system for navigation.

Figure 3.1 A water resources system. (*Source: President's Water Resources Policy Commission, 1950.*)

Several types of conveyance systems are shown. There are several canals, and levees are shown to protect the city from flooding. From the large dam is shown a set of outlet works that discharge to the municipal water plant. Several other pipeline structures are shown as part of the municipal system and the sprinkler irrigation system shown on the left side of the diagram.

Several treatment plants are in view. These include the municipal water treatment plant, the city and industrial waste treatment plant, and farther downstream a community treatment plant.

The lockage system enables navigation as far upstream as the diversion dam, presumably to enable the delivery of commodities to the city or perhaps to haul away agricultural products from the region.

Pumping stations are in evidence below the city waste treatment plant (to supply the equalizing reservoir) and at the farm at the bottom of the diagram.

To control the flows, there will be numerous valves, gates, and spillways, but these are mostly too small to be seen. Spillways can be seen at the dams.

The city will have water supply, wastewater, and stormwater systems, mostly out of view with underground facilities.

Most of the system shown is of surface water, but a well is shown on the right side serving a farming operation. If the watershed is near a coastal area, there may be barrier systems to prevent saltwater intrusion.

Figure 3.2 shows a better view of an irrigated system. From it you can see supplies from surface water and groundwater as well as delivery and return water systems. Seepage from the canal systems illustrates how irrigation systems recharge groundwater tables.

From Fig. 3.1 the reader can appreciate the disaggregation of a complex water resources system into subsystems and how the whole system is made up of the sum of subsystems, interacting together. Some of the subsystems are urban water supply systems, wastewater and water quality management systems, stormwater and flood control systems, irrigation and drainage systems, reservoir and navigation systems, and the natural water flow system superimposed on the man-made systems.

Another view of a system is shown on Fig. 3.3 (Federal Energy Regulatory Commission, 1994; Shen et al., 1985). The view is of a portion of the Platte River which is the subject of a case study in Chap. 19. It illustrates hydraulic structures as components of a system that includes dams and reservoirs, powerhouses, canals, diversion structures, natural riverine systems, and surface–groundwater systems.

Components of Water Resources Systems

The watershed as the organizing framework

One of the most important principles of water resources management is the use of the watershed or river basin as the framework for planning and organizing systems. For this reason it is mentioned at the beginning of the discussion of components of systems. While an individual watershed is not a hydraulic structure, it can be a component of a larger watershed system. Chapter 16 discusses watersheds in detail.

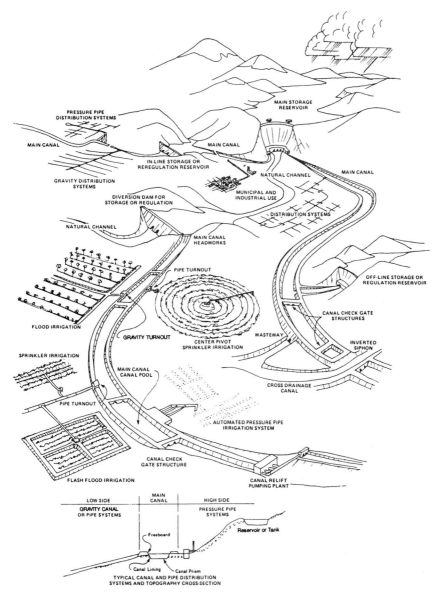

Figure 3.2 Schematic of an irrigated river basin. (*Source: Buyalski, 1991.*)

Dams and reservoirs

Dams are a special class of hydraulic structure, basically a barrier to the flow to enable a reservoir to be created to store water. Dams have a number of special components, including service and emergency spillways, outlet structures, and drains.

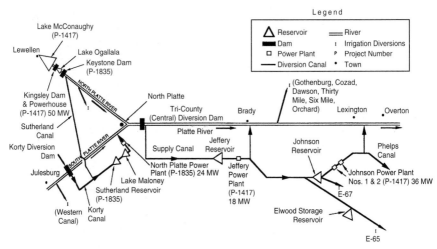

Figure 3.3 Schematic of Platte River. (*Source: Shen et al., 1985.*)

A reservoir is a lake where water is stored, either natural or dammed artificially. Reservoirs require special management attention and are discussed in Chap. 13.

As discussed in Chap. 13, dams and reservoirs are controversial. Nevertheless, they have traditionally been the main tools for river basin management.

Dams and reservoirs figure prominently throughout this book because they are so central to water resources management. Figure 3.4 shows one example of a dam and reservoir: the Jordan Dam and Reservoir system, located on the Coosa River near Wetumpka, Alabama. The dam is part of the Alabama Power Company's facilities; the figure shows the powerhouse and the gated spillways. The dam was built in 1928, early in the private-power era in the United States.

Figure 3.5 shows the massive Guri Dam, located on the Caroni River in Venezuela, one of the largest hydroelectric projects in the world. The facility is owned by CVG-Electrification del Caroni C.A. (EDELCA), and the dam was planned by the Harza Engineering Company, with a principal part of the work being under the direction of Victor A. Koelzer, who is mentioned in this book's acknowledgments. At its final stage, with water elevation of 270 m, Guri Dam will have a useful storage of 86,000 MCM (69.7 MAF) of water and a maximum generating capacity of 10,060 MW (Palacios and Chen, 1979).

Both Jordan Dam and Guri Dam are located in warm-weather climates; Fig. 3.6 shows Bradley Dam in Alaska. Figure 2.7 showed snowpack that feeds the watershed for Bradley Dam.

Figure 3.4 View of Jordan Dam. (*Courtesy Alabama Power Company.*)

Figure 3.5 Guri Dam. (*Source: Victor A. Koelzer.*)

Figure 3.6 Bradley Dam in Alaska. (*Courtesy Bechtel Corporation.*)

Diversion structures

Diversion structures are devices interposed in a stream to change its direction or flow patterns. They may include intakes, boat chutes, river training structures, or fish ladders.

Diverting water from streams was an early feature of irrigation in the western United States. A surface-water diversion of this kind can be seen in Fig. 3.1, which illustrates a diversion dam and a high-line canal. Sometimes farmers went long distances upstream on rivers to divert water and then carried it by flat canals for many miles downstream, perhaps for distances of up to 50 miles in some cases. Denver's High Line Canal is an example of this approach. It brings water over 50 miles from the South Platte River to the city.

The fish ladder is a special kind of diversion structure, meant to provide migrating fish species with a means of swimming upstream. A fish ladder is shown on the reregulating reservoir at the bottom of Fig. 3.1.

Conveyance systems

The structures and components that are associated with conveyance systems include natural open channels and canals, pipelines, pipe networks and sewers, bridges, and levees.

Open channels are either natural or man-made. A river is an example of a natural open channel. Left alone, a river develops a complex pattern of flow and sustains a complex ecological system. Actually, a river is much more than just a channel. It contains the main channel, the tributaries, the floodplain, the full riparian ecological zone (the zone where the ecology depends on the river), and the alluvium that conveys water under the stream. Rivers require in-stream flows of certain magnitudes and quality to sustain the life they support.

A canal or lined drainage ditch is an example of a man-made channel. Man-made channels can convey water from place to place efficiently, but they interfere with natural rivers and are sometimes regarded as unwanted. A lined irrigation or flood control canal might be regarded by some as an eyesore, but if water conservation is the goal, it might be considered an asset. Figure 3.7 shows a drop inlet, a

Figure 3.7 View of inlet structure, Colorado Big Thompson Project.

Figure 3.8 California Aqueduct, East Branch. (*Source: California Department of Water Resources.*)

transition structure between an open channel and a tunnel. This structure is part of the Colorado Big Thompson Project, discussed in Chap. 12. Figure 3.8 shows a lined canal that is part of the California Aqueduct system, also described in Chap. 12.

Analyzing open channels is a frequent problem faced by engineers. Designing a canal to convey water without silting up or overtopping is one example. Another is to analyze how high a flood will rise in a river channel (Chap. 11). Still another is to compute the hydraulic regimes in a channel to determine if fish habitat is adequate (Chap. 16).

Pipes, or closed conduits, can be classified as tunnels, transmission pipelines, pressure pipe networks, or sewer networks. A tunnel may operate under pressure or as an open channel. A transmission line usually involves a single pipe to convey water from one place to another, sometimes great distances. Pressure pipe networks, as in urban water distribution systems with grid patterns, move water

Figure 3.9 Pipeline view. (*Courtesy CertainTeed Corp.*)

from place to place in a network. Sewer networks funnel small collection sewers into a large collector sewer at the lower end and are followed by interceptor and outfall sewers. Figure 3.9 illustrates conceptually the installation process for a medium-sized pipeline.

Pipelines often form part of extensive networks and systems. Figure 3.10 shows a water supply system operated by a regional water supply organization in Germany called the Zweckverband Bodensee Wasserversorgung (Bodensee Regional Water Supply Cooperative). The main source of the water is the Bodensee (Lake Constance), and the treated water network extends northward nearly to the Main River. Figure 3.11 illustrates the water treatment setup.

Figure 3.10 Regional water supply system for Bodensee region. (*Courtesy Zweckverband Bodensee Wasserversorgung.*)

A bridge (Fig. 3.12) is part of a conveyance system in the sense that it provides a method to separate a streambed from a road, a rail line, or some other structure crossing a stream. A bridge affects the flow in the stream in the same manner as a pipe conduit might; that is, it constricts the flow and causes resistance and backwater. Bridges are expensive structures requiring considerable care and maintenance.

A culvert has a function somewhat like a bridge in that it enables a separation of grades between a small stream and a roadway or other

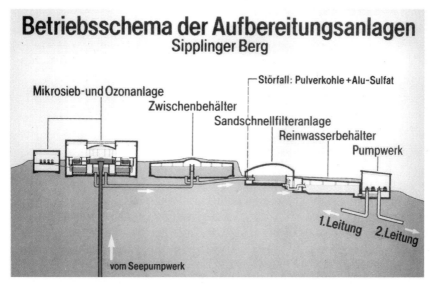

Betriebsschema der Aufbereitungsanlagen
Sipplinger Berg

Mikrosieb-und Ozonanlage

┌ Störfall: Pulverkohle + Alu-Sulfat

Zwischenbehälter

Sandschnellfilteranlage

Reinwasserbehälter

Pumpwerk

1.Leitung   2.Leitung

vom Seepumpwerk

Figure 3.11 Treated water setup for Bodensee Regional Water Supply Cooperative. (*Courtesy Zweckverband Bodensee Wasserversorgung.*)

Figure 3.12 Bridge over South Platte River. (*Source: David W. Hendricks.*)

embankment above. Culverts make up a large portion of the expenditure on urban drainage and roads.

Levees are part of conveyance systems in that they form the banks of channels to protect land areas from flooding. Levee failures happened during the Great Mississippi River Flood of 1993 (Chap. 11).

Locks

A lock is a device to raise or lower boats and ships up or down a river. Petersen (1986) provides details on lock planning design. A lock can be seen at the bottom of Fig. 3.1.

Basically, a lock works by providing a small lake for a vessel to rise up or be lowered, depending on its direction of travel. If a vessel is moving upstream, it enters the lock, which is then closed. Water is allowed to flow into the lock from upstream to bring the water level to the upper pool level. Then the vessel departs. Downstream navigation is the reverse. The vessel enters at the upper pool level, then water is discharged until the pool is at the downstream level. In either case, water is released, providing a way to flush fish and water downstream.

Lock operation can discharge large quantities of water. According to Linsley et al. (1992), the highest lift in the United States is the John Day Lock on the Columbia River, at 113 ft. Most locks on the Mississippi River and Tennessee River systems are 110 ft wide by 600 ft long, although some are larger and some smaller. Suppose a lock has a lift of 50 ft and is 100 ft wide by 500 ft long. If the lock fills and empties once per hour, the discharge required to operate it averages out at 694 cfs.

Hydroelectric plants

Hydroelectric plants are facilities to generate electric energy from water discharge. The basic components of a hydro facility are shown in Fig. 3.13. Water from the reservoir flows through the penstock to the powerhouse, which contains a turbine-driven generator. Water flowing through the turbine at high pressure turns the blades and generates power; then the water is released through the draft tube to the tailwater. Hydro plants are useful for meeting demands for peak power. When used in conjunction with conventional steam or natural gas plants which operate at a more constant load, the hydro plant can be turned on to add power quickly to a system.

In water resources systems analysis and planning, a key issue in hydroelectricity is the quantity of power generated. A basic relationship between discharge and head determines it:

$$P = \frac{\eta\gamma QH}{550}$$

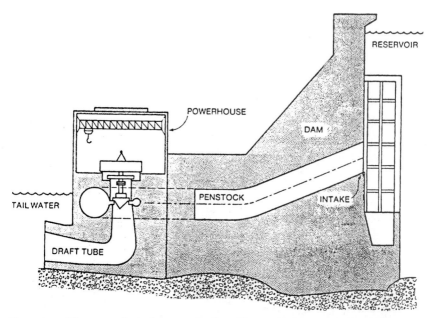

Figure 3.13 Elements of a hydropower facility. (*Source: U.S. Army Corps of Engineers, EM 1110-2-1701, 1985.*)

where η = efficiency of the turbine
γ = specific weight of water (normally 62.4 lb/ft^3)
Q = discharge in ft^3/sec
H = head of water in ft

This yields power in horsepower. To convert to kilowatts, multiply by 0.746. Note that as efficiency decreases, the power yield from a given combination of discharge and head also decreases.

With the power in kilowatts, one can determine the total power potential by considering the head and discharge available, determining the duration of the generation, then computing kilowatt-hours, the standard unit of electrical energy.

There are four basic types of hydro plants: run-of-river, pondage, storage, and reregulating. They differ by amount of storage and function. The run-of-river plant lacks any useful storage and generates power depending on the streamflow available. The pondage system has enough storage to meet the daily peak variation, and its reservoir can fluctuate somewhat. The storage facility has enough storage to carry water from season to season. The reregulating reservoir provides enough storage to smooth out the large fluctuations that arise from peak power generation.

Figure 3.14 Bad Creek pumped storage facility.
(*Source: Duke Power Co.*)

Pumped storage is a special type of storage facility that allows the storage of electrical energy. In it water is pumped to a reservoir at a higher elevation, to be released when the energy is needed. Figure 3.14 shows Duke Power's Bad Creek pumped storage plant, located in Oconee County, South Carolina. The facility moves water from the lower reservoir at an elevation of 1110 ft to the upper reservoir at an elevation of 2310 ft, and has a capacity of 1,065,000 kW.

In the early years of industrialization in the United States, water power was a key advantage for small towns and plant sites. Today, small hydroelectric plants are still used to generate energy, and can be important for villages and small towns that lack access to larger sources. Figure 3.15 shows a small hydro plant of the Tennessee Valley Authority, the Nolichucky Dam, which was taken offline due to severe sedimentation caused by abandoned mine runoff.

Figure 3.16 shows the construction of hydro facilities inside a dam, in this case the Bradley Dam in Alaska.

Figure 3.15 TVA's Nolichucky Dam. (*Courtesy Tennessee Valley Authority.*)

Pumping stations

Pumping stations impart energy to water and raise it in elevation or pressure. There are several different types of pumps, mainly built around the centrifugal or turbine pump format, and they have numerous applications. In general, centrifugal pumps are useful to add pressure head to systems, and turbine pumps are useful to pump large quantities of water at relatively lower heads. The relationship between discharge and head for a pump is similar to that of a turbine,

$$P = \frac{\gamma QH}{550\eta}$$

where the symbols are the same, but in this case the efficiency term changes places to denote that the power required to pump given quantities of discharge and head increases as the efficiency decreases.

Figure 3.17 shows a large pumping station, the Farr Pump Plant, part of the Colorado Big Thompson Project, described in Chap. 12.

Valves, gates, and spillways

Valves, gates, and spillways are control devices for conveyance systems or dams.

Figure 3.16 Construction of Bradley Dam hydropower units. (*Courtesy Bechtel Corporation.*)

Valves are found in all pipelines and pipe networks, and are normally used when pressure must be controlled or when flow must be shut off completely. The most common valve is, of course, the spigot in a household. Its larger cousin, the gate valve, is the most common control device in urban water supply systems. Several other types of valves are useful for different applications such as pressure reduction, quick shutoff, and close control of discharge. Examples of such valves are the needle valve and the butterfly valve.

Gates are used in both pipes and channels, but normally are not used when the pressure is high. A common type of gate is the slide gate, which is a flat plate that slides over the opening in a pipe. Larger versions of slide gates may be seen at the tops of dams.

Spillways are used as emergency overflow devices to protect dams.

Figure 3.17 Farr Pump Plant, Colorado Big Thompson Project. (*Source: NCWCD.*)

There are a number of types, normally classified as service spillways or emergency spillways. Water flowing over a spillway normally flows into some type of energy-dissipation device to avoid erosion downstream. Figures 3.4 and 3.5 illustrate spillways.

Measurement devices

Different types of measurement devices are found in open channels and closed conduits. They are essential for metering water and for establishing systems of charges. In a small pipe, such as a household service line, an inexpensive propeller meter will probably be found. There are millions of these in service in the United States, and they form the basis for meter reading and monthly water charges. Chapter 17 describes a controversy that occurred over these meters.

In larger pipelines different types of metering devices are found. Orifice meters and magnetic flow meters are common.

In open channels flow measurement is more difficult. The Parshall flume was developed at Colorado State University about 1915 and has served around the world to meter water in irrigation canals. Different types of weirs serve similar purposes. Figure 3.18 shows the

Figure 3.18 Ralph Parshall testing his flume. (*Courtesy Photographic Archives of Colorado State University.*)

developer of the Parshall Flume, Ralph Parshall, in an early field test situation.

One difficult problem is measurement of flow in a sewer that flows as an open channel. With water containing debris and with fluctuating depths, it is not easy to install a low-cost, reliable metering system.

Integrated urban water systems

The concept of the "integrated urban water system" was promoted by McPherson (1970). Figure 3.19 shows the connections between the urban water supply, wastewater, and stormwater systems. In a practical sense, the connections occur to the extent that water managers make them happen. For example, stormwater can be used to recharge

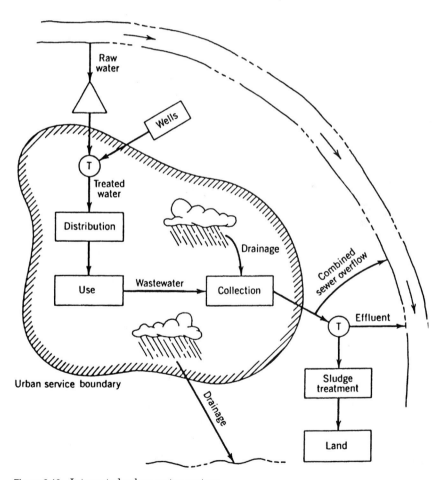

Figure 3.19 Integrated urban water system.

a water supply source, and the organizational and financial aspects of utilities can be combined. For the most part, however, water supply, wastewater, and stormwater are separate systems.

Urban water supply systems

Urban water supply systems provide for acquisition, treatment, and delivery to domestic, commercial, and industrial customers. The four basic parts as shown in Fig. 3.20 are the source of supply, treatment plant, distribution system, and point of use.

The source of supply may be either surface or ground water. Surface sources are either streams or lakes. If streams are used, then a diversion is necessary and in-stream flow rules apply.

Treatment plants may range from simple setups, involving little more than chlorination, to elaborate and expensive processing systems. Most large cities have elaborate plants, but an exception is New York City, which for many years furnished unfiltered surface water from the Catskill Mountains. In rural areas and in many small and medium-sized cities, groundwater is often supplied with just chlorination.

Distribution systems represent much of the "hidden capital" of water supply systems, involving around two-thirds of the system investment and a major share of the maintenance problems. They involve pipes, valves, pumps, storage tanks, and associated structures.

The point of use represents the privately owned part of the water supply system, but one which affects drinking water quality.

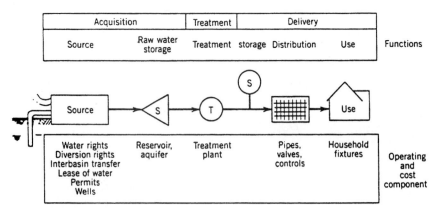

Figure 3.20 Urban water supply system.

Urban wastewater systems

The wastewater system begins where the urban water supply system leaves off. The collection system takes the residuals of domestic, industrial, commercial, and public uses. Figure 3.21 illustrates the main components of a wastewater management system.

A number of different terms are used to describe wastewater systems. "Wastewater" and "sewage" are synonymous terms. "Sewerage" refers to the pipes, pumping stations, and facilities that handle the wastewater. "Sanitary sewage" is regular wastewater, usually municipal. This is sometimes called "foul sewage" in England. "Combined sewers" transport both sanitary sewage and storm drainage. This translates to "mixed wastewater" in Germany. "Separate sewers" transport only sanitary sewage except for some infiltration and inflow. A "lift station" is a pumping station to overcome the lack of a gravity route. A "force main" is a sewer that is under pressure, and may be called a "pressure sewer." A "main" or "trunk sewer" is one of the principal sewers of a system, one that collects the lateral sewers, and an "interceptor sewer" is one that intercepts main sewers and carries the wastewater to a treatment plant. An "outfall sewer" conveys wastewater to a point of disposal. Treatment plants are classified as primary, secondary, or tertiary, depending on the degree of treatment, but other terms include "advanced wastewater treatment" (AWT).

The treatment plant is a twentieth-century addition to the wastewater management system. Sludge handling is an important part of the program. Usually, no one wants sludge, but agricultural uses for

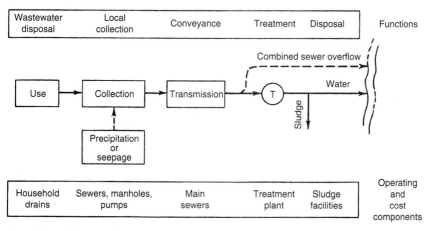

Figure 3.21 Urban wastewater management system.

processed sludge are increasing. In any case, sludge is expensive and complex to handle, and it may be a hazard to the environment.

Treatment plants

The two basic types of treatment plants for water have already been mentioned: one to process raw water for domestic or industrial use (water treatment plant), and another to process wastewater before it is discharged into a stream (wastewater treatment plant). Chapter 14 presents diagrams that show typical layouts of both types of treatment plants.

These treatment plants are actually complex processing systems that become more complex and expensive with each new regulation. The water treatment plant must comply with the Safe Drinking Water Act, and the wastewater plant with the Clean Water Act (see Chaps. 8 and 14). Depending on the nature of the raw water source, the water treatment plant will need different components, and the components of the wastewater plant depend on the strength and composition of the raw wastewater and on the ability of the receiving water to assimilate waste.

Urban stormwater systems

The configuration of a stormwater system is shown in Fig. 3.22. It can be thought of as two separate systems: a "minor" system for storm drainage and a "major" system to handle emergency flows.

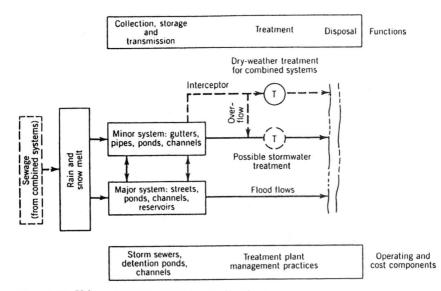

Figure 3.22 Urban stormwater management system.

The minor drainage system is also called the "initial" system or the "convenience" system. Minor systems include gutters, small ditches, culverts and storm drains, detention ponds, and small channels.

The major system involves streets and urban streams and should include the floodway and the flood fringe areas. It is often not planned, it just happens. Areas of frequent rain are normally ready to pass flood flows due to experience with previous flood flows. In some areas, however, floods may hit when there is activity in the floodplain, resulting in heavy damage, loss of life and property, and social trauma. (See Chap. 11.) The major system is like the emergency spillway on a dam. Major systems must take into account both physical facilities and floodplain management. By leaving floodplains free of buildings and structures, they remain available as a conveyance channel for floodwaters.

A combined sewer system is a hybrid between a stormwater system and a sanitary sewer system, with the addition of regulator and treatment facilities.

The water quality subsystem is superimposed on top of the minor and major systems. Water quality problems arise from the washoff of surface pollutants, from combined sewer overflows, or from the erosion of pollutants from the inside of sewers. (See Chap. 14.)

Wells

The basic control element for groundwater systems is the well (see Chap. 18 for more about groundwater). There are, of course, many different types and purposes of wells. The well might be considered as the point of diversion from the stream of water flowing through the aquifer. Sometimes a well is also used as a recharge point, but the withdrawal well is by far more common.

Figure 2.8 shows wells in the two basic types of aquifers, confined and unconfined. Two wells are shown in the confined aquifer: one that flows, an artesian well; and one that does not flow and requires pumping. The unconfined aquifer has one well, which requires pumping from the water table.

A recharge area is shown where water may infiltrate into the aquifer. Where it infiltrates, it initially becomes an unconfined aquifer, then it becomes confined due to the presence of the confining layer.

Figure 2.9 shows the basic elements of the well: the casing to support the walls of the drill hole, and the screen at the part of the aquifer where water is withdrawn. In addition, for nonflowing wells a pump and motor are required to provide the energy to lift the water from the well.

Questions

1. Give an equation to relate hydropower to head and discharge of a stream. How does this equation show the possibilities of low-head hydro?

2. Bridges are normally thought of as part of highway features rather than as conveyance structures for waterways. How are bridges vulnerable to water flow and hydraulic forces, and what are their effects on flood backwater?

3. Levees are key features of flood protection systems in low-lying areas. Who should be responsible for maintaining them?

4. Sketch a lock and explain how it works.

5. Sketch a pumped storage system and explain how it works.

6. Automation of water resources systems requires accurate meters for control decisions. If wastewater flow is difficult to measure, how can automatic control systems be implemented?

7. The chapter describes the "integrated urban water system." What are the organizational barriers to integrating systems in this manner? See also Chap. 21.

8. Of the parts of the urban water distribution system, which is most capital-intensive? What are its maintenance requirements?

9. Do you agree that permits should be required for stormwater discharges? Why or why not? See also Chaps. 6 and 14.

10. Under what circumstances might discharge of wastewater be used to recharge aquifers? What, in your opinion, should be the restrictions on such disposal and recharge?

References

Buyalski, C. P., et al., *Canal Systems Automation Manual,* U.S. Bureau of Reclamation, Denver, 1991.

Federal Energy Regulatory Commission, Kingsley Dam and North Platte/Keystone Diversion Dam Projects, Nebraska, Revised Draft Environmental Impact Statement, April 1994.

Linsley, Ray K., Joseph B. Franzini, David L. Freyberg, and George Tchobanoglous, *Water-Resources Engineering,* McGraw-Hill, New York, 1992.

McPherson, M. B., *Prospects for Metropolitan Water Management,* ASCE Urban Water Resources Research Program, ASCE, New York, 1970.

Palacios, Pedro, and Henry H. Chen, Planning, Symposium on the Guri Hydroelectric Complex, Proceedings of the American Power Conference, 1979.

Petersen, Margaret, *River Engineering,* Prentice-Hall, Englewood Cliffs, NJ, 1986.

President's Water Resources Policy Commission, *A Water Policy for the American People,* 1950.

Shen, Hsieh Wen, Kim Loi Hiew, and E. Loubser, The Potential Flow Release Rules for Kingsley Dam in Meeting Crane Habitat Requirements—Platte River, Nebraska, Colorado Water Resources Research Institute, Ft. Collins, 1985.

U.S. Army Corps of Engineers, *Hydropower,* EM 1110-2-1701, Washington, DC, December 31, 1985.

4

Planning and Decision-Making Processes

Introduction

After listening to problems of water resources managers, one industry leader said, "in the past we focused on *projects,* but now we focus on *process.*" This chapter is about this process—*a planning and decision-making process.* It is the set of processes for planning and gaining approvals for water projects and management actions. The process involves everything from planning and implementing small water system improvements to resolving complex, interstate disputes with high political content. It includes both normal and straightforward decision-making processes and situations with complex political and seemingly irrational outcomes. Experience is the best teacher, but hopefully the principles in this chapter, along with the case studies in Part 2 of the book, will enlighten water managers and engineers about the complexities of the process.

Capital investments

Table 3.1 outlines a list of typical capital investments that are required for water management systems. Planning of such capital improvements requires engineering work and securing approvals from permit, political, and financial authorities. Often capital projects are undertaken by single agencies with full authority to implement them, if permits and financing can be secured. Increasingly, however, joint action is favored, and coordinated approaches become part of the planning process. Sometimes special agencies are formed just to undertake projects. While the examples in the text are mostly from the United States, projects in other countries experience the same types of processes, although with different players. Due to the nature of the processes, political systems are key factors. When international development banks are involved, they have great influence.

Management actions

Another set of planning and decision-making processes involves nonstructural management actions such as regulation, coordination, allocation, reallocation, pricing, financial planning, land use control, political action, regional cooperation, agreements, public education, and emergency response (Table 4.1). For these, permits and financing may or may not be required, but the key feature will usually be getting the actors together and coordinating win–win strategies. Due to this heavy emphasis on coordination, political systems take on more significance, and differences between countries become more evident.

Planning and Decision-Making Scenarios

Planners use many "jargon" words, such as strategic plans, long-range plans, action plans, standing plans, policy plans, master plans, contingency plans, sector plans, budget plans, financial plans, area-wide plans, engineering plans, implementation plans, and facility plans (Grigg, 1985). One reason for so many terms is that planning is a creative activity, and it involves innovative people who think up these terms. In actuality, the terms represent mixtures of plans for capital investment and management actions, short range and long range.

TABLE 4.1   Management Actions and Scenarios

Action	Requirements
Water quality management	Combines regulation and coordination
Surface water allocation	Combines allocation law and coordination
Drinking water regulation	Health issue with regulatory control
Groundwater regulation	Complex regulation and land use issue
Reallocation of storage	Legal issue requiring coordination
Floodplain management	Land use issue with nonstructural action
Financial plans and actions	Involves capital investment and pricing
Regional conflicts over environment	Requires coordination and political action
Regional investments and projects	Opportunity for joint savings and economies
Regional water use control	Requires coordination and agreements
Drought contingency actions	Have strong land use and political links
Fish and wildlife enhancement	Public interest issue requiring joint action
Watershed management	Land use issue

Basically, planning is classified by who is in charge, what the coverage is, and what stage the planning is in. For example, the U.S. National Water Commission's (1972) report on water resources planning classified scenarios according to jurisdiction, scope, and stage of planning. The jurisdictions included federal, interstate regional, state, intrastate regional, and local; and I would add organizational, and include levels within organizations. That makes eight different jurisdictional levels. Scopes included multisectorial planning, sectorial planning, functional planning, and stages included policy planning, framework planning, general appraisal planning, and implementation planning. The stage of planning could also be described by the time dimension (policy, strategic, near term or long term).

These scenarios of planning are applied to both capital investments in facilities (Table 3.1) and management actions other than capital investment (Table 4.1).

Evolution of Water Resources Planning in the United States

The search for workable processes for planning has been difficult (Holmes, 1979). Water affects most sectors of the economy and the environment, and the goal has been to make planning "comprehensive." As it became clear that multiple purposes and players were involved, the goal became more ambitious: comprehensive, coordinated, joint planning" (CCJP).

In earlier years, the term "water resources planning" mostly meant planning for facilities that met economic goals such as hydropower and/or irrigation. Goals were focused on single purposes—power, navigation, irrigation, or flood control—and there was more push for economic development than environmental needs. As the nation developed, it became clear that water involves many industries, geographic areas, and public-interest viewpoints, and the possibilities of multiple-purpose development became apparent. The Flood Control Act of 1917 called for "a comprehensive study of the watershed," including the study of power possibilities, although the Corps of Engineers, the nation's main flood control agency, was resistant to the concept of multiple-purpose planning (Holmes, 1972).

The National Industrial Recovery Act of 1933 called for a "comprehensive" program of public works to consider the full spectrum of water resources uses, which included, according to Holmes (1972), control, utilization, and purification of waters; prevention of soil and coastal erosion; development of water power; transmission of electric energy; river and harbor improvements; flood control; etc.

Disputes grew after World War II, and in the 1950s after national water policy studies, a Senate Select Committee on Water Resources was appointed. In 1961, the committee called for the federal government, in cooperation with the states, to prepare comprehensive plans for the development and management of major river basins. The Select Committee led to the Water Resources Planning Act, first passed in 1962, which served to institutionalize the term "comprehensive planning." In its statement of policy, the act states that it is the "policy of the Congress to encourage the conservation, development, and utilization of water and related land resources of the United States on a comprehensive and coordinated basis by the Federal Government, States, localities, and private enterprise...."

The Water Resources Planning Act

The Water Resources Planning Act provided support for state planning programs, for the establishment of a National Water Resources Council, and for river basin (level B) studies, and considerable activity was initiated under it in the 1960s and 1970s.

In implementing the act, the Water Resources Council recognized three levels of plans: A, B, and C. Framework studies are examples of level A plans, river basin plans are examples of level B, and a level C plan is at the implementation level. In theory, a level C plan should be consistent with level A and level B plans, providing for coordination in the process.

The period 1965–1980 was an active one in water resources planning. One might say that in this period the New Deal concepts of activist government involvement in water resources were tested and failed. By 1981 the concepts of the Water Resources Planning Act were pretty much dead, due to decisions by the Reagan administration. The Carter "hit list," which dealt with both reform in government and with environmental issues, was part of the reason for the demise. While there was considerable support for government reform and environmental preservation, the Carter administration was thrown out because of problems such as high inflation. Today, the climate for the planning and decision process is more challenging than ever. Recent emphasis has been on tax cutting, regulatory reform, privatization, and other nongovernmental initiatives.

Comprehensive planning today

Chapter 9 describes how a "comprehensive framework" seems to have replaced "comprehensive planning" as a concept. The challenge is to coordinate and harmonize the process. Rather than following a rationalized "comprehensive planning process," proposals today must pass a series of feasibility hurdles including technical, financial, legal, environmental, and political, and judicial review is often required.

Environmental interest groups use national legislation and the courts to pursue their goals; states sometimes provide forums for coordination; national and state legislation set policy for decision making; and interstate issues are often involved.

With new emphasis on fish and wildlife and on species protection, the scope of "comprehensive planning" has changed. The overall process is much more complex than was originally thought by those who coined the term "comprehensive planning." In the United States the process is worked out in the messy arena of participatory democracy, and no single person or agency has complete control of it, even for a single problem or issue.

The ASCE's Water Resources Planning and Management Division

As the complexities of water resources planning increased, the American Society of Civil Engineers (ASCE) recognized that more attention to planning and policy was needed (Committee on Water Resources Planning, 1962), and in 1973 the Water Resources Planning and Management Division was formed. The division celebrated its twentieth anniversary at its convention in Seattle in 1993.

Planning Processes

Planning processes have well-defined structures and tasks. In 1972, the U.S. National Water Commission's Panel on Water Resources Planning (1972) presented this definition: "Planning is the creative and analytical process of: (1) hypothesizing sets of possible goals, (2) assembling needed information to develop and systematically analyze alternative courses of action for attainment of such goals, (3) displaying the information and consequences of alternative actions in an authoritative manner, (4) devising detailed procedures for carrying out the actions, and (5) recommending courses of action as an aid to the decision-makers in deciding a set of goals and courses of action to pursue."

This is an excellent definition, covering the traditional view of planning. It includes the basic steps of identifying problems, setting goals, devising plans, and carrying them out. Notice the separation between "planners" and "decision makers," and that the public is assumed to be represented by the decision makers. One might say that this is a "representative government" model. An alternative definition might be "the steps necessary to develop a plan, program, or strategy, and all of the governmental processes necessary to gain the permits, authorizations, and financing needed to implement the proposed action."

In general, the planning process deals with the questions: Given a goal for water management, what is the best way to accomplish it, and can approval and support be gained? Finding the best plan is technical in the sense that financial, institutional, economic, legal, and engineering aspects are all called "technical." Gaining approval requires dealing with the public, politicians, and regulatory processes.

Planning requires finding the best alternatives, given society's goals, objectives, and preferences. The Committee on Social and Environmental Objectives of ASCE's Water Resources Planning and Management Division (1984) clarified these terms this way: A goal is a general aim or desired end; an objective is a statement of purposes that is more limited and specific than a goal; a policy is a statement of intent that is broader than goals and objectives; a program is a statement of ongoing or proposed actions that lead to implementation of policies; and a constraint is a boundary condition which limits what may be done.

Two big changes in planning are apparent from 1972 to today. First, the emphasis is on less capital-intensive approaches to water management, and planning and decision making are increasingly directed at management actions that do not involve projects. Also, the importance of environmental and social impact analysis has increased. All of this has greatly increased the technical complexity of planning and the number of experts that become involved.

Second, public involvement has become a paramount issue. This is, to some extent, the result of increased emphasis on worldwide interest in democratization, but in the United States, it goes further, into a process of unraveling representative government and implementing a new form of "direct democracy," in which the media have much more influence.

In spite of both these changes, the traditional process is still valid. In fact, the National Water Commission acknowledged that management issues need to be planned too, and that the public needs to be involved. The differences are in the degree to which management issues and public involvement dominate.

The traditional process applies to management issues as well as project planning, but management actions may get less attention than projects because the investment levels, environmental impacts, and public awareness are less. Often, the hidden stakes in management actions are enormous. For example, the resources invested in planning conflicts and the potential shifts in wealth due to water rights changes can be quite significant.

Because water resources is a multiple-purpose, multiple-means activity, and because it involves multiple interest groups and players, planning must be done collaboratively. If it is not, excessive conflict results. It is this search for collaboration that has led to the use of the term "comprehensive, coordinated joint planning" (CCJP).

Another trend has been the increased influence of social scientists

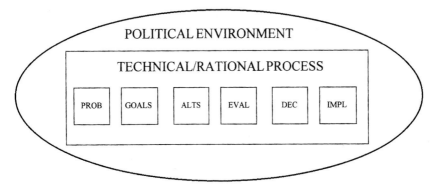

Figure 4.1 The planning process in a political environment.

in water resources policy analysis, as opposed to engineers. Reuss (1992) explains it as due to the engineer's penchant to gain control over uncertainty, while social scientists concentrate on analyzing uncertainty and its consequences. The result is that engineers become frustrated with longer, less certain processes, while social scientists are content to engage (and extend) the planning process.

Planning for project capital investments deals with implementing and financing the best plan for structural measures. Planning for management actions focuses on problem solving and policy implementation. Both involve conflicts over multiple purposes, multiple jurisdictions, and different interest groups.

To deal with any of these situations, I favor a model of a technical process operating inside a political environment. The concept for the rational-political model of planning and decision making is illustrated in Fig. 4.1, which shows the rational process taking place within a political environment that involves the public and all of the players.

Quantitative techniques are useful in management planning, but final decisions on high-level policy problems are too complex to be solved completely by them.

Rational model of planning

Managers like the rational way of doing business: Identify the problem, assess the options, and make a decision. They sometimes become frustrated with the political process, which is unpredictable, messy, and hard to understand. The rational process involves a set of steps which include:

- Problem recognition and identification
- Goal setting

- Establishment of criteria and measures of goal achievement
- Formulation of alternatives
- Evaluation of alternatives, assessment of impacts
- Decision on preferred alternative
- Implementation

This process embodies most of the traditional engineering functions as well as legal, economic, and financial work, and if environmental impacts are involved, it includes environmental impact analysis. This process can include steps up through final design of a project or program, and might cost up to 20 percent of a project. In the Two Forks project (see Chap. 10), most of the $40 million expenditure was for steps involved in the rational model.

The tools of the planner are especially important in the technical process. These include engineering planning and design tools, modeling and systems analysis (Chap. 5), techniques for economic and multi-objective evaluation, including benefit–cost analysis, financial analysis, legal opinions, and environmental and social impact analysis.

Political model of planning

In the water industry, problems are usually too complex and unstructured to formulate only in a rational planning format. Final resolutions of water conflicts often lie in the legal, financial, and political arenas, rather than in the technical arena. Elections, court battles, bureaucratic rule making and decision making, and water right purchases are involved. To succeed in this environment, managers working primarily on technical problems will take back seats in water conflicts unless they combine technical expertise with in-depth knowledge of policy, planning, communications, finance, and public involvement.

Political models involve intangible factors such as identification of players and interest groups, identification of trade-offs and negotiating strategies, public participation, establishment of incremental alternatives as well as far-reaching solutions, consideration of individual and group preferences, analysis of voting behavior, and other political science concepts.

Water resources managers are presented with many complex problems. Examples include whether to champion one project or another, how to unravel conflicts over water allocation, and how to deal with multiple stakeholders over management issues within a certain geographic area or across water management functions. In today's complex society, many managers are unable to solve problems such as these, and they stick to simpler problems, ignoring the tougher ones.

This is one of the reasons that engineers have lost some of the leadership in water management.

One cause of political problems is the divergent agendas of the players in the water industry. Sometimes the political problems are caused by the agendas of agency staffers. Water service providers have different goals from regulators. Planning and coordinating organizations work to harmonize water management, but sometimes they have bias toward one goal or another. Public-interest organizations have diverse goals.

Both elected and appointed officials work in politics. Agencies and bureaucrats work to drum up outside support for their programs, sometimes whether the program is needed or not. The goal may be to enhance images by enlarging programs and budgets. These larger programs result in higher employment, more powerful positions, and higher pay. These are the kind of bureaucratic forces that led to "Parkinson's law," which states that bureaucracies grow whether there is a need for them or not, due to human needs for power and influence.

Standards such as water quality standards or water conservation limits are political in that they involve the interests and opinions of engineers and interest groups. In storm sewer standards, the city engineer prefers larger systems since they reduce maintenance and local nuisance flooding, but the developer prefers smaller ones or none at all since profit is involved. The result is often a politically set standard.

Public administrators must understand political tradeoffs. A key role for them is to convey information about options to elected officials. Budget politics are of great importance to the development and maintenance of water systems, and the manager needs to give careful attention to them.

Department-versus-bureau conflicts have to do with the relative needs of a subunit within a larger department. A rate increase for electricity, for example, could lead to a decision by the utility manager not to seek one for water.

Water provides a lever with which to control land development and use, a source of wealth and influence. Land developers and industrial leaders support water management actions that are favorable to them. Thus, the water decision-making process attracts powerful people and business interests. This is also true of other public services, such as transportation. Sanders (1984) summarizes the political results.

In politics, there is a tendency to defer certain capital and maintenance needs, especially those that are not very visible, such as pipeline repairs. Large, visible pork-barrel water projects are another matter. Solving regional problems requires enlightened leadership highlighted by cooperation. Water problems require this today more than they require the "push it through" style.

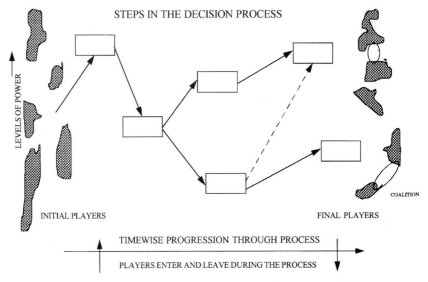

WATER RESOURCES PLANNING PROCESS

Figure 4.2 Political aspects of the planning process.

Whereas the rational model recognizes problems, sets goals, identifies alternatives, and evaluates impacts, costs, and benefits, the political model has additional features: coalition building among stakeholders, identification of critical decision points, adoption of strategies, shifts in goals, alternatives, strategies, and coalitions, and eventual outcomes that depend more on what can be done than on what might be the best under an unconstrained planning process. Because we live in an irrational world, the above process may not always be clear, and the uncertainty may bother some managers. Figure 4.2 shows this process.

A problem is initially identified through some process of management or politics. The next step by management is to determine if there is a commitment by the authorities to solve the problem. If there is a commitment, a goal can be established.

The manager will want to know who the stakeholders and decision makers are. The stakeholders need to be involved up front in a substantive way. Some water managers and members of the public are cynical about public participation because they do not think it is a sincere process. I knew a 1970s manager in a planning study who quit and said, "I can't mislead the public anymore."

Going through the planning process may take a great deal of time. The process unfolds over time, maybe years. Figure 4.2 shows this, with the steps of the process taking place in the view of the stakeholders, with provisions for feedback to adapt any aspect of the rational model to accommodate what the stakeholders want.

The figure shows the problem-solving process as it would apply to a single issue evolving over time. The features of the process are the problem to be solved, stakeholders, coalitions, goals, strategies, study processes, decision points, and possible outcomes.

The stakeholders are arrayed as shown in levels of power or influence in the decision process. The positions of the various actors in this array may vary over time, introducing another dimension to the dynamism of the process. The stakeholders array themselves in coalitions or interest groups. These rise and fall in influence and interest during the process. Stakeholders enter and leave the overall process during the long time period of many water resources problems and projects. We may not like the fact that not all stakeholders are equal, but it is a reality.

At the far right of the diagram lies a set of possible outcomes. The possible outcomes have variable characteristics, including technical alternatives, institutional alternatives, alternative goal achievement, alternative management arrangements, alternative timing, and alternative location dimensions. Sometimes the alternative outcomes can be related to other outcomes, as in intersectorial planning problems.

Along the route to the decision lie numerous crucial decision subpoints that involve some or all of the stakeholders. These may be meetings, reviews, completion of studies, new developments and surprises, changed attitudes. In between these nodal decision process subpoints lie decision subprocesses. Sometimes these seem quiet and inactive, but the committed groups know that crucial matters are under way. For example, influence and power are shifting, and knowledge is building. Where organizations are involved, it is in the decision subprocesses that decision support is needed. The decision subpoints may be the steps of the planning process, such as identifying alternatives, but these steps are really complex exercises in themselves.

It is important to realize that the processes referred to are carried out in the absence of perfect information. There is a lack of organized information and intelligence about what is going on with allies, neutral parties, and opponents in water resources problem solving.

Some stakeholders will influence outcomes from the beginning. They need clear goals early on, so that political strategies can be formulated. They have advantages over others who may decide to "go along" or participate less actively in the decision process. To be effective, coalitions of stakeholders are needed and maximum influence is sought. Environmental organizations have gained the reputation of being focused in their goals in opposing water resources development.

Partnerships and Alliances

"Partnering" has become popular as a way to avoid litigation and disputes in business. In water management, the concept can be applied as a way to develop collaborative solutions. There are several examples of partnering and collaboration in problem solving in the case studies. One place where these approaches is essential is in estuary management (see Chap. 22). An example of partnering for estuaries is Coastal America.

Coastal America is a "collaborative partnership process for action" that joins the forces of federal, state, and local agencies with private interests to address environmental problems along the nation's shorelines (Coastal America, 1993). Coastal America is implemented by a Memorandum of Understanding that relies on existing statutory authorities. It is based on a team approach, and utilizes a National Implementation Team (NIT), seven Regional Implementation Teams (RITs), and a Coastal America office at the President's Council on Environmental Quality.

The Tennessee Valley Authority has undertaken another experiment in partnering through "River Action Teams." This initiative is part of TVA's clean water action plan, meant to make the Tennessee River the "cleanest and most productive commercial river system in the United States by the year 2000" (TVA, 1993). The TVA sees the "integrator role" as critical in problems like this. The TVA sees their leadership role as providing insightful information, developing creative solutions, building effective coalitions, and tracking the progress of cleanup efforts.

Conflict Resolution

Cooperative approaches and partnering have the goal of reducing conflict and increasing the likelihood of finding the best plans and getting them accepted. However, the political aspects of planning and decision making often result in conflict. Techniques to deal with such conflict include negotiation, conflict resolution, and alternative dispute resolution (ADR).

Awareness of the need for win–win approaches in water management is increasing. The ASCE organized a symposium in 1989 on the engineer's role in resolving water-related conflicts. The editors, Warren Viessman and Ernest Smerdon (1990), wrote that "engineers must be society-wise as well as technology-wise," and that "curricula related to water resources planning and engineering should be proposed to more fully address subjects such as conflict management, dealing with critical events, decision making processes, working with the public and governmental bodies, policy analysis, and vehicles for fostering interactive (public-agency–decision maker) approaches to water resources problem solving."

Negotiation, as well as being a good strategy to achieve agreement, is one of the techniques of ADR. It is not, according to Beyea (1993), a process of compromise, but a method for finding solutions that give both parties 80 to 90 percent of what they need.

Although attempts at conflict resolution are still the exception rather than the rule, some are impressive. Montana has evolved a collaborative, consensus-building approach to a state water plan (Moy, 1989). It has six objectives: to transcend jurisdictional boundaries and involve all players (inclusive public involvement); agreement of parties on what issues are (agreement on issues); pursuit of consensus solutions (pursuit of consensus); balancing of competing water uses (balance uses); enhancement of coordination (enhance coordination); and continuous updating (flexible). It is intended that state-wide and basin-wide issues will go through the process. During 1988–1989 the process was used on an experimental basis, and Moy expects that it may lead to a successful state water plan.

McKinney (1988) analyzed the Montana experience and compared it to other literature about collaborative, consensus-building processes. He concluded that there are seven conditions for success: all affected parties represented in process; joint identification and evaluation of issues and alternatives; adequate time for consensus to develop; clear delineation of roles and responsibilities of decision-making bodies; general public must understand and support process; selection of issues that are amenable to collaborative problem solving, avoidance of issues that are not; and mandatory implementation of solutions.

Public Involvement in Planning and Decision Making

Perhaps the most important part of planning is gaining consensus and public support for plans and actions. However, the transition from a society with simpler decision processes to today's "pure democracy" has been bumpy with regard to the public's role in water resources decision making.

Public involvement became institutionalized in the 1960s when the antipoverty program required what was called "maximum feasible participation." However, public involvement is not meant to be a capitalized, proper noun. It is meant to be a two-way process that captures the spirit of effective democratic government. It is meant to be straight communication, honest communication, based on mutual respect and a sense of interdependence (Puget Sound Water Quality Authority, 1986).

Prior to 1970, there was little "public involvement" in water resources decision making. By 1970, however, it was recognized that the public needed to become an organic part of the planning process. In 1972, the U.S. National Water Commission Consulting Panel on Water Resources Planning (1972) stated that: "Public participation in the planning process is generally recognized as desirable. Yet an effective method of achieving public participation has not been devised." The panel went on to explain that the public should be part of the decision process, but their full involvement is best expressed by local government representatives.

According to Priscoli, public involvement and conflict management techniques help engineers gain agreement on facts, alternatives, and solutions (Priscoli, 1989). Public involvement and conflict management are seen as techniques along a spectrum that includes public information, task forces and advisory groups, public meetings, workshops and problem-solving meetings, conferences, mediation sessions, collaborative problem solving, negotiation, and arbitration. The spectrum goes from passing out information about a decision (public information) to agreeing about the decision (negotiation and arbitration). Priscoli makes seven observations about why public involvement and conflict management help incorporate environmental and social objectives into water management: They define the relationship between environmental quality and other social values; they clarify relationships between social values (equity, freedom, justice) and social structure; they help reclaim the "civil" in engineering; they help deal with conflicting visions about the world; they help us adapt to changes in society; they recognize that "process communicates intent"; and they are keys to defining the main dimensions of social acceptability.

Public involvement is illustrated by release policies on the TVA system (Ungate, 1992). In September 1987 the TVA board authorized a study of how to operate their system of 30 reservoirs. The TVA's study process, done without the supervision of the Federal Energy Regulatory Commission (FERC) because it is exempt from FERC requirements, was carried out under the provisions of the National Environmental Policy Act (NEPA) and involved all public and private interests that would have had a say under FERC relicensing proceedings. The TVA used two complementary approaches: decision analysis (a straightforward planning process); and the NEPA process to provide a framework for communicating with the public and all interest groups. Four phases of decision analysis were used: clarification of major issues, identification of options, evaluation, and "bullet-proofing." The NEPA process operated alongside the decision analysis and provided public review and scoping with wide participation by the public and interest-group players such as the TVA board and staff, power distributors, the navigation industry, local elected officials concerned with flood control, environmental groups, local economic development officials, and state and federal environmental, resource, and energy agencies. Ungate, the TVA projects manager for water resources, concluded that the key to the success of the process was the focus on the most important steps in decision making: the alternatives, the application of relevant information and logic to evaluation, and consideration of the values of all parties. This resulted in a focus on creative solutions rather than fighting an adversarial review process.

The American Water Works Association (AWWA) recently devoted an entire journal issue to public involvement (November 1993). They advocated that water supply utilities no longer rely just on technical integrity in today's social and political environment (Dent, 1993). Management must commit to long-term public affairs programs which focus on employee communication, customer service, public information, media and community relations, youth and public education, public affairs, and public involvement. The author believes that the public involvement process requires digging under specific positions to discover mutual interests, and that the public affairs programs must undertake problem-solving and alternative dispute resolution techniques.

In the same issue of the *AWWA Journal,* Diester and Tice (1993) discuss how to make the public a partner in project development, and give an example of siting a new reservoir for reclaimed water in California. They recommend that effective public involvement will include a sincere commitment by the agency and a complete acceptance of the principle that public involvement cannot be turned off when the heat is turned on. Their 12-step process places the decision process within the public involvement process; in other words, the public involvement process is the organizing principle for the decision process.

No matter how well it is treated, the public may sometimes seem unreasonable. People respect credibility and honesty, however, and if they are given straight answers, and they believe that management has their interests in mind, they will cooperate. To provide guidance in dealing with the public, the American Public Works Association (1984) produced a short, popularized guidebook which contains useful suggestions for communication and public involvement.

Multiobjective Evaluation of Projects and Actions

Water resources projects and management actions involve multiple purposes, multiple means, and multiple constituencies. For this reason, finding the best course of action is somewhat like performing a

juggling act. Compared to the simpler case of private business decisions—oriented toward the bottom line—decision criteria for water
actions are more complex.

Economists have responded with theories and techniques for multiobjective evaluation. It is a good test of making government work to
respond to all of the issues involved in society. Unfortunately, individuals do not always agree with the best common actions, so conflict
management is required.

Multiple-purpose project evaluation

The simplest technique is a matrix that shows the benefits provided
by alternative actions as they relate to each purpose and as they impact each affected group. One decision support matrix contains five
categories of information: actions, purposes, groups, evaluation criteria, and benefits. Figure 4.3 shows a decision matrix that illustrates a
sample project evaluation scheme. It shows different projects being
compared on the basis of how much each contributes to the achievement of a particular goal.

The economics profession began to give attention to multiobjective
evaluation in the 1930s. Benefit–cost analysis traces its roots to the
Flood Control Act of 1936. World War II interrupted the progress of
water resources planning, but by 1950 the Inter-Agency Committee
on Water Resources had published guidelines for project evaluation.
This booklet on evaluation (the "Green Book") was a comparison of
the evaluation procedures used by several federal agencies:
Agriculture, the Army, Commerce, Health, Education and Welfare,

	PROJECT A		PROJECT B		PROJECT C	
	+	-	+	-	+	-
GOAL I Group A Group B						
GOAL II Group A Group B						
GOAL III Group A Group B						
GOAL IV Group A Group B						

Figure 4.3 Goals achievement matrix.

Interior, Labor, and the Federal Power Commission. It set forth details of how to measure benefits and costs, standardizing issues such as setting prices for commodities, and generally sought to bring standardization to a field that was pretty unorganized. Also, cost allocation was covered in the manual.

Over the next 30 years, improvements and standardization were introduced to project evaluation. Of particular note was the economic analysis made under the Principles and Standards that were required by the Water Resources Planning Act. By 1983, principles and guidelines for economic and environmental studies were published by the Water Resources Council (1983).

The Principles and Guidelines, approved by the Secretary of the Interior and the President, begin with a description of the planning process. The process includes identification of problems and opportunities; inventory, forecast, and analysis; formulation of alternative plans; comparison of alternative plans; and selection of recommended plan.

The Principles and Guidelines prescribe four accounts for the evaluation: national economic development (NED), environmental quality (EQ), regional economic development (RED), and other social effects (OSE). These are intended to show the effects on the national economy; on ecological, cultural, and esthetic values that cannot be measured quantitatively; on regions with regard to incidence of NED effects, income transfers, and employment effects; and on urban and community impacts and effects on life, health, and safety. This is indeed a broad array of evaluation criteria, going far beyond simple benefit–cost analysis.

The Principles and Guidelines set up detailed criteria for displaying project effects. These respond to the need to establish goals achievement matrix formats for the display of decision information.

Cost allocation procedures are set up on the basis of joint and separable costs. Separable costs are the reduction in financial cost that would result if a project purpose was excluded from the plan. Joint costs are those that remain after all of the separable costs have been deducted. Joint costs may be allocated over the project purposes in proportion to project benefits or project uses.

Cost effectiveness and incremental cost analysis are additional tools to guide the decision maker to the best project or plan. Orth (1994) explains them as they have been developed within the U.S. Army Corps of Engineers to help analysts evaluate environmental mitigation and restoration plans. Cost effectiveness analysis enables the determination of the lowest-cost approach to obtain a given environmental output, and incremental cost analysis examines the added cost for an additional unit of environmental output. Environmental outputs can be measured in terms of "Habitat Units" which result from U.S. Fish and Wildlife Service (1980) procedures and include items such as acres of a particular type of resource, species population count, productivity of vegetation or trees, or ecological diversity.

In addition to the goals achievement matrix and displays such as those of the Principles and Guidelines, other methods are the cross-impact matrix and various techniques for multiobjective trade-off analysis.

Economics as a tool in planning

Economics is a main tool area for water resources planning. Economics deals with the production, distribution, and consumption of commodities. In that sense, it applies to water development, distribution, and consumption, and covers subjects such as the utility of a project, how to distribute water equitably, and issues dealing with consumption, such as water conservation.

Two basic issues arise over and over again: the economic efficiency of schemes, and the equity issues involved in water development, distribution, and consumption. Economic efficiency is measured in terms of national income. This means that a particular water management scheme will result in a certain amount of production of income in terms of products, services, wages, and the traditional measures of economic progress. Equity deals with the distribution of the income to different groups and regions. Both efficiency and equity analysis arise in multiobjective evaluation. Efficiency is measured by national economic development, and equity arises in environmental analysis, regional development, and social welfare. Equity can also arise in any analysis of the distribution of benefits to different groups.

Applications of economics

Economics is useful to quantify social benefits and costs of water actions. The production function that is widely used in systems analysis and management science is an economic function, usually net benefits. Another place where economics is useful is in economic development studies. There is an important difference between economic criteria and financial criteria. Economics deals with the public interest—benefits to all of us and to us as individuals—but usually the benefits are not tied directly to our pocketbooks, as in flood control benefits which might not be needed for many years. Financial issues hit us directly, however, and they deal with who pays.

Because water decisions are so political in nature, they are not always based on pure analysis. Even federal projects that had benefit–cost analysis applied to them were not selected purely on the basis of the economics. The pork-barrel effect can always override economic analysis.

Economics seemed more influential during the project-building era than it is now, because the benefit–cost technique was used to compare projects. Today, with more emphasis on financial, legal, and political issues, economics has taken somewhat of a back seat, but it is still important.

Benefit–cost analysis

The most directly usable tool of multiple-purpose evaluation in economics is benefit–cost analysis (BCA). BCA has been around for over 50 years. It was created in response to a requirement in the 1936 Flood Control Act that stated that projects would be authorized if the benefits exceeded the costs, regardless of whom they accrued to. This was a New Deal concept, one that sought to provide a mechanism of public investment to benefit the nation at a time of economic hardship. At its core, BCA is a way to compare different categories of benefits and costs for water projects or actions. Benefits make sense only if they line up with the goals of a project, however.

The use of BCA is made clear in the Principles and Guidelines as it applies to analysis of the NED objective. It also relates to the RED objective in terms of how the NED benefits are distributed to regions, but it does not apply to the EQ or OSE categories. In fact, environmentalists do not like BCA, because it is hard to quantify environmental benefits.

BCA basically requires summing up the benefits and costs, and then comparing them. The Principles and Guidelines describe how to compute them for the following project purposes: municipal and industrial water supply; agriculture; urban flood damage; hydropower; inland navigation; deep-draft navigation; recreation; commercial fishing; other direct benefits such as those increases in goods and services that result from project purposes; and effects on employment.

Integrated Resource Planning

Based on reports of success in the electric utility industry and with some water utilities, the American Water Works Research Foundation (AWWARF) has sponsored a research grant on how to apply "integrated resource planning" (IRP) to water utility planning. As of this writing, a guidance manual is being prepared, to be published by AWWARF in 1996 (American Water Works Research Foundation, 1995). According to the AWWARF, utilities say that IRP is "a continuous process with ever changing variables, public involvement is critical to the success of IRP, and the water industry must embrace IRP to be successful. An IRP program involves several components:

- Defining the overall goals and objectives and establishing milestones
- Identifying all of the stakeholders and their concerns and involving them throughout the process
- Determining the problems, critical planning issues, and potential conflicts to be addressed during the process
- Identifying and managing risks and uncertainties
- Implementing the IRP
- Evaluating the effectiveness of the process and making appropriate adjustments

From the above, it appears that IRP is an organized collection of the techniques discussed in this chapter, and it may be well to obtain the AWWARF's guidance manual to learn about details of applying it.

Integrated, or comprehensive, planning is evolving as a technique in the United States. Wagner (1995), in a keynote address to the ASCE's Water Resources Planning and Management Division, described attributes of the evolving technique:

- Extensive stakeholder involvement
- Locally defined, measurable goals, defined by stakeholders
- Comprehensive, all-inclusive planning scope
- Use of risk assessment, with priorities and schedules
- A plan of action, with provisions for adapting

Wagner also stated a caution about adaptive management: Unless there is a sense of *urgency for action,* the process can degenerate into a perpetual cycle of planning without progress. He showed how adaptive management was applied in Jamaica Bay, where initial cost estimates were $2 billion; with the adaptive approach about $1 billion has apparently been saved.

Paradigm for Planning Based on Coordination

A new paradigm stressing cooperation and coordination is needed for planning and decision making in the water industry (Grigg, 1993). This idea is described in more detail in Chap. 9. It is particularly true in interjurisdictional situations with many participants (as most water actions have). A new paradigm could enhance coordination in the rational planning process working within the political environment.

Although there are examples of successes in planning, several examples of unsolved conflicts are cited in the case-study chapters. These failures might be helped with better coordination, and maybe some will be helped in the future. Issues such as those cited above seem to result when one or another of the water management partners has not fulfilled an essential responsibility. But whose responsibility is it to fix problems that fall between the roles of service organizations and regulators? The answer is that no one knows, so when something goes wrong, the response is often a lawsuit or a new law, increasing the costs to society or the regulatory gridlock.

Consider the incentive structures of water agencies, regulators, planner/coordinators, and support organizations. With the possible exception of regulators, who have a narrow focus, only planner/coordinators have the incentive to identify and solve cross-cutting problems. This lack of proper incentives is the fundamental issue driving the need for planning and coordination in the U.S. water industry.

Two examples illustrate the costs when proper incentives and coordination are missing. Chesapeake Bay is an example of a large, interjurisdictional problem requiring joint, coordinated action (see Chap. 22). At the signing ceremony for an intergovernmental agreement in 1983, Jacques Costeau warned the celebrants that there were enormous political disincentives to collective action, and that the real difficulties, maintaining momentum and interjurisdictional cooperation, were still in the future (Chinchill, 1988).

Two Forks is an example of difficulty in finding a collective formula for action in a metropolitan area and in a river basin. Another great figure in water management, Abel Wolman, said about river basins, "basin approaches come into criticism by some on the score that basins are essentially non-economic or social units. Viewed by themselves, they represent artificial spheres of action irrelevant to societies needs. The engineer-planner finds them convenient, because he sees them as continuous hydrologic worlds" (Wolman, 1980).

These two wise men of water management, Jacques Costeau and Abel Wolman, saw the same problems with coordinated, cooperative, collective actions. They are on the one hand badly needed, and on the other, extremely difficult. Chapter 9 reflects more deeply on this dilemma in water management.

Summary and Conclusions

Water resources management is an important *planning and decision-making process*. The process includes many scenarios, ranging from simple to complex.

Both capital investments and management actions must be

planned. To approach them, I recommend considering the process as including a technical process inside a political environment.

The United States has searched for many years for a paradigm for comprehensive water resources planning. Today, the role of the federal government has changed from planner/investor to regulator, and it is very difficult to implement new projects. In the United States, planning takes place in a participatory democracy, and no single agency has complete control.

The political environment of water, and the reality that there rarely exists an agency with the full authority needed to solve water problems, show the need for stronger coordination. The main challenge is integrating water management (see Chap. 1). An example is Holland, where Kuijpera (1988) noted: "Integrated water management is the main future aim of Dutch water management. However, it's not yet an operational concept but still a rather abstract concept." Integrating the social and technical systems of water is a formidable task.

Regardless of the situation, the well-known processes of planning are helpful, especially with technical tools such as models and databases. Collaboration through partnerships and alliances helps a lot, along with public involvement to build support for actions. Conflict resolution is a key tool for coordination and reaching agreement.

Multiobjective evaluation, use of the Principles and Standards, and application of economics helps in the analysis phases. Benefit–cost analysis is a helpful tool for some applications. Applied research projects can be a tool to aid in planning. They might involve a group of water managers working with a central research team.

A new planning algorithm that pursues problem solving in the public interest is needed. Chapter 9 discusses the concept in more detail. I have offered a model framework with 15 elements. The main element is that a coordinated framework for action exists. Three process elements require that the framework be comprehensive, collaborative, and include stakeholder involvement. Two control elements require local control within a national policy framework. To promote action, three process requirements include having identifiable processes, an action orientation, and the quality of being adaptive. Six further requirements are that the framework promote sustainable development, be integrated, promote good management practices, be science-based and risk-based, and build capacity (Grigg, 1996).

Water resources planning affects many interest groups, and is both a technical exercise and a balancing of interests. The technical exercise deals with the complexity, and the balancing of interests deals with the conflict. Even when management plans are carefully laid, the technical analyses are perfect, and the public is involved, a perfectly logical plan may fail because powerful political forces disagree with the plan. Involving them in the early stages of problem definition, along with brainstorming and team building in multijurisdictional problems, can help promote a successful solution.

Questions

1. One of the important steps in the planning "process" is to be sure there is a commitment to solve the problem at hand. Explain why commitment is so important before effort and money are invested in planning.

2. Name the principal provisions of the U.S. Water Resources Planning Act. How did these work out, as compared to the goals of the policy makers who passed the act? What lessons can be drawn from this experience?

3. Explain and give examples of what is meant by level A, level B, and level C planning.

4. Imagine you go to a foreign country to explain the "rise of environmentalism" as it relates to U.S. water resources management. How would you explain it? Do you think the advantage in water conflicts is with water developers or environmentalists? Suggest three techniques or laws that environmentalists might use to stop a water project, and describe how water managers might respond to each to lead to a better project.

5. Explain how the environmental impact statement (EIS) relates to the planning process for water resources development.

6. China's 3-Gorges dam project is of great significance. Discuss the environmental impact, the social impact and the decision process that international lenders should follow to consider lending money for the project.

7. Define briefly each of the following six tests of feasibility: economic, financial, political, environmental, social, technical.

8. Explain the difference between tangible and intangible benefits of water resources projects.

9. What is the difference between the inflation rate and the discount rate in economic analysis?

10. In the Principles and Standards for water resources planning set up by the Water Resources Planning Act, what were the four "accounts" for analysis of costs and benefits?

11. What is the difference between financial feasibility and economic feasibility?

12. The text contains a number of cases about water management in the state of Colorado. Do you believe that Colorado should have a state water plan? Why doesn't it? What is a state water plan, and what should be included? On an overall basis, what should be the role of state government in water resources planning? Which agency should develop a state water plan? What process should be followed to develop a state water plan?

13. Explain why benefit–cost analysis is not totally adequate for project evaluation, leading the United States to develop the Principles and Standards for water planning and project evaluation.

References

American Public Works Association, *Better Communication: The Key to Public Works Progress,* APWA, Chicago, 1984.

American Society of Civil Engineers, Water Resources Planning and Management Division, Committee on Social and Environmental Objectives, *Social and Environmental Objectives in Water Resources Planning and Management,* ASCE, New York, 1984.

American Water Works Research Foundation, Project Update, Drinking Water Research, January/February 1995.

Beyea, Jan, Beyond the Politics of Blame, *EPRI Journal,* July/August 1993.

Chinchill, J., Chesapeake Bay Restoration Program: Is an Integrated Approach Possible?, in William R. Walker ed., *Water Policy Issues Related to the Chesapeake Bay,* Virginia Water Resources Center, Blacksburg, VA, 1988.

Coastal America, *Building Alliances to Restore Coastal Environments,* Washington, DC, January 1993.

Committee on Water Resources Planning, Basic Considerations in Water Resources Planning, *Journal of the Hydraulics Division, American Society of Civil Engineers,* HY 5, September 1962.

Diester, Ann D., and Catherine A. Tice, Making the Public a Partner in Project Development, *Journal of the American Water Works Association,* November 1993.

Dent, Joan, Public Affairs Programs: The Critical Link to the Public, *Journal of the American Water Works Association,* November 1993.

Grigg, Neil S., *Water Resources Planning,* McGraw-Hill, New York, 1985.

Grigg, N., New Paradigm for Coordination in Water Industry, *American Society of Civil Engineers Journal of Water Resources Planning and Management,* Vol. 119, No. 5, September/October 1993, pp. 572–587.

Grigg, Neil S., *A Coordinated Framework for Large Scale Water Management Actions,* American Society of Civil Engineers, New York, in press, 1996.

Holmes, Beatrice Hort, *A History of Federal Water Resources Programs, 1800–1960,* U.S. Department of Agriculture, Economic Research Service, Washington, DC, June 1972.

Holmes, Beatrice Hort, *History of Federal Water Resources Programs and Policies, 1961–1970,* U.S. Department of Agriculture, Economics, Statistics and Cooperatives Service, Miscellaneous Publication No. 1379, U.S. Government Printing Office, Washington DC, September 1979.

Inter-Agency Committee on Water Resources, Subcommittee on Evaluation Standards, Proposed Practices for Economic Analysis of River Basin Projects, Washington, DC, May 1950, revision, May 1958.

Kuijpera, C. B. F., Towards Integrated Water Management in the Netherlands, International Workshop on Water Awareness, Skokloster, Stockholm Region, Sweden, June 27, 1988.

McKinney, Matthew J., Water Resources Planning: A Collaborative, Consensus-Building Approach, *Society and Natural Resources,* Vol. I, No. 4, 1988.

Moy, Richard M., *Montana's Water Plan: An Incremental Approach,* Department of Natural Resources and Conservation, Helena, MT, 1989.

Orth, Kenneth D., Cost Effectiveness Analysis for Environmental Planning: Nine Easy Steps, IWR Report 94-PS-2, U.S. Army Corps of Engineers, Institute for Water Resources, Washington, DC, October 1994.

Priscoli, Jerome Delli, Public Involvement, Conflict Management: Means to EQ and Social Objectives, *Journal of Water Resources Planning and Management,* ASCE, Vol. 115, No. 1, January 1989, pp. 31–41.

Puget Sound Water Quality Authority, *Public Involvement in Water Quality Policy Making,* Seattle, WA, June 1986.

Reuss, Martin, Coping with Uncertainty: Social Scientists, Engineers, and Federal Water Resources Planning, *Natural Resources Journal,* Winter 1992.

Sanders, Heywood T., Politics and Urban Public Facilities, in Royce Hanson, ed., *Perspectives on Urban Infrastructure,* National Academy Press, Washington, DC, 1984.

Tennessee Valley Authority, Announcement of TVA River Action Teams, Communication Plan, Board of Directors, January 20, 1993.

Ungate, Christopher D., Equal Consideration at TVA: Changing System Operations to Meet Societal Needs, *Hydro Review,* July 1992.

U.S. National Water Commission, Consulting Panel on Water Resources Planning, *Water Resources Planning,* Washington, DC, 1972.

Viessman, Warren, Jr., Water Management: Challenge and Opportunity, *Journal of the Water Resources Planning and Management Division, ASCE,* Vol. 116, No. 2, March/April 1990, pp. 155–169.

Viessman, Warren, Jr., and Ernest T. Smerdon, eds., *Managing Water-Related Conflicts: The Engineer's Role,* American Society of Civil Engineers, New York, 1990.

Wagner, Edward O., Integrated Water Resources Planning Approaches the 21st Century, Keynote Address at the 22nd Annual Conference of the Water Resources Planning and Management Division, Cambridge, MA, May 8, 1995.

Water Resources Council, *Economic and Environmental Principles and Guidelines for Water and Related Land Resources Implementation Studies,* Washington, DC, March 10, 1983.

Wolman, A., Some Reflections on River Basin Management, Proceedings, International Association for Water Pollution Research Specialized Conference on New Developments in River Basin Management, Cincinnati, OH, 1980.

5

Systems Analysis, Models, and Decision Support Systems

Introduction

Although one can never understand completely the full complexities of water resources systems—including structural, environmental, and social aspects—one can learn much about them by using systems tools. This chapter outlines how quantitative analysis using models and decision support systems can aid the *planning and decision-making process* by helping to find the best plans and assessing the impacts, thus helping to gain administrative approval and financial support for water resources actions.

Years ago, when computers and models first came on the scene, their usefulness was to some extent overpromised. When they could not do as much as promised, engineers and managers did not embrace them. Now, however, the tools can do much more, and with computers on each desk, their use is increasing rapidly. In fact, I believe that the "systems approach" will be more meaningful in the future due to the greater use of computers. Not only will it involve more emphasis on models and data, it will also foster greater appreciation for "systems thinking."

This chapter presents conceptual explanations of the tools and explains how they can be used for management purposes. The reader should consult other texts, such as the one by Mays and Tung (1992), for technical details.

Concepts and Definitions

There is a lot of jargon in the fields of computer modeling, systems analysis, database management, and decision support systems, so we shall begin by defining the basic concepts as they relate to water resources management.

First is the concept of the *water resources system.* We defined it in Chap. 1 as a combination of water control facilities and environmental elements that work together to achieve water management purposes. There are a number of examples in this book; Fig. 5.1 shows one in the form of a schematic of the water distribution system in the Jordan Valley.

Remember that a water resources system also includes social elements. Figure 5.2 illustrates how people interact with an irrigation system.

The next concept we consider is *systems analysis,* which has many definitions. The definition we shall use is:

Water resources systems analysis: the application of computer-based models and databases to analyze water resources systems on a holistic basis, in order to show how the elements of the systems interact with each other and with their external environment.

Notice that this definition focuses on the analysis of a whole system. Other definitions mix this aspect with project planning and management procedures, which I do not consider to be part of systems analysis. At any rate, here are three more definitions:

"The study of an activity by mathematical means to determine its desired end and the most efficient method of obtaining it" (*Webster's II,* 1984)

"A coordinated set of procedures that can be used to address issues of project planning, engineering design, and management" (Ossenbruggen, 1984)

"An inquiry to aid a decision-maker choose a course of action by systematically investigating his proper objectives, comparing quantitatively where possible the cost, effectiveness, and risks associated with alternative policies or strategies for achieving them, and formulating additional alternatives if those examined are found wanting" (Rudwick, 1973)

Other, related terms are as follows.

Systems approach to water resources management: a systematic approach to conceptualizing the water resources "system" and using the tools of systems analysis (databases, models, geographical information systems) to identify and evaluate management strategies

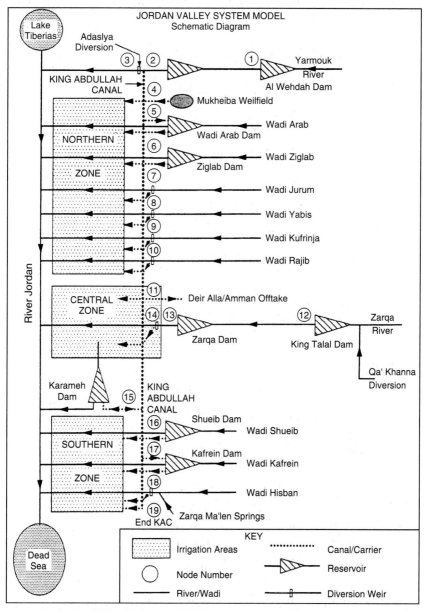

Figure 5.1 Jordan Valley decision model. (*Source: Jordan Valley Authority.*)

Water resources system: a combination of water control facilities and environmental elements that work together to achieve water management purposes

Sociotechnical system: a combination of a technical system (water resources system) with its sociopolitical environment

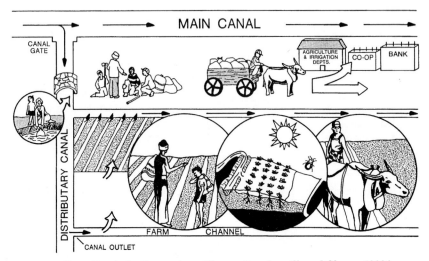

Figure 5.2 Operating irrigation system. (*Source: Lowdermilk and Clyma, 1983.*)

Systems engineering: "the art and science of selecting from a large number of feasible alternatives, involving substantial engineering content, that particular set of actions which will best accomplish the overall objectives of the decision makers, within the constraints of law, morality, economics, resources, political and social pressures, and laws governing the physical, life and other natural sciences" (Hall and Dracup, 1970)

Decision support system (DSS): an advisory system for management, usually computer-based, that utilizes databases, models, and communication/dialog systems to provide decision makers with management information

The term "decision support systems" (DSS) has evolved to replace and succeed older terms such as "management information system" (MIS). Mainly, terms such as these describe collections of databases, computer simulation and optimization models, graphical interfaces, mapping and geographical information systems (GIS), and various analysis and presentation techniques. They are a convenient way to describe how the analysis tools are used in management, for "decision support."

Comparing international approaches to systems analysis, UNESCO's International Hydrologic Programme developed a report, "The Process of Water Resources Project Planning" (UNESCO, 1986), and outlined five stages: (1) plan initiation and preliminary planning; (2) data collection and processing; (3) formulation and screening of project alternatives; (4) development of final study results; and (5) design. They proposed conflict resolution between stages 3 and 4, and a political "go–no go" step at the beginning of stage 4. If the project was given the "go" sign, then stage 4 would include modeling, including impact analysis, risk assessment, costs and benefit analysis, and development of operational models. After stage 4, there would be another political process to see if the project would be funded or not. Stage 3 would also involve modeling, but to screen alternatives. The working group included members from the United States, East Germany, West Germany, Australia, Poland, Greece, Denmark, and an observer from Israel. My personal opinion is that they tried to include too much in the concept.

As you see, there is overlap and "jargon" in these terms. Let us try to put the terms in perspective. We can apply the tools at specific levels when conditions are known and controlled and decisions are "structured." We can also apply them at general levels which deal with less structured, political problem solving, often at the policy level. At this level, the tools generate information and knowledge that can be used in political decision processes, but they are not the "final answer" by any means.

Rudwick (1973) explained the dichotomy between the specific and general levels this way:

> At one end of the spectrum are the mathematically oriented analysts who wish to apply a set of optimization techniques to highly structured problems. . . . On the other end of the work spectrum . . . are those analysts whose starting point is the unstructured problem of the decision-maker. Their major objective is to build a proper structure to the problem, including uncovering the true goals of the decision-maker.

In my opinion, a way to clarify the terms is to focus on use of the term "systems analysis" to mean what it says: the analysis of systems. Then, when a system has been defined in terms of interacting components, systems analysis is the analysis of how these components interact and what the results are. Terms such as the "planning process," the "problem-solving process," or the "decision process" can be used for the broader decision process. These terms are less academic, easier to understand, and connote the political aspects of the decision process.

While systems analysis involves mathematical modeling, there is more to it. For example, defining the system's boundaries can be a substantive issue. Viessman and Welty (1985) devote a chapter to the "totality of water resources systems, the concern with how the components interact, and how they can be combined into efficient systems for meeting prescribed objectives."

Rogers and Fiering (1986) adopt a broad definition. They state that "systems analysis" includes operations research, cybernetics, management science, and systems analysis. They do not regard the use of a simulation model as systems analysis, but merely as an electronic computation of a routing study. To qualify as systems analysis, they require that alternatives be chosen in accordance with some computational algorithm, and that hydrologic inputs be varied stochastically.

Djordjevic (1993) developed a concept for the application of cybernetics to water resources management. He uses the term "cybernetics" to mean an essentially new approach to systems planning, operation, and control—one that focuses on using information as a resource in decision making. Like many "systems" terms, one may not necessarily understand its use from a dictionary definition, such as "cybernetics is the theoretical study of control processes in electronic, mechanical, and biological systems, especially mathematical analysis of the flow of data in such systems" (*Webster's II,* 1984).

Djordjevic's thinking about applying cybernetics lines up with the concepts of "systems thinking." He sees the systems approach as generalized through systems thinking, which analyzes systems in interaction with their environments. He sees cybernetics as treating problems with clearly defined goals (this would place it more in the "mechanistic" camp). Then, he sees information and mathematical modeling as resources for decision making. The structure of cybernetic disciplines includes a mathematical basis for systems definition, probabilistic methods, optimization methods, and special applied disciplines such as water resources.

Evolution of Water Resources Systems Analysis

Although planning techniques have been around for a long time, systems analysis has evolved since about the 1950s. It was made possible by the availability of the computer; one could, in fact, say that the foundation of the planning process is politics, and the foundation of systems analysis is mathematics.

Early research on water resources systems analysis was carried out at Harvard University. A paper by Maass and Hufschmidt (1959), "In Search of New Methods for River Basin Planning," refers to September 1956 as the starting date of a three-year research program at the Littauer Graduate School of Public Administration at Harvard. The goal of the activity was "to improve methodology for the planning and design (used in its broad sense) of multi-unit, multi-purpose water resource systems," and "to compare many more alternatives than they [planners and engineers] do under present practice," in order to "yield significantly greater profits or benefits."

My first contact with systems analysis was when Warren Hall was invited to Colorado State University to lecture on water resources systems analysis. This was in 1968, and we used the draft of his book, *Water Resources Systems Engineering,* written with John Dracup (Hall and Dracup, 1970). They used the concept of water resources systems *engineering* to formulate problems in broader terms than systems analysis alone.

Hall and Dracup acknowledged that groups at Harvard, Cornell, Berkeley, Stanford, Illinois, and other individuals had all reached the same conclusion: The practice in water resources of analyzing a single alternative plan and submitting it to lawmakers for a yes-or-no decision was no longer in the best interests of the public. Their goal was to adapt the modern tools of operations research and systems analysis to assist in the development of a number of alternatives for water management. In those years, the focus was on finding the best plan for water development, but in later years systems analysis has been found to be equally as useful to assess management issues such as demand reduction.

Hall and Dracup recognized that the tools of systems analysis can only assist; they cannot replace the water resources decision-making process. This observation, developed as a philosophy by Hall in the 1950s and written in the late 1960s, rings true today. In fact, the observation drives some of the organization of this book. Systems analysis gets one chapter to show how it can be used as a tool of management. The decision-making process gets another chapter. The rest of the book is devoted to the other aspects of problem solving, issues such as law and finance.

Hall had keen insight into the complexities of water resources systems. His book shows this, and the many conversations we had underlined this combined theoretical-practical insight. Thus, to me, the most interesting parts of his book were the chapters about real-world problem solving. The rest of the book explained the tools of systems analysis and gave examples of objective functions, investment timing, water supply, large-scale systems, groundwater, and water quality.

In the 40 years since the work by Hall and the Harvard group, a great deal has been written about the tools of water resources systems analysis. To cite just two examples, Chaturvedi (1987) and Mays and Tung (1992) present excellent complete texts covering the tools of systems analysis. Chaturvedi credits his origins with the Harvard Group, and Mays was a student in Ven T. Chow's group at Illinois, both being excellent starting points for work in systems analysis.

So far, the promise of systems analysis has fallen short of its goals. This was made clear by Rogers and Fiering (1986). However, quantitative analysis must still be the foundation for water resources management. We must have a balance between systems analysis and practical politics of water resources. The future holds good promise for systems analysis, both the application of computers and of "systems thinking."

Systems Thinking and Problem Solving Applied to Water Problems

Systems thinking

A concept that is closely related to systems analysis is "systems thinking." There are a number of different views of this concept, and it is useful as a way to look at problems holistically. To me, "systems thinking" means to think about whole systems at once, obviously a necessary way to approach water resources management. Actually, that is what systems analysis is, but with a different approach.

In the early 1990s, a business text on "systems thinking" became a best seller (Senge, 1990). Senge's approach focuses on "learning organizations," following his theory that organizations that "learn" will "continually enhance their capacity to realize their highest aspirations." He identifies five "component technologies" that characterize the learning organization: (1) systems thinking, (2) personal mastery (competence), (3) mental models, (4) building shared values, and (5) team learning. These relate to several different parts of water resources management, but we will focus on systems thinking at this point. To Senge, systems thinking is the "fifth discipline" that ties together the other four.

Senge believes that "today, systems thinking is needed more than ever because we are being overwhelmed with complexity." Systems thinking, according to Senge, is "a conceptual framework, a body of knowledge and tools that has been developed over the past fifty years, to make the full patterns clearer, and to help us to see how to change them effectively." He believes that it is "a discipline for seeing wholes. It is a framework for seeing interrelationships rather than things, for seeing patterns of change rather than static 'snapshots.'" These capabilities are badly needed for a "comprehensive framework" for water resources management (see Chap. 9).

Senge's version of system thinking can be applied quantitatively through application of the "systems dynamics" modeling technique developed at MIT by Jay Forrester. It is an interesting set of methods to simulate complex processes that involve the flow of both materials and information.

Systems thinking and problem solving

Regardless of the situation, each problem has certain characteristics: What is the problem? How do different groups see it? What are the alternatives for solving it? How do the alternatives shape up in terms of the decision criteria? What will be the impacts on different groups of the different possible decisions? What should the timing of the solution be? How should the solution be implemented?

To illustrate how "systems thinking" and "problem solving" represent higher forms of systems analysis, I will mention work on the "art of problem solving" by a master of systems analysis, Russell Ackoff (1978). Ackoff was a professor at the University of Pennsylvania, and he wrote a popular book on operations research.

Ackoff liked being referred to as a problem solver; he admitted that solving problems had been his main occupation as an adult. He went through three phases: an early period with a philosophy-based approach (logic and rational thinking); a middle period of following a scientific approach (systems analysis); and a later period that emphasized art (creativity). As a result of this experience, Ackoff believes that only art can produce exciting solutions that are really "beautiful." This is what we want in creative problem solving, of course.

Ackoff envisions a "problem" as having five components: (1) the ones faced by the problem (the decision makers), (2) the controllable variables, (3) the uncontrolled variables, (4) the constraints, and (5) the possible outcomes. The uncontrolled variables make up the problem environment. The value of the outcome is a specified relationship between the controlled and uncontrolled variables. Ackoff goes on to present many interesting aspects of problem solving, including science, philosophy, and art. Clearly, his approach to problem solving presents "systems analysis" in the broadest possible framework. I like Ackoff's list of the essential properties of good management: competence, communicativeness, concern, courage, creativity—all "C's."

Systems thinking helps to explain the difference between the structured and unstructured problems of water resources systems. It illustrates the differences between lower-level, operational water problems (structured, rational, mechanistic) and higher-level, water policy issues (unstructured, political, systemic). It is a challenge to apply systems thinking to the higher-level policy and systemic problems (Cunningham and Farquharson, 1989).

Systemic problems result from organizations adjusting to change, acting like organisms that grow and maintain themselves, seek equilibrium in reaction to stresses and strains, and try to survive. This is closer to the political view of water decision making than to the rational view, where the organization is assumed to seek a set of well-defined goals; players in the water industry seek to survive; and those outside the closed system have different agendas than those inside, causing conflict.

In systems thinking, the tasks of an organization are adaptive/regulative, coordinative, productive, and maintenance. Adaptive/regulative tasks deal with the organization's environment and regulators. Coordinative tasks involve directing and coordinating inputs to production. Production deals with the products of the organization, and maintenance repairs and helps the organization survive. These are also the tasks of managing water systems, but it is the overall water industry that does them, not individual players, and without effective coordination, conflict and problems result.

The social system involves values, beliefs, and interests of participants. Systemic problems arise when parts of the system disrupt the equilibrium of other parts, as when values clash. An example is the conflict of scientific management (time and motion studies) with the need for humane treatment of workers. In water resources, one conflict is between water suppliers and environmentalists; another is the clash between an economic objective to divert water from a valley and a social goal to maintain irrigated agriculture in the valley.

In systems problem solving, one must understand the relationships between subsystems and their environments. We seek to dissolve rather than resolve problems, that is, to work out conflicts between subsystems and their environments. In practical terms, this means finding win–win water resources plans.

Systems problems can be divided into mechanistic and systemic types. *Mechanistic problems* involve relatively closed systems, which are not open to interference from other problems. *Systemic problems* relate to problems outside the closed system, and arise from the complexity of situations that require a systemic, not mechanistic approach to problem solving (Cunningham and Farquharson, 1989).

Mechanistic problem solving seeks to resolve problems—this being possible when problems are well structured or involve closed systems. Mechanistic problem solving isolates the broken components of systems and repairs each one. Systems problem solving seeks to understand relationships and the general situation, then to repair relationships.

Mechanistic problem solving focuses on scientific laws and procedures, whereas systems problem solving focuses on relationships. In mechanistic situations, problems are defined by the problem solver, whereas in the systemic approach, the problem is not defined, but evolves from force fields and the dynamics of the components. In systems problem solving, solutions respond to both technical and social systems, and to strategic and tactical issues.

Cunningham and Farquharson give an example of applying the systems method to a hospital environment. The steps they describe are: assemble key decision makers; define each subsystem's critical requirements; formulate the "mess" (that is, the facts, total picture of the problem, assumptions); examine the underlying forces, that is, do a force field analysis that isolates the forces helping or hindering; develop action plans to restrain negatives and enhance positives; and select action plans by consensus. As you see, this is quite close to versions of water planning/systems processes.

Systems thinking and water problems

Combining the rational and the political models of water resources planning requires that we examine the *whole system*—the water system and the social system within which it operates. It helps if we use the "systems viewpoint."

Water systems are "sociotechnical" systems. This characteristic have been studied by management scientists who present principles of systems thinking. They take a social theory of organizations, not the production-oriented view that is inherent in the rational model. Sociotechnical systems are technical systems with strong links to society. They can be mechanistic (the technological system, such as a telephone), or systemic (the telephone and its impact on society). Social thinkers criticize simple concepts (a telephone that works) and see the big picture (telephones isolate people, they do not work face-to-face, and society feels isolated). These are important philosophical issues; in fact, the National Academy of Engineering devoted a symposium to them (Sladovich, 1991). Water systems share sociotechnical problems like this with other technological systems.

The social theory applies to all organizations, but water resources decision making is even more complex because of the interdependence of organizations. The complexity of water decision making is greater because in organizations the players involved in production, maintenance, adaptation, and management can be coordinated by a single authority, but in water systems they cannot.

Although water resources problems are more complex than the hospital example, we can learn about them from systems thinking. Complexity is reduced by studying the interaction of complex systems. Water systems are sociotechnical, physical management systems, including organizations, which consist of activities of individuals, groups, and physical systems acting to produce output to fulfill needs such as production, maintenance, adaptation, and management. Since water systems are sociotechnical, we see that the social system, made up of the values, beliefs, and interests of participants, is the essence of the political environment in which water management takes place.

The usual goal in water resources management, "to accomplish the objectives of the decision makers," assumes that we know the decision makers and their objectives. In the political model, however, this is not always the case. In the final analysis, systems thinking seems most promising as applied to water resources by public involvement, partnering, and techniques of alternative dispute resolution. Public involvement identifies the players, keeps them informed and involved in the decision process, and seeks to integrate the technical and social aspects of solutions. Partnering is a way to develop cooperation. Alternative dispute resolution techniques allow identification of the agendas of the social subsystems, and what are win–win plans. These were the subjects of the preceding chapter.

Application of Decision Support Systems

In a practical sense, the systems approach is applied within the broad framework of a decision support system (DSS). As discussed earlier, a DSS involves models, data, and communication systems such as graphical interfaces and perhaps other graphical aids such as mapping and geographical information systems (GIS). A view of these is given in Fig. 5.3.

The use of a DSS should be straightforward. A water resources problem must be solved, a decision maker must be involved, data about the actual situation must be collected, analysis must be con-

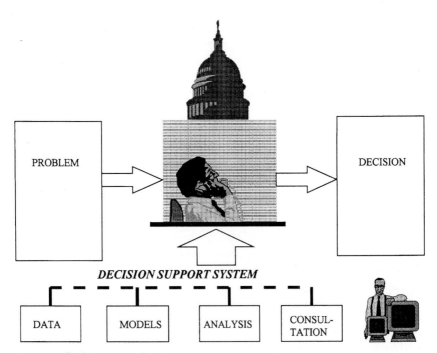

Figure 5.3 Decision support system.

ducted to determine the options and consequences, then decisions must be made about management options. The analysis may or may not involve a mathematical model.

In practice, things do not always work so smoothly. The decision maker asks for a study of a certain situation. The modeler tries to carry out a valid simulation, and may or may not produce useful work. In the final analysis, the decision maker, or the modeler's supervisor, has to evaluate the quality of the work.

Most problems involve the basic parameters of systems analysis: state variables, decision variables, inputs, and outputs.

State variables characterize the status of a system at any time. An example is the quantity of storage of water in a reservoir at any time.

Decision variables are the variables that can be controlled, for example, the rate of release of water from a reservoir at any time.

Inputs to a system are the externally applied inputs, for example, the inflow hydrograph. Other inputs might be available infrastructure, energy, money, labor, and ideas.

Outputs are the results of decisions, for example, the outflow hydrograph. Outputs then become the water supplies for municipal and industrial (M&I) water, irrigation water, wastewater management results, water quality parameters, stormwater management and flood control outcomes, energy production, transportation, and environmental enhancement of natural systems.

For a water resources issue, decision information will be required on these and other parameters. These will vary from scenario to scenario, but the number of different types will not be all that many. The decision information will respond to the elements of the problem as outlined in the previous section: the nature of the problem, the views of different groups, alternatives to solve it, how alternatives shape up in terms of decision criteria, impacts on different groups of possible decisions, timing of solutions, and how to implement the solution. These questions play out differently in scenarios such as those outlined in Chap. 1 (service delivery, planning and coordination, organizational, water operations management, regulation, capital investments, or policy development). To illustrate, let us consider two of the most common examples, operational problems and capital investment. Both of these lend themselves to decision support systems with models and data. In the case of operational decisions, much of the DSS effort goes into how reservoirs should be operated. In the case of the capital investment, the main quantity to focus on is return on financial investment.

System configurations

Hall and Dracup (1970) state that a system is "a set of objects which interact in a regular, interdependent manner." Their concept is to isolate the system, the principal interacting objects, and define their relationship with the system's environment by inputs and outputs. Defining the boundaries or configuring the system is the first step.

A sociotechnical system will have more parts than a purely physical system because of the social elements. It will simplify things if we think about systems in levels—the technical system and the sociotechnical system.

For a technical system, systems analysis will simulate physical system behavior (*state variable*) under different conditions of supply and demand (*inputs*) that yield given levels of reliability of *outputs* with certain price tags and return on investment (*decision variables*). Some variables will be controllable and others uncontrollable. Also, there will be constraints.

Figure 5.4 shows the concept of a technical system and its interacting components, with the external environment shown around the technical system. This diagram, which is quite conceptual, illustrates many different aspects of a system. For example, one of the components inside the system could be the water system of the Jordan Valley shown in Fig. 5.1, and the remaining elements could be the agricultural, urban, and industrial development subsystems, with the external influences being relationships with neighboring countries. When an external environment is placed around the system, we begin to gain the sociotechnical view, which includes the values, beliefs, goals, agendas, and interests of the players. Chapter 4 deals with these issues.

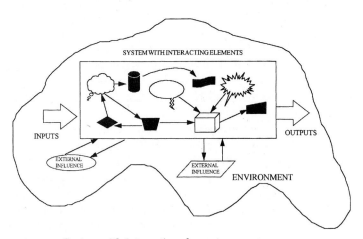

Figure 5.4 System with interacting elements.

Now, let us look at the components of the DSS to see how they work together to support decisions.

Data systems

Water resources data are required for inputs, demands, system states, and information items needed to run the models or to make decisions. Types of water resources data include hydrologic variables, system characteristics and states, and many others. Data includes water quantity, quality, and climate information. Data must be available and reliable to be useful.

At a simple level, a decision support system will include a model which involves some hydrologic and hydraulic data and some system characteristics. The model should predict how the system will perform under different hydrologic loadings. Ideally, all of the data will be entered into the model from external sources, and the data will not be "hard-wired" into the model. That way, the model is generic and can predict what will happen if the data change.

In fact, models vary from the simple to the complex and do not always give reliable answers, but data are there for anyone to see and evaluate. Thus, many believe that data give better answers to system questions than models do. For this reason, the "data-centered" approach to modeling and decision support systems is gaining acceptance (Woodring, 1994).

Woodring (1994) shows how a DSS can be constructed with a focus on the data. He identifies the database management systems (DBMS) as a key feature of the DSS. According to Woodring, a DBMS is "a systemized collection of computer processes that operate together to define logical and physical organization of, all interactions with, and control operations on data in data sets." Using concepts from systems engineering, Woodring described structured methodology, data flow diagrams, and entity relationship diagrams as building blocks for a data-centered DSS.

The data-centered approach and the management of data in modeling are evolving. The Corps of Engineers' Hydrologic Engineering Center (HEC) has developed a data storage system (another DSS) for their HEC family of models. The system enables efficient organization and transfer of data between HEC programs.

Database management is an important part of water resources management. With the advent of improved database management packages, more organization of data has come about. Standardization in data has come about by a set of ANSI defined commands known as Structured Query Language (SQL), and new relational database programs that enable the user to describe relationships between the different types of data are giving new power to database management in the water resources field.

Models

The basic purpose of models in water resources systems analysis is to simulate the behavior of physical systems. There are many different kinds, and it is hard to generalize about them. Wurbs (1994) describes the different kinds of water models and includes the following categories: general-purpose software; demand forecasting and balancing supply with demand; water distribution system models; groundwater models; watershed runoff models; stream hydraulics models; river and reservoir water quality models; and reservoir/river system models.

For river basin planning, early models were mainly hydrologic, and many hydrologic models are now available (Chap. 2 discusses hydrologic modeling). For example, research to develop the Stanford Watershed Model and similar watershed models such as those by the U.S. Geological Survey began in the 1960s. These models can simulate the systems effects of hydrologic processes such as infiltration, runoff, groundwater flow, interception, evapotranspiration, and storage, and they compute water balances of complex watersheds.

Hydraulic models deal with the movement of water in open channels or pipe networks. Open-channel models such as HEC-2 have proved useful for years for floodplain and flood insurance studies. They generally compute water surface profiles along channels, in floodplains, and through bridges. River system models, including sedimentation, are increasingly necessary to analyze environmental issues. The more sophisticated hydraulic models simulate unsteady flow, sometimes in two or three dimensions. An unsteady flow model would be necessary, for example, to simulate a dam break scenario. The HEC manges a program called DAMBRK, which was developed by the U.S. National Weather Service to simulate the creation and movement of dam-break flood waves. The HEC also has a program called UNET to simulate unsteady flow through a full network of open channels.

Also in the category of hydraulic models are those that simulate flows in water distribution systems. These are essential to manage pressures in urban distribution networks.

Flood prediction programs are both hydrologic and hydraulic in nature. The Corps of Engineers has developed a model called HEC-1. Reservoir operations models might be considered as hydrologic-hydraulic in nature, because they simulate the operation of reservoirs to predict the transformation of an inflow hydrograph to an outflow hydrograph. The Corps has developed a sophisticated package for reservoir simulation (HEC-5).

Groundwater models use a special kind of simulation to show movement of groundwater in different kinds of aquifers. In addition to groundwater quantity, models can also simulate groundwater quality.

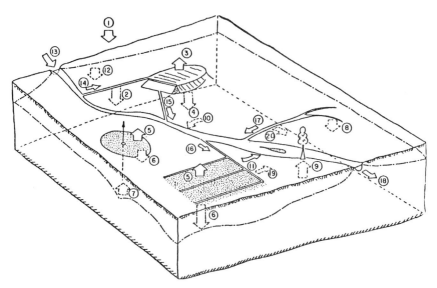

Figure 5.5 Block diagram of SAMSON model. (*Source: Grigg, 1984; diagram courtesy of Hubert Morel-Seytoux.*)

An example of a surface-groundwater hydrologic model is SAMSON (Stream-Aquifer Simulation Model), which was developed by Hubert Morel-Seytoux (Colorado Water Resources Research Institute, 1985) at Colorado State University in the 1980s to simulate the interaction of surface water and groundwater in the South Platte River Basin (see Chap. 19). Figure 5.5 illustrates details of the hydrologic balance used by the model.

Water quality models can deal with hydrologic issues (nonpoint sources and sediment) or with hydraulics (transport models). They are discussed in Chap. 14.

River basin simulation models can include network models like MODSIM or water balance simulations that rely on other types of logic or computational algorithms. MODSIM, developed at Colorado State University by John Labadie, has been widely used to aid cities in their water accounting tasks. For example, the city of Ft. Collins, Colorado, used MODSIM to do the drought study described in Chap. 20. Figure 5.6 illustrates the complex link-node setup of the model for a local system. This is described in more detail later in the chapter.

Integrated model packages include the Stormwater Management Model (SWMM) which was initiated about 1970, leading to a working model that is still very much in use in the mid-1990s. This model simulates the runoff to inlets, the transport in sewers, the change in water quality from runoff, in-system storage, and the water quality in receiving waters.

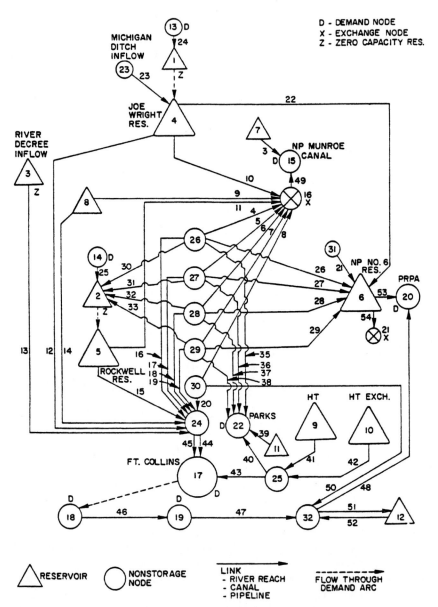

Figure 5.6 MODSIM link-node configuration. (*Source: Grigg, 1984; diagram courtesy of John Labadie.*)

For capital investments, economic-financial modeling is used. Financial models are particularly useful because one can examine alternative scenarios and run sensitivity analyses for different interest rates, payback plans, and other potential management decisions. Chapter 7 discusses the basic elements of financial modeling.

The U.S. Office of Technology Assessment (1982) of the U.S. Congress surveyed model use and reported that models have expanded the nation's ability to understand and manage water; increased accuracy of estimates; made possible analyses that were not possible without models; and have the capability to provide greater benefit for decision making in the future. They caution, however, that models are complex and require skilled personnel as well as budget support. This optimistic finding was disputed by Rogers and Fiering (1986), and as one who responded to the Office of Technology Assessment survey from a post in state government, I thought that although models are useful, the report was too favorable. One problem that occurs, in my opinion, is that surveys such as this are filled out by modelers.

Models have many useful applications, and their usefulness is increasing. They are used in many applications, such as modeling reservoir operational scenarios, modeling stream water quality issues for permit checking, and modeling the performance of urban water distribution systems.

Sensitivity analysis of model results is also quite useful. Models are, realistically speaking, not very accurate. Problems include lack of data, difficulty in characterizing water resources systems, errors and lack of skill in model operation, unfamiliarity with computers, and just poor engineering. Sensitivity analysis allows one to compensate somewhat for these problems by examining how the system behaves under different assumptions, such as poor data availability.

I have found that the use of models in court is somewhat questionable because of the accuracy factor. I do believe that they will be used more in the future, however. An example was reported in the *Denver Post* on November 21, 1991: "Computer Models Squaring Off over Water." This article referred to the use of surface-groundwater models to try to illustrate that groundwater pumping would harm wetlands. Each side had a model that showed a different outcome from the pumping.

In Chap. 2, I outlined the importance of water accounting as a basis for basin models. The Stockholm Environment Institute and the Tellus Institute (1993) have produced an accounting model called "A Computerized Water Evaluation and Planning System" (WEAP) which accounts for most basin components. WEAP includes the components shown in Fig. 5.6. According to William Johnson (1993), WEAP is based on water accounting principles (see Chap. 2), and it "gives a holistic, integrated picture of the supply and demand system of the study area at any point in time, and under different user-specified sets of conditions. This picture includes rivers, creeks, reservoirs, and groundwater as sources of supply and water withdrawals, discharges, and instream flow requirements as demands." Johnson illustrates the application of the WEAP model to the Upper Chattahoochee Basin, Georgia (see Chap. 19).

Ultimately, we need models of large-scale, complex systems. These models will give us improved capability to deal with river basin issues, for example. Techniques include hierarchical analysis and decomposition of large-scale systems into subsystems. Although we are using many complex models, we should not forget the fundamental water-accounting nature of models and that simple spreadsheet models are useful in many applications because of their simplicity and ease of use.

Communication systems and dialog

The dialog aspect of the decision support system can mean different things. If you are the decision maker, you are interested in asking questions and getting answers about your problem. To the analyst, the dialog can also mean moving data back and forth between the database and the models. The communications and dialog aspects of DSS are very important, and much of the effort of developing DSS systems is in graphical user interfaces (GUI) that improve the user friendliness of computing.

Risk Assessment

One of the important uses of models and data is to assess risk of different management alternatives. Risk assessment is an important tool in water resources management because many hydrologic phenomena involve uncertainty and must be analyzed using statistics.

Risk assessment in water resources is a subset of the same problems associated with the environment in general. Colorado State University organized a program called "Ecological Risk Assessment and Management" to address this area. Quoting the U.S. Environmental Protection Agency, the program defines ecological risk assessment as a "process that evaluates the likelihood that undesirable ecological effects may occur or are occurring as a result of exposure to one or more stressors."

The ecological risk assessment process is aimed at environmental restoration or natural resources management scenarios. It seeks to combine technical input with regulatory and political considerations to determine the best risk-based management alternatives.

Hydrologic models provide good insight into the risk of failures in water resources. Probabilities of extreme events are assessed using hydrologic statistics. Examples of the applications of these technologies can be seen in Chap. 11 on floods and Chap. 20 on droughts.

Example: Water Supply System

We can apply systems analysis to water supply systems, such as that of the city of Ft. Collins, Colorado. We seek in this case to study alternatives for increasing the quantity and reliability of the supply. Ft. Collins approached this problem in the 1980s by developing a water supply policy that was based on a drought study.

For a study like this, we first need to characterize the system, that is, the supply, treatment, and distribution systems. We restrict the system to water supply, and do not include the wastewater side. For the purpose of this analysis, we restrict the system further to water supply upstream of the treatment plant and distribution system. Figure 5.6 illustrates a characterization of the Ft. Collins water system and shows that the basic components to be modeled are the hydrologic inputs, the conveyance channels and pipes, and the storage points.

To know how to exercise the system model, we must know the questions to be answered and who will participate in the planning and decision process. For this example, we restrict the question to the statistics of the system's water supply yield. With an analysis of the yield, we can determine the reliability of the system to meet customer demands in dry and wet years.

The sequence of the modeling study and resulting decision was:

1. Configure the system and assemble the simulation model.
2. Compile historical data for all water rights and sources.
3. Calibrate the model to make sure it is predicting the total system yield correctly.
4. Using hydrologic techniques, generate synthetic sequences of inflows.
5. Perform statistical analyses of total system yields.
6. Compare system yields to present demand and projected growth in demand.
7. Assemble risk factors in terms of probability of system failure.
8. Issue decision advice on measures to increase the security of the water supply.

Example: Operation of Reservoirs under the Appropriation Doctrine

John Eckhardt (1991), an engineer in the State Engineer's Office of Colorado, prepared a dissertation at Colorado State University about using a decision support system for the operation of reservoirs under the appropriation doctrine of water law. The research illustrates systems analysis in a practical way.

The problem is how to operate reservoirs which contain water belonging to different water rights owners under a complex system of natural water and stored water that results from the unique features of the appropriation doctrine, both how it is theoretically supposed to work, and how it really works, given all of the real-world constraints.

Eckhardt analyzed the problem and developed a framework for a real-time reservoir-operation DSS as shown in Fig. 5.7. Note that the

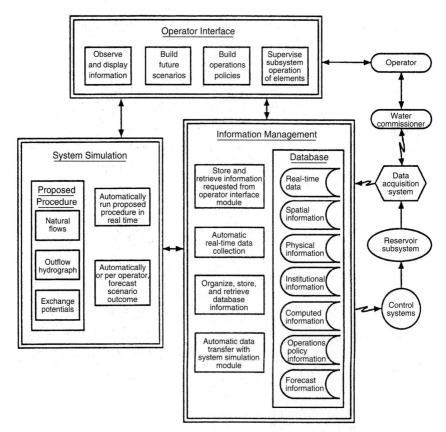

Figure 5.7   Framework for reservoir decision support system. (*Source: Eckhardt, 1991.*)

framework contains an operator interface, a provision for system simulation, and an information management subsystem. It is a good example of a needed DSS for this system. At this writing (1995), the state engineer's office is preparing a system with some of the features shown.

Colorado River Decision Support System

The Colorado River Decision Support System (CRDSS; see Chap. 19) is a good example of current thinking about DSSs. A feasibility study identified three general uses for the CRDSS: interstate-compact policy analysis (see Chap. 15), determination of water for development, and water resources administration (see Chap. 15). Some of the anticipated problems, such as interstate compact analysis and evaluation of in-stream flows for endangered species, are at the policy level. Other problems, such as optimization of water rights administration and on-line sharing of information, might be more mechanistic.

Five components of the CRDSS were envisioned: (1) a total river basin model, (2) water rights planning and subbasin models, (3) a consumptive use model, (4) databases, and (5) hardware and networking configurations.

The goal for the CRDSS was to "provide a dynamic, efficient, and effective information base for understanding and solving the problems at hand." The feasibility report listed the following required characteristics:

- Unambiguous treatment of data and information
- Efficient and effective visualization of information
- Data analysis techniques allowing full use of information
- Formal techniques for plan evaluation
- Rapid evaluations of policy alternatives
- Dynamic modeling of system state and updating of information

CRDSS models and databases are to be valuable elements in future water planning and water rights administration in Colorado. They will require an effective, sustainable management strategy to ensure long-term maintainability of the models developed. The CRDSS must perform tasks in a realistic way that recognizes Colorado's water rights administration system. The Big River model must give identical simulations of the entire river basin to enable Colorado to check its policies against those of the bureau and states, and vice versa.

Chapter 19 gives more details about the CRDSS. The system illustrates current thinking about how a state government can work with local water users and the federal government to develop a state-of-the-art tool for river basin management. We shall use the CRDSS to focus on some issues about managing DSS efforts.

Managing Decision Support Systems

People sometimes become almost mesmerized by computing systems and then become disillusioned when they begin to understand the limitations of such systems. To avoid this trap, I believe that certain principles should be followed to make sure that investments in DSS systems remain productive. If the principles are violated, a lot of money may be wasted on useless models and databases that are later abandoned. To illustrate these principles, a few suggestions that arose from the CRDSS study follow.

With the CRDSS, Colorado paid for an advanced information system featuring databases, models, and decision support features. In effect, the system is a capital asset that requires the same care as any other capital asset, except that it has two distinctive features: continuously evolving technology, and a revolving door of users. Dealing with these features requires a special approach to CRDSS development and management.

As CRDSS is to be data-driven, its basic platform is data. Eventually, this data will be available in geographical information system maps and in digital numerical form. While we have a good idea of some of it, such as the consumptive use inputs and stream gauging data, the organization and frameworks for data management will evolve on a collaborative basis. Eventually, a single management entity will be required to manage the data.

Next, the CRDSS features simulation models. First, the Big River Model is required to give identical simulations of the entire river basin to enable Colorado to check its policies against those of the bureau and states, and vice versa. This being the case, the Big River Model must be developed in parallel with bureau efforts, although an interim version may be developed while the bureau plans its program.

Next, the Water Rights Planning Model (WRPM) is to be a simulation model of water balances within Colorado's subbasins of the river. These include the Gunnison, Yampa, Upper Colorado and San Juan, and other subdivisions. Initially the WRPM will be required to undertake standard modeling tasks such as reservoir operation accounting, water rights and exchange modeling, and stream channel routing. Most important, the WRPM must perform these tasks in a realistic way that recognizes Colorado's water rights administration system.

As the CRDSS evolves, each of its components (databases, Big River Model, WRPM) must come under unified management or the capital asset will be dispersed. A long-term management plan to provide the unified management is required. This management plan must be worked out within the responsible agency, Colorado's Department of Natural Resources. Suggestions for the plan follow.

1. *Databases.* Several databases are under development, and they generally fit one of three types: data that can be managed within the State Engineer's Office (SEO) water rights administration databases; data that are unique to and necessary for the operation of the models; and other data, such as those used in policy studies but not falling into the other categories. As the CRDSS is developed, the SEO data should be captured in the SEO database, the model input data should remain with the CRDSS models, and the other data should be cataloged in reports and on disks for later retrieval on an as-needed basis.

2. *The Big River Model.* The Big River Model will eventually require a joint management approach with the bureau, or it will not be possible to maintain identical versions.

3. *The Water Rights Planning Model.* Development and management of the WRPM is actually a very significant program for Colorado's water agencies and planning programs, because the models will serve as the technological organizing concept for basin management plans and water rights administration. Due to its importance, development of a protocol to select a model technology, apply it to sub-basins, and make it available for water planning and water rights administration is a critical activity requiring utmost care.

To approach model selection and development, a management framework is needed for how the model will be maintained, improved, and distributed, and how users will be trained to use it. Quality control will be a very important aspect of the program.

The variables in the management framework must be identified and determined. An initial view of them, with tentative determinations, might be as follows:

- *Overall management responsibility* could be with an individual state agency, an interagency committee, or an independent board/staff. Right now, the mechanism is a project management team operating under the combined authority of the SEO and the Colorado Water Conservation Board (CWCB). How this will work post-CRDSS development has not been announced. This function would include overall policy development on model use, improvement, and management.

- *Model maintenance* includes maintaining the code, issuing updated versions, assuring that model versions retain integrity, maintaining a model audit process to check any version for validity, and generally being the responsible party, technically speaking. This function could either be kept under the overall manager or contracted out to an independent, technically competent entity.

- *User support* includes communication with users, training, model distribution, and related functions. Normally, it would be expected to reside with model maintenance, but it could be separate.

Clearly, model maintenance and user support will have a cost and require technical excellence. The actual success of these functions will determine the long-term success of the CRDSS and the payoff from the capital investment. Moreover, given that the CRDSS models will be critical to the future of Colorado's water planning and water rights administration, it is imperative that decisions be made early on as to how these functions will be handled. These decisions will then lead to decisions about the required nature of the WRPM.

To get a view of variables affecting the WRPM, let us look at a short story about the use of the WRPM in the future. The story is intended to illustrate key variables and decisions.

> September 12, 2009, 3 P.M., Denver. Angela Hernandez of the Consolidated Colorado Water Authority (CCWA) calls Frank Chen, a model maintenance specialist in the joint water DSS center of the CWCB and SEO.
>
> "Frank, we have version 4.2 of the Colorado River WRPM running on our network, and we just ordered a new server and upgraded distributed graphics processors, so we believe it's time to upgrade to version 5.0. Can you E-mail us the software and set up a level one training session? While we are at it, we'd like to teleconference the training to our sub-basin stations on the West Slope."
>
> "Sure, Angela. The CCWA has a blanket license agreement and it provides for regular upgrading and for training at the flat rates specified in the license agreement. I'll get the E-mail to you this afternoon, and you can relay it via the Internet to your stations. We'll schedule the teleconference via Colorado Telcon for next week."
>
> "Remember, Angela, in version 5.0 we have finally removed the last hard-wired data from the WRPM, and it is now truly a generic, sustainable, defect-free model technology, fully portable to all platforms and cleansed of all the computational biases and blind spots of the original developers. We know this because the model audit team convened last year by the Comprehensive Colorado River Collaborative Consensus Coordinating Body (CCRCCCB or C2RC3B) gave us a rating of A+ for the model."
>
> "That's great news, Frank, and it shows the wisdom of those engineers and managers who set up the first guidelines for CRDSS in the early '90's. Wow, what a return on the taxpayers' investment!"

This story illustrates several key points:

- A model maintenance office with competent technical staff

- Legal instruments to control models and user access
- Methods to self-finance model maintenance and improvement
- Software transmittal methods
- Collaborative use of models
- A process of audit and continuous model improvement.

These functions illustrate two essential elements for success: control by the model management office and a process of continuous improvement. If these are to succeed, the source code must belong to the state, must be well organized and clearly verifiable, subject to modular improvement or conversion to new, object-oriented formats, and must generally be a code that makes a good foundation for later workers to build on. With this in mind, the following additional requirements are suggested:

1. Code is 100 percent owned by state (absolute requirement).
2. Code is well organized and adaptable to software improvements.
3. Code can be understood by future programmers.
4. As graphics improve, code is adaptable to partial graphic displays, and visual outputs that instill confidence in users.

Conclusions

Systems analysis provides a way to deal with the complexity and conflict of water resources within the framework of quantitative analysis. Clearly, it will be directed toward more computer-based models and data systems to aid decision makers.

While Rogers and Fiering (1986) found that systems analysis had not lived up to its promise, this author is convinced that the barriers will fall in the face of better technology and improved understanding. Deficiencies in databases and inadequacies in modeling will be overcome by research, and institutional resistance will give way to demonstrations of usefulness. The issue of insensitivity of systems to variations in design choices is a generic problem, and will indicate that systems analysis is not always needed.

Advanced computing is using graphical interfaces and aspects of artificial intelligence. Models are being assembled under the umbrellas of shell programs that can call up different system components and examine their interfaces.

While it cannot resolve all conflicts, computer-based decision information can be used productively in water resources management. For water resources issues, certain decision information will be required.

This will vary from scenario to scenario, but the number of different types will not be all that many. The decision information will respond to elements of the problem such as nature of problem, views of different groups, alternatives to solve it, how alternatives shape up in terms of decision criteria, impacts on different groups of possible decisions, timing of solutions, and how to implement the solution. These questions play out differently in scenarios such as those outlined in Chap. 1 (service delivery, planning and coordination, organizational, water operations management, regulation, capital investments, or policy development).

For operational decisions, much of the DSS effort goes into how reservoirs should be operated. Problems are mostly structured, and the main challenge is hydrologic modeling.

For capital investments, the main quantity to focus on is return on financial investment. A systems analysis is required to simulate physical system behavior (state variable) under different conditions of supply and demand (inputs) that yield given levels of reliability (outputs) with certain price tags and return on investment (decision variables).

At this writing (1996), tools of systems analysis have come a long way from their overpromised but underdelivered start-up, and now they are poised to take an even more important role in water resources management. They cannot solve the political problems, but they can help by helping to deal with some of the complexities.

Questions

1. What is your philosophy of the "systems approach" and "systems analysis"? Do you believe that they are being used to their full potential in water resources management?

2. In what detail do managers need to understand models and database management tools?

3. According to Peter Senge, the author of *The Fifth Discipline* (1990) systems thinking is "a conceptual framework, a body of knowledge and tools that has been developed over the past fifty years, to make the full patterns clearer, and to help us to see how to change them effectively." How can this approach help water resources managers?

4. What is a "sociotechnical system," and how would an irrigation system fit this description?

5. Do you believe that better use of risk assessment could help water resources managers? Can you give some examples?

6. Explain the components of decision support systems and how they are used.

7. Discuss how data, information, and knowledge can help resolve the conflicts of water resources management.

8. The following equation can be used to generate synthetic data for water supply planning:

$$Q_i = \overline{Q} + r(Q_{i-1} - \overline{Q}) + t_i s(1 - r^2)^{1/2}$$

where Q_i = monthly flow in month i
$\overline{Q}$ = average monthly flow
r = correlation coefficient
Q_{i-1} = flow in month $i - 1$
t_i = random component
s = standard deviation

Explain how this equation can be used in reservoir or water supply planning.

References

Ackoff, Russell L., *The Art of Problem Solving,* John Wiley, New York, 1978.

Chaturvedi, M. C., *Water Resources Systems Planning and Management,* Tata McGraw-Hill, New Delhi, 1987.

Colorado Water Resources Research Institute, South Platte Team, Voluntary Integrated Water Management: South Platte River Basin, Colorado, CWRRI Report, September 1985.

Cunningham, J. Barton, and John Farquharson, Systems Problem-Solving: Unravelling the "Mess," *Management Decision,* Vol. 27, No. 1, 1989.

Djordjevic, Branislav, *Cybernetics in Water Resources Management,* Water Resources Publications, Littleton, CO, 1993.

Eckhardt, John R., Real-Time Reservoir Operation Decision Support under the Appropriation Doctrine, Ph.D. dissertation, Colorado State University, Spring 1991.

Grigg, N. S., et al., Voluntary Basinwide Water Management: South Platte River Basin, Colorado, Completion Report 133, Colorado Water Resources Research Institute, October 1984.

Hall, Warren A., and John Dracup, *Water Resources Systems Engineering,* McGraw-Hill, New York, 1970.

Johnson, William K., Accounting for Water Supply and Demand: An Application of Computer Program WEAP to the Upper Chattahoochee River Basin, Georgia, U.S. Army Corps of Engineers, HEC-TD-34, Davis, CA, August 1993 (draft).

Lowdermilk, M. K., and Wayne Clyma, *Diagnostic Analysis of Irrigation Systems, Volume I: Concepts and Methodology,* Water Management Synthesis Project, Colorado State University, 1983.

Maass, Arthur, and Maynard M. Hufschmidt, In Search of New Methods for River Basin Planning, *Journal of the Boston Society of Civil Engineers,* Vol. XLVI, No. 2, April 1959.

Mays, Larry W., and Yeou-Koung Tung, *Hydrosystems Engineering and Management,* McGraw-Hill, New York, 1992.

Ossenbruggen, Paul, *Systems Analysis for Civil Engineers,* John Wiley, New York, 1984.

Rogers, Peter P., and Myron B. Fiering, Use of Systems Analysis in Water Management, *Water Resources Research,* Vol. 22, No. 9, pp. 146S–158S, August 1986.

Rudwick, Bernard H., *Systems Analysis for Effective Planning,* John Wiley, New York, 1973.

Senge, Peter M., *The Fifth Discipline: The Art and Practice of the Learning Organization,* Doubleday Currency, New York, 1990.

Sladovich, Hedy E., *Engineering as a Social Enterprise,* National Academy Press, Washington, DC, 1991.

Stockholm Environment Institute and Tellus Institute, A Computerized Water Evaluation and Planning System (WEAP), Boston (from a description by the Hydrologic Engineering Center, Davis, CA, December 1993).

UNESCO, The Process of Water Resources Project Planning, IHP Working Group A.4.3.1., Paris, July 7, 1986 (draft).

U.S. Office of Technology Assessment, *Use of Models for Water Resources Management, Planning and Policy,* OTA, Washington, DC, 1982.

Viessman, Warren, Jr., and Claire Welty, *Water Management: Technology and Institutions,* Harper & Row, New York, 1985.

Webster's II, New Riverside University Dictionary, Riverside Publishing Company, Boston, 1984.

Woodring, Richard Craig, A Data-Centered Paradigm for Enhancing Water Resources Decision Support Systems, Ph.D. dissertation, Colorado State University, Fall 1994.

Wurbs, Ralph A., Computer Models for Water Resources Planning and Management, U.S. Army Corps of Engineers, Institute for Water Resources, IWR Report 94-NDS-7, July 1994.

6

Water and Environmental Law, Regulation, and Administration

Ideally, collaborative decision making with public involvement should be the mechanism for water management, but realistically, law and regulations provide the major coordinating mechanism. To function effectively, water managers must understand administrative and regulatory procedures as well as law.

Years ago, main topics in water law were the riparian and appropriation doctrines. Today, water law extends to water quality and environmental law, and covers eastern states as well as western. This chapter presents an outline of water and environmental law and administration, and extends the concepts into the regulatory realm to show how the law is translated into management actions. The companion to this chapter is Chap. 15, which describes details and cases about water administration systems, including permit systems, transfers, and interstate compacts.

A Matrix of Water Laws

When we speak of "water law," we often think only of statutes, but the spectrum of water law includes three levels and three branches of government. In addition, there is activity in the international arena, but it normally does not affect the management of water inside of national boundaries. The following is a matrix of where water laws are found. In addition, there are executive orders.

	Constitutions	Statutes	Regulations	Case law
Federal	Constitution	Federal laws	Agency regulations	Federal cases
State	Constitution	State laws	Agency regulations	State cases
Local	Charters	City codes	Agency regulations	Local cases

Water Quantity Law

Water quantity law within states (intrastate law) is largely a state matter. Between states, interstate compacts and other federal laws are involved, and the U.S. Supreme Court has been actively involved in settling disputes.

Getches (1990) presents an overview of the main systems of state water quantity law: riparian, appropriation, and hybrid systems. The eastern states mainly follow the riparian doctrine. Nine states follow the appropriation doctrine (Alaska, Arizona, Colorado, Idaho, Montana, Nevada, New Mexico, Utah, and Wyoming). Another 10 follow hybrid systems (California, Kansas, Mississippi, Nebraska, North Dakota, Oklahoma, Oregon, South Dakota, Texas, Washington). Hawaii follows a system based on historical precedents of the kingdom of Hawaii, and Louisiana follows a system based on the French civil code. The other 29 states follow the riparian doctrine.

On the surface, the riparian doctrine is used mainly in the humid eastern United States, and the appropriation doctrine is used mainly in the semiarid western United States. Hybrid systems evolved in different places, and involve the use of permits and other devices as surrogates for water rights and as mechanisms to allocate water and settle disputes. When we go deeper, however, aspects of the water law systems become much more complex.

Because water allocations to water users are a matter of state interest, states must develop administrative systems to control water rights. In Colorado and other western states this is done through the State Engineer's Office. In eastern states, systems for administering the riparian and hybrid systems are not as well organized as in the West, but they are evolving. Chapter 15 presents additional details about water administration and describes several case studies, including eastern and western state approaches and an interstate compact.

Riparian doctrine

In the riparian doctrine a person whose land abuts the water is said to be a *riparian* land owner. "Riparian" means on or relating to the bank of a water course.

According to Getches (1990), the riparian landowner's rights are: to the flow of the stream; to make a reasonable use of the water body, as long as other riparians are not damaged; to have access; to wharf out; to prevent erosion of the banks; to purify the water; and to claim title to the beds of nonnavigable lakes and streams.

The origins of the riparian doctrine are probably in Europe, as both England and France have versions of it. However, there is some dispute over its exact roots. Its basic version in the United States is common law, and it has not been enacted in statutes or state constitutions, but has been used as the basis for court decisions. The pure form has been extensively modified, so that there exist a number of variations.

One aspect of pure riparianism is the natural flow rule, which entitles the landowner to the flow of the stream undiminished in quantity or quality. This concept is based on a situation where, for example, a millwheel might be involved, and the diversions, if any, would be small. This is not practical today, when there is so much pressure for water development. Thus, the riparian doctrine must give way to a *reasonable use* doctrine.

Consider the case where a city or an industry diverts water from a stream and takes it great distances, perhaps returning it to a wastewater treatment plant that discharges into another drainage basin. This would be an *interbasin* transfer, and technically speaking, it would violate the principle of riparian rights.

The reasonable use doctrine is intended to deal with the impracticalities of the riparian doctrine. It allows landowners to use waters if they do not interfere with reasonable uses of other landowners. These kind of adjustments to pure riparianism have evolved into administrative systems, which are essentially patchworks of riparian doctrine and practical, politically acceptable methods for allocating and managing water.

Whereas the appropriation doctrine provides a definite method of administration, the riparian doctrine does not. However, court cases have been based on it. Also, the doctrine influenced state statutory systems as they developed hybrid methods and procedures for water administration. Today, many disputes remain to be settled. The hybrid systems and the reasonable use doctrine do not answer all of the questions, such as what to do when shortages occur, how to provide for in-stream flows, and how to deal with interbasin transfers.

Regulation of water quantity by permit systems

To deal with the impracticalities of the riparian doctrine, some states have developed permit systems. These superimpose permit requirements over riparian and other rights. Getches (1990) listed 17 states with permit systems (Arkansas, Delaware, Florida, Georgia, Illinois, Indiana, Iowa, Kentucky, Maryland, Massachusetts, Minnesota, New Jersey, New York, North Carolina, Pennsylvania, South Carolina, and Wisconsin).

Permit laws vary in nature. Basically, a permit is like a water right: It entitles the holder to use the water. However, it is not a property right; it is a permit. Permits may be for the withdrawal of a particular quantity of water for urban use, such as, for example, a permit to withdraw water for a city of 50,000. Conditions on such a permit would be negotiated between the administrative agency and the diverter.

What is the security for the water user when he or she has a permit? This is an important legal question. If an industry makes a large investment on the basis of a permit, what control does it have if a competitor wants a share of the water? This and many other questions remain to be answered about permits.

Permit authorities, normally state agencies, must make decisions on legal questions about allocating water among users. This can be a tough challenge, because there are many competitors for these rights. In the West, legal systems have evolved to avoid having an administrative agency become a "water czar" in this way.

Prior appropriation doctrine

In the western United States, the prior appropriation doctrine is followed. This doctrine arose to deal with the fact that there was not enough water to satisfy all users, so a system of allocation was needed early on.

Getches (1990) describes the doctrine as having been developed in the nineteenth-century West to provide for miners' uses of public lands, then being extended to farm uses on private lands. It found its way into state constitutions, and now is the main principle for water allocation in 19 western states, although 10 of these states follow hybrid approaches.

Strictly speaking, the water belongs to the public, but the doctrine provides for the right to use it, in order of priority, as long as it is being applied to a beneficial use. Traditionally, a valid appropriation would contain three elements: the intent to apply the water to a beneficial use, an actual diversion, and demonstration of application to the beneficial use.

Burger (1989) provides additional insight into the development of water allocation laws in arid regions. He sees irrigation as the fundamental element of development in arid countries, being necessary before urbanization and industrialization can proceed. Thus, water law developed first to accommodate irrigation, and water was closely tied to land. Then priorities developed, first domestic water, then agricultural water, and then urban and industrial uses. Burger sees Roman water law as the basis for Western systems, because the Romans provided much of the legal systems for Western civilization. Thus, Roman water law principles found their way into many of the arid countries in Europe and the Middle East.

Roman law classified things according to ownership: public, private, and community ownership. Examples would be private water rights, public natural resources such as air, and community rights to some resources such as grazing. These concepts still generate disputes about water law and ownership.

Security of title to water has been necessary in water law in arid regions because without security of title, no one would invest in irrigation. This is the same issue discussed above, which still needs attention in permit states.

The prior appropriation doctrine has given rise to an extensive system of water adjudication and administration. Rice and White (1987) describe the fundamental concepts of the doctrine as including the appropriation systems (mandate and permit systems); the fact that beneficial use is the measure; and that a water right is an interest in real property.

A water right must be initiated and perfected by appropriation and adjudication. Then water rights are administered by a system of rules and regulations, including calls on the river. Water rights can be lost through forfeiture or abandonment. Systems are necessary to transfer water rights to new owners or new uses. The Colorado system for administering water rights is described in Chaps. 13 and 15, and transfers are discussed in Chap. 15.

Practical issues in the appropriation doctrine

In administering the appropriation doctrine, numerous practical problems arise. Imagine trying to determine precisely each water right owner's entitlement in a stream that rises and falls according to hydrologic variation, with uncertain routing of flows from one point to another, unknown return flows, variable weather, and everyone diverting and releasing water according to schedules not under control of the administrators. It is certainly a challenge, especially in states such as Colorado, where there is a lot of development.

Administration under Colorado water law is presented as a case study in Chap. 15. One of the biggest issues is: Given that the state follows the appropriation doctrine, should it have a "state water plan"? On one side, some say that water is the most important resource the state has, and the state should certainly have a plan to manage it. On the other side, some say that Colorado has a state water plan—it is called the appropriation doctrine. Which is correct? Actually, the appropriation doctrine is embodied in the state's constitution of 1876, which states that "the right to appropriate water shall never be denied."

This question—Should the state have a plan?—involves numerous issues, including geographic, value-based, and institutional conflicts; public trust doctrine versus appropriation doctrine; third-party effects; in-stream flows and recreational water; interbasin transfer; scientific complexity in the appropriation doctrine; and interstate compacts. These issues also affect other western states, and they are presented in some detail in Chap. 24.

John Eckhardt, a manager with the State Engineer's Office in Colorado, presented a thesis at Colorado State University with a chapter entitled "Real-Time Reservoir Operations in Colorado: Background and Problems" (1991). He described problems of water accounting and reporting under the appropriation doctrine: problems of on-stream reservoirs that require the use of forecasting versus hindsight to estimate quantities of available water; and problems of multiple water users, all with their own demands and schedules, that must be accommodated within a system of administration that is plagued with imperfect data and decision support systems. The result is a system of "gentleman's agreements" among the parties that require sophisticated interpretation. On top of these problems is the specter that water quality issues may arise in the future in water rights administration.

Water Quality Law

Water quality management is discussed is some detail in Chap. 14. The centerpiece of the U.S. system is the Federal Water Pollution Control Act of 1972 (the Clean Water Act), which created the permit-based system of water quality management we have today. It involves a national permit system for point-source dischargers, a uniform system of technology-based effluent standards for industrial dischargers, and federal financial assistance. Evolution of the provisions of the act is discussed in Chap. 14, and this chapter provides an overview of the act to fit into the discussion of the framework of water and environmental law.

The Safe Drinking Water Act

The Safe Drinking Water Act (SDWA) is the successor to regulatory programs of the U.S. Public Health Service which began in 1914. It sets primary and secondary drinking water standards that apply to all public water supply systems. The primary standards govern contaminants that threaten health, and the secondary standards address those that threaten "welfare."

The American Water Works Association (AWWA) tracks trends in the SDWA and publishes a review article from time to time. Three of these are summarized here (Pontius, 1990, 1992, 1994).

As with the other major federal regulatory programs, the SDWA assigns the U.S. Environmental Protection Agency (EPA) to set regulations and oversee implementation of the act. State governments are to take primary enforcement responsibility (called "primacy") for the act. Public water systems are obliged to meet the regulations. This is a "command and control" system.

Under the 1974 act, the EPA was to adopt revised Primary Drinking Water Regulations by 1977, but the goal was not met. The actual schedule was: by 1977, inorganics, organics, microbiological contaminants, turbidity, and radionuclides; and by 1982, monitoring requirements for corrosion and sodium. Trihalomethanes (THM) were addressed on a variable schedule, depending on system size, and by 1983 THM compliance methods were addressed. Secondary drinking water standards were set in 1979.

The 1986 amendments to the SDWA were extensive. Some highlights included the requirement to set maximum contaminant levels (MCL) for 83 contaminants, to set MCLs for a list of priority pollutants updated every three years, to establish criteria for filtration of surface waters, and to require disinfection for all public water supplies.

By 1990, the EPA had promulgated regulations for volatile organic chemicals, fluoride, surface water treatment, total coliform bacteria, synthetic organic and inorganic chemicals, lead, and copper. By the time of the AWWA's 1993 article, regulations under the SDWA had about tripled.

Clearly, the SDWA will have strong and permanent effects on the U.S. drinking water industry. It drives the industry's regulatory structure, finance, research, and product development. These strong effects illustrate the key role of regulation in the water industry as a whole.

Environmental Statutes

There has been an explosive growth in environmental statutes since about 1970. Figure 6.1 shows a conceptual view of this growth. Some diagrams like this show no environmental statutes before about 1945, but they neglect the predecessor laws to today's legislation. For example, the beginning of water quality and public health laws were in place before 1945.

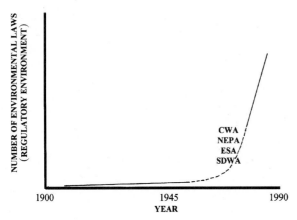

Figure 6.1 Growth of environmental laws.

One of the most far-reaching statutes is the National Environmental Policy Act (NEPA). It established goals and objectives for environmental policy and put into place the requirement for environmental impact statements (EIS). The EIS has had a major impact on the cost and results of water planning. For two examples, Chap. 10 describes the EIS process for the Two Forks project, and Chap. 15 describes the Virginia Beach Water Supply problem and EIS.

The Endangered Species Act provides substantial authority to the Fish and Wildlife Service to designate species for protection. It is discussed in more detail in the next section.

The Fish and Wildlife Coordination Act provides for wildlife resources and requires that wildlife conservation will receive equal consideration and be coordinated with other features of water resource development programs.

The Wild and Scenic Rivers Act provides that selected rivers of the nation which possess outstandingly remarkable scenic, recreational, geologic, fish and wildlife, historic, cultural, or other similar values shall be preserved in free-flowing condition.

The Coastal Zone Management Act provides broad regulatory authority for coastal zone states.

The Resource Conservation and Recovery Act provides tools for the management of groundwater.

Wetlands protection is a growing arena of environmental law, with most of the authority coming from Section 404 of the Clean Water Act, but with growing recognition in its own right (see Chap. 16).

Other statutes include the Comprehensive Environmental Response, Compensation, and Liability Act (CERCLA), or Superfund; the Federal

Insecticide, Fungicide, and Rodenticide Act (FIFRA); Forest Service legislation; farm bills; and the Federal Power Act.

The Endangered Species Act

The Endangered Species Act (ESA) has attracted much attention because it gives the Fish and Wildlife Service so much authority. Basically, the service can stop a project by listing a species as endangered. The U.S. Fish and Wildlife Service is responsible for all species except ocean species, which are looked after by the National Marine Fisheries Service. The agencies determine if a species is near extinction, and if so, they list it as "endangered" or "threatened." The agency is then required to devise recovery plans for the species. Exemptions can be provided by a "god squad," which is made up of cabinet-level officials (Robinson, 1992).

When the ESA turned 20 years of age in 1993, there was a flurry of publicity. A review stated that in its first years the act concentrated on win–win campaigns to save species such as whales, American bald eagles, falcons, pelicans, turtles, and manatees. However, the later years saw some questionable campaigns (Miller, 1993). In late 1992 the Bush administration reached a settlement with a group of environmental organizations to protect 400 more species by 1997, raising the number of protected plants and animals from 750 to 1,150 (Schneider, 1992).

Related to water resources planning and coordination, Section 7 of the ESA is one of the most powerful administrative authorities available. Section 7 is "the section of the Endangered Species Act of 1973, as amended, outlining procedures for interagency coordination to conserve Federally listed species and designated critical habitat. Section 7(a)(1) requires Federal agencies to use their authorities to further the conservation of listed species," and a Section 7 consultation involves "the various section 7 processes, including both consultation and conference if proposed species are involved" (U.S. Fish and Wildlife Service, 1994).

If, under the ESA, the Fish and Wildlife Service issues a "jeopardy opinion," stating that a species is in jeopardy, then a "recovery plan" is necessary, with elements as required by federal agencies. Such plans can be quite expensive, and are avoided as much as possible by water management agencies.

As an example, whooping cranes make up one of the most visible cases of endangered species; see the Platte River case, Chap. 19. Figure 6.2 illustrates how one "action site" (the proposed Two Forks Dam, see Chap. 10), is related to the extent of the critical habitat for an endangered species, the whooping crane.

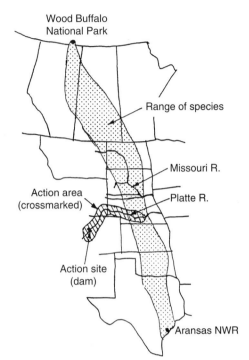

A dam on the Platte River in Colorado (action site) also may affect the water regime for a whooping crane critical habitat (action area) 150 miles downstream in Nebraska.

Figure 6.2   Range of endangered species action area. (*Source: U.S. Fish and Wildlife Service, 1994.*)

Transboundary Issues

One problem causing much conflict in water laws is the allocation of water across boundaries, a "transboundary conflict."

Recognizing the importance of water conflicts, the International Water Resources Association (IWRA) devoted an entire issue of their journal, *Water International,* to water, peace, and conflict management (Vlachos, 1990). The issue placed the spotlight on numerous conflicts, and Vlachos reported that conflicts will increase because nearly 40 percent of the world's population lives in river basins shared by two or more nations.

Utton (1987) stated that the issue is how to handle "migratory" natural resources such as water, air, and living things in divided political environments such as the International Boundary Water Commission between the United States and Mexico; the Colorado River, involving seven states and Mexico; international groundwater management; the Great Lakes; the Pecos River between New Mexico and Texas; Pomeranian Bay between East Germany and Poland; the Tigris-Euphrates in Turkey, Syria, and Iraq; the Rio Grande; the Nile; India–Pakistan conflicts; and others.

Interbasin transfer

In an interbasin transfer (see Chap. 15), water is removed from its natural watershed and transferred to another basin. Interbasin transfer can be viewed from the framework of at least three legal theories, riparianism, appropriation doctrine, or interstate law.

In the riparian view, Heath (1989) examines the issue of interbasin transfer as it applies in North Carolina. Heath sees interbasin transfer as an issue of riparian/permit systems of water law. He finds that riparian landowners commonly assume that they alone are entitled to use the waters running along their properties.

Interbasin transfers are always controversial. In the East they are handled on an ad-hoc basis, usually under the rules of the permit system or authority of the state government. In the West, they are handled more formally under state law.

Interstate compacts

Water flowing across state or national lines produces problems of water quantity and quality. How can the issues be resolved? The answer to this difficult question lies in making treaties, and in the United States these are called interstate compacts. The earliest of these go back to about 1920 (see Colorado River case, Chap. 19), and some are being considered even today (see ACF case, Chap. 19).

The interstate compact is a way to negotiate between political stakeholders in a river basin. Chapter 19 outlines how the real problem of water management in river basins is the lack of correspondence between political and hydrologic lines. A compact is a way to work out differences among the political groups. More details are provided in Chap. 15.

International Water Law and Law of the Oceans

In the United States the Constitution applies to disputes between states, but no such instrument is available to resolve disputes between nations. There is really no such thing as "international water law" because each country has its own system, some more developed than others. In this chapter I have reviewed U.S. systems of water law, for example, but other countries have different systems. Islamic countries, for example, follow religious approaches to water law based on the Koran.

McCaffrey (1993) reviews a number of ongoing international disputes over water: in the Middle East, the Jordan River, the Tigris-Euphrates Rivers, the "Peace Pipeline" project, and the Nile River;

the Indus and Ganges Rivers in the Asian Subcontinent; the United States, Mexico, and Canada; and the Parana River in Latin America. Each of these has a long story to tell of disputes over water.

According to McCaffrey, international law is a decentralized system relying mostly on self-help and the opinion of the world community. It lacks compulsory jurisdiction and centralized enforcement, necessary features of national systems. He says that treaties and international custom are the two forcing mechanisms to make international water law work, and quite a few treaties are in place. McCaffrey quotes the U.N. sources that there are some 2,000 of them in place, some more than 1,000 years old.

For international custom, the work of "recognized experts" is an important element. That is, when experts make recommendations for the resolution of disputes, such customs are suggested. Also, customs can be promulgated by international groups such as the Institut de Droit International (Institute of International Law, IDI), the International Law Association (ILA), and the International Law Commission (ILC) of the United Nations. These groups have promulgated a number of documents with agreed-on principles related to water law. They include the 1961 Salzburg Resolution on the Use of International Non-Maritime Waters (IDI), the 1979 Athens Resolution on the Pollution of Rivers and Lakes and International Law (IDI), the 1966 Helsinki Rules on the Uses of the Waters of International Rivers (ILA), the 1982 Montreal Rules on Water Pollution in an International River Basin (ILA), and the 1991 draft Law of the Non-Navigational Uses of International Water Courses (ILC).

In the final analysis, international disputes are much like those in the United States in that they ultimately depend on the willingness of the parties to come to terms. McCaffrey (1993) concludes: "the key to both peaceful relations with regard to shared water resources and sound management thereof is ongoing communication between the states concerned, preferably on the technical level . . . Experience has shown that such communication can most effectively occur through some form of joint mechanism such as a commission composed of experts from all basin states. Unfortunately, political frictions often prevent the formation of such bodies precisely in the cases they are needed most."

Drainage and Flood Law

In an engineering sense, there is a distinction between drainage and flood control problems. "Flood control" generally refers to larger, riverine problems, whereas "drainage" refers to diffuse waters and to more local problems.

In the law of drainage, there are three basic doctrines: the common enemy rule, the natural flow rule, and the reasonable use rule (Goldfarb, 1988). The common enemy rule basically says that you can do anything you like to protect your property, and it does not matter how you affect your neighbor. The natural flow rule is essentially antidevelopment; that is, you cannot change anything that will affect natural flows. Obviously, that is an impractical rule.

The reasonable use rule is the more common approach today. States that have precedents either in common enemy or natural flow roots are moving toward compromise. Under the reasonable use approach, you can modify your land somewhat, even if you affect your neighbor, but there would be a test of reasonableness. This rule recognizes that development will occur, but that there is a community obligation to work together to accommodate it.

Regulations such as detention storage to hold flood flows to historical levels are examples of reasonable use doctrine approaches. Say a city requires developers to detain stormwater up to the two-year flow. Anything greater than that will be an altered flow, but there will be a community responsibility to deal with it.

The legal basis for governmental involvement and regulation in stormwater and flood control is in state constitutions and local charters that authorize cities to improve the health and welfare of citizens. At the federal level, the authority is in various statutes that enable agencies such as the U.S. Army Corps of Engineers to undertake flood control work.

Stormwater programs cut across water and land use issues, and their authorities are not always clear. Typical local government program elements include city stormwater standards; subdivision regulations; stormwater quality control programs; erosion control and land quality programs; and programs for the control and beautification of urban areas, such as stream restoration, greenbelt construction, recreation, and environmental education.

Stormwater standards might involve a city ordinance that sets the return periods and levels of service required for stormwater systems. Subdivision regulations might impose standards and requirements on developers, and include items such as sidewalk requirements, street crowns, and inlet sizes. Related to them might be development standards that impose requirements for amenities such as greenbelts, walkways, ponds, and other stormwater-related facilities. However, today it is stormwater quality that is attracting the greatest attention in the regulatory agenda.

The 1972 Clean Water Act provided the basic authority to regulate stormwater, and in the 1970s there was a flurry of activity in studies and data collection, but no regulations were imposed. The Environmental Protection Agency began to move in 1984 to clarify its procedures for regulating stormwater dischargers. The rules issued September 26, 1984, in the *Federal Register* stated that stormwater discharges that come from urbanized commercial or industrial areas, or those "designated by the Director," had to apply for permits, and the *Federal Register* of March 7, 1985, stated that the deadline for applying for the permits would be extended to December 31, 1985.

This 1980s attempt to require permits failed, but Section 405 of the Water Quality Act of 1987 Amendments and Reauthorization mandated new controls and deadlines under the NPDES. The Natural Resources Defense Council filed several lawsuits, and on November 16, 1990, the EPA issued final stormwater discharge rules that required a complex two-phase application procedure for municipalities over 100,000 with separate stormwater discharges (about 225 cities and counties), for industrial dischargers (about 100,000), and for construction sites disturbing more than 5 acres.

By November 1991, cities over 250,000 were to have completed the Part 1 application with inventories of stormwater systems, screening for illegal connections, and plans for more extensive sampling. By May 1993, all cities over 100,000 were to have completed the Part 2 application with a separate list of industrial dischargers, sampling results, a stormwater management program, and a financial plan (Rubin, 1992). Rules for industrial plants affected manufacturing, power plants, airports, and municipal wastewater treatment plants.

In 1994, scrambling was on to rationalize the rules, which could be too expensive for the benefits received. There was quite a controversy about the potential cost. On the one hand, the EPA was quoted as believing that industrial applications would cost about $1000 per facility and cities about $50,000 to $75,000 (Beurket, 1990), but cities were calling their costs higher by an order of magnitude, with compliance costs much higher and benefits uncertain (Tucker, 1991). Tucker showed that Denver, Aurora, and Lakewood, Colorado, working with the Urban Drainage and Flood Control District, would spend about $2 million just on the application process. He also stated that Sacramento, California, estimated its compliance costs to be about $2 billion to meet standards for certain heavy metals.

Flood control programs by local governments have focused increasingly on floodplain management, and away from structural projects. See Chap. 11 for a discussion. This being the case, local land use authority is used to enforce the provisions. Over the years, a long list of legislative authorities has been passed for various aspects of flood control.

Groundwater Law

Getches (1990) describes the complexities of groundwater law. It may be based on land ownership doctrines or on the notion that water is a shared public resource. States take different approaches, and might recognize the ownership of water by the overlying land, but also recognize the need to limit use to reasonable levels, as well as to enable the landowner to protect the groundwater from contamination.

Savage (1986) notes that while groundwater is the principal source for drinking water for more than 50 percent of the American population, there is no comprehensive national groundwater legislation. There are, however, seven major statutes with some groundwater requirements, a number of EPA regulations dealing with groundwater contamination control, legislative proposals that duplicate existing authorities, and groundwater strategies that exist or are being developed in 46 states, which focus on existing authorities and deal effectively with groundwater problems. This explains why, in her opinion, there is no need for a national groundwater legislation. Chapter 23 discusses some of the thinking about the need for a national program to which Savage was responding at the time.

Some of the state preventative strategies include permitting and surveillance of potential contamination sources; siting requirements for industrial and municipal waste-generating treatment and other source facilities; promotion of proper pesticides and fertilizer use and animal waste management; public education; and prohibition of certain activities such as land disposal of toxic sludge and hazardous waste. Also, a number of states have groundwater quality standards or use classification systems either relating to drinking water quality or to manage industrial activity impacts on groundwater quality. All states monitor potable groundwater supplies, and most states conduct additional monitoring near potential polluters or in vulnerable areas. A significant number of preventative programs have been implemented, including common groundwater quality goals through use classification systems; expanded regulatory authority; improved septic system management and agriculture management; permit programs for groundwater withdrawal; delegation of authority to municipalities for inspection; and strategies for public education.

Savage believes that a new law will only duplicate existing statutory authority and will deluge state and local governments with new regulations. Also, it would not carry adequate funding for implementation.

Hydro Relicensing

The Federal Energy Regulatory Commission (FERC) has responsibility for relicensing of nonfederal hydroelectric power projects. Before the year 2000, about 200 projects are scheduled for relicensing, and about 150 are expected to apply for original licenses (Lamb, 1992).

The commission's authority stems from the Federal Power Act, which requires that it give equal consideration to environmental issues. The Electric Consumers Protection Act also requires that a project be best adapted to a comprehensive plan for a waterway, and requires license conditions to "equitably protect, mitigate damages to, and enhance, fish and wildlife, including related spawning grounds and habitat."

The Water Resources Development Act of 1986

The Water Resources Development Act of 1986 introduced a number of new reforms, especially in project planning and cost sharing. First, project planning became a two-phase project, with the first phase being brief and fully federally funded. In the second phase, a lengthier procedure is jointly managed and funded, resulting in recommendations for implementation. Next, the act provides that the sponsors' shares of project costs have increased and are based on percentages of total cost. Apparently, sponsors now provide financing during the construction period. This means that more complex intergovernmental-local cooperation agreements are needed between the core and sponsors (Vlachos, 1988).

Regulation in the Water Industry

Regulation is aimed at controlling activities to protect the public interest where private markets do not. Regulation in the water industry deals with health and safety, water quality, fish and wildlife, quantity allocation, finance, and service quality. Regulators are an important part of the water industry (see Chap. 8).

Environmental regulation is a critical political and economic issue. On the one hand, regulatory controls protect the environment, but controls can be so tight and arbitrary that resentment and political backlashes occur. As a result, some are searching for better models than the current statute-by-statute approach to regulation. Ecosystem protection through more comprehensive approaches may be one answer.

Figure 6.3 illustrates the regulatory dilemma faced by water agency and industrial dischargers. The public's interest in water quality is reflected in pressure from both the volunteer side (environmental groups) and the elected side (Congress and state officials). The results eventually come to bear in the form of permits, monitoring, and enforcement of regulations. The press is an important factor, and

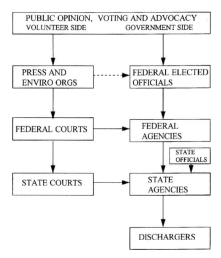

PUBLIC OPINION, VOTING AND ADVOCACY
VOLUNTEER SIDE   GOVERNMENT SIDE

PRESS AND
ENVIRO ORGS   FEDERAL ELECTED
                     OFFICIALS

FEDERAL COURTS   FEDERAL
                     AGENCIES

                     STATE
                     OFFICIALS

STATE COURTS   STATE
                     AGENCIES

                     DISCHARGERS

Figure 6.3   Regulatory and political controls on dischargers.

courts apply pressure, sometimes in response to public opinion. The diagram illustrates all "four" branches of American government: executive (agencies), legislative (Congress and state legislatures), judicial (courts), and the press (the unofficial fourth branch that influences public opinion).

Regulation of water management is necessary because of the interrelated and shared nature of the water resource. One person's waste affects another person's drinking water. Moreover, water utilities have monopoly franchises, either because they are public agencies or because their franchises are granted by state agencies. Thus, it would be impossible to have a water industry that was completely unregulated.

A saying that captures this principle is "You can't have the fox guarding the chicken coop." It is a version of the "separation of powers" principle in government. On the other hand, it appears that separation of powers is sometimes violated by the regulators themselves, because the same agencies that write the rules enforce them.

Interest groups, especially environmental groups, push their agendas through regulations and laws. The regulatory arena is where conflicts over business versus environment are worked out. In this sense, regulation is a "coordinating mechanism" for the water industry.

Each sector of the water industry has its own regulatory programs (the Safe Drinking Water Act; the Clean Water Act; floodplain regulation; in-stream flow laws; dam safety; the Federal Power Act; the National Environmental Policy Act; and others). Anyone working for the U.S. Environmental Protection Agency is a regulator, although some are more so than others. By the same token, the state EPAs or water quality agencies are regulators. The U.S. Army Corps of Engineers has a regulatory component by delegation of the authority to implement Section 404 of the Clean Water Act. The U.S. Fish and Wildlife Service and the Forest Service have entered the regulatory arenas with the Endangered Species Act and enforcement of federal water-related rules in National Forests.

In the West, state engineer offices are regulators in the sense that they control the diversion of water from streams and wells. Eastern states are increasing their activity in this area. State natural resources departments with dam safety missions regulate aspects of safety in the West. Similar functions have developed in the East.

State public service commissions regulate costs of water service for some utilities. Electric, gas, and telecommunications utilities rates are made public, and comparisons of costs are now easier for the public to make (National Regulatory Research Institute, 1983).

Environmental organizations are key players in regulation (Brimelow and Spencer, 1992).

A regulatory program must have an enforcement mechanism to be taken seriously. Law enforcement is a police function which everyone understands. This is very important in water management, but there are degrees of enforcement, just as there are in criminal justice systems.

Most of the experience in the water field is from the Clean Water Act, which gives authority to the EPA to enforce its provisions. This authority includes to enter and inspect premises, review records, test monitoring equipment, and take samples (Eizenstat and Garrett, 1984).

There have been a number of cases against corporate officials by the U.S. Department of Justice. Case histories of Clean Water Act enforcement actions, with court decisions against the polluters, include *United States v. Allied Chemical Co.* Allied Chemical Co. was fined $13.2 million for polluting the James River with toxic chemicals. This was later reduced to $5 million and an $8 million contribution to a fund to mitigate damage. In another case, *United States v. Ouellette,* a city sewage superintendent was convicted for falsifying monthly discharge permit reports (Eizenstat and Garrett, p. 2-36).

As you see, there is no comprehensive water industry regulatory policy. The total picture is a melange of federal, state, and local laws and regulations that govern water service providers and individual water users. We often hear calls for "regulatory relief" and "regulatory reform" because people and businesses do not like being "regulated." However, some regulation is a price we pay to live together in a civilized society.

Water Administration

The method of administering the law is through regulatory programs implemented mostly by state agencies. This administrative work forms a special set of tasks that might be called "water administration." Different aspects of it are described in Chap. 15 for water quantity and in Chap. 14 for water quality. To summarize these tasks, I offer a framework for water administration that consists of:

- Identification and response to problems
- Formulation of laws and rules
- Development of programs to administer the laws and rules
- Staffing, budgeting, and implementation of programs
- Monitoring and enforcement programs
- Systems for appeal of penalties and rulings
- Arrangements to review and modify laws and rules

Several examples of this framework are given in the case studies. For water quality, several issues related to enforcement are described (Chap. 14). Also, Chap. 22 illustrates how rules apply to estuaries. Chap. 15 illustrates how water quantity is administered.

Roles of Courts

Unfortunately, the third branch of government, justice, is an integral part of water resources management. I say "unfortunately" because when an action gets to court, it means that the voluntary, coordinated approach has broken down, but courts are necessary parts of the system. Also unfortunately, courts are not always able to deal with the level of complexity of water issues that professional engineers or scientists deal with.

The justice system can involve federal, state, or local courts. Courts that have been involved in cases cited in this book include the U.S. Supreme Court (see Pecos River case, Chap. 15); a U.S. District Court (see ACF case, Chap. 19, and Holt Reservoir flood case, Chap. 11); a state district court (see Colorado water administration case, Chap. 15); and an administrative law judgment (see Albemarle-Pamlico case, Chap. 22).

The main part of water law is statutory, but much of it is "case law," under which complex situations have been tried and precedents have been set. Attorneys search hard for cases to prove their points and to build arguments based on precedents.

Lawsuits are used often as management measures to gain decisions about complex issues. For example, in the ACF case (Chap. 19), Alabama filed a lawsuit against the Corps of Engineers, and that forced the initiation of a comprehensive study involving three states as well as the Corps.

Civic Environmentalism

The tight grip of regulation on business is a hot political issue, and the public seems to want less government. However, the public also wants a clean environment. This is the "front line" of sustainable development.

This author's opinion is that there will be no rollback of environmental laws and rules, and society must find ways to make progress within these constraints. However, sometimes coordination simply does not work; then, agency rules or the courts call the shots.

I have found that many others call for more cooperative, coordinated approaches, but the real issue is how to make them work. DeWitt John, in a book entitled *Civil Environmentalism: Alternatives to Regulation in States and Communities* (1994), presents an interesting concept for this dilemma. John identifies three features of the current "command-and-control" approach to environmental regulation: federal preemption of state and local authority, fragmentation, and a combination of tough procedural requirements with ambitious goals. In this system, the federal government is the "gorilla in the closet," to be employed when state and local programs fail. Anyone who has participated in this system senses the discomfort it brings to those who are regulated. John advocates the approach called "civil environmentalism" as an alternative to command-and-control regulation. This is a more collaborative and integrative approach to environmental policy and involves more bargaining among the players.

John presents five features of civic environmentalism: (1) a focus on the unfinished business of nonpoint problems, pollution prevention, and protecting ecosystems; (2) extensive use of nonregulatory tools; (3) interagency and intergovernmental cooperation; (4) a search for alternatives to political confrontation; and (5) a new role for the federal government as a participant in decisions made at the state or local level.

Questions

1. Early water law in the United States dealt mainly with control of water withdrawals and uses, but now water law is much broader. Explain what is meant by this and give examples.

2. What is meant by a "regulatory program" in water resources management? Give two examples of regulatory programs that are common in the United States.

3. Explain the differences among the appropriation doctrine, the riparian doctrine, and the permit system of water law.

4. Moslem countries have a unique doctrine of water law. How would you expect this doctrine to be related to religion?

5. If two states in the United States are unable to resolve their problems with shared waters, what venue and actions are available to them to resolve their problems? How would similar problems between two sovereign nations be resolved?

6. What is the basic doctrine of water withdrawal and use law in the western United States? Explain briefly how this doctrine works. Is this doctrine fully workable now? Why or why not?

7. Compare the appropriation doctrine and the riparian doctrine (permit system) for efficiency and equity, and explain how each does or does not protect the "public interest."

8. Explain the "public trust doctrine" in water law and management and how it affects state water policy.

9. Explain the following terms as they relate to water law: appropriation, adjudication, riparian, exchange plan, and augmentation.

10. Give three examples of national-level laws in the United States that affect water resources planning.

References

Beurket, Raymond T., Jr., US EPA Issues Stormwater Regulations, *APWA Reporter,* December 1990.

Brimelow, Peter, and Leslie Spencer, You Can't Get There from Here, *Forbes,* July 6, 1992.

Burger, Alewyn, Water Law, Seminar at Colorado State University, Spring 1989.

Eckhardt, John R., Real-Time Reservoir Operation Decision Support under the Appropriation Doctrine, Ph.D. dissertation, Colorado State University, Spring 1991.

Eizenstat, Stuart E., and David C. Garrett III, Clean Water Act, in L. Lee Harrison, ed., *McGraw-Hill Environmental Auditing Handbook,* McGraw-Hill, New York, 1984.

Getches, David H., *Water Law in a Nutshell,* West Publishing, St. Paul, MN, 1990.

Goldfarb, William, *Water Law,* Lewis Publishers, Chelsea, MI, 1988.

Heath, Milton S., Jr., Interbasin Transfers and Other Diversions: Legal Issues Involved in Diverting Water, *Popular Government,* University of North Carolina, Fall 1989.

John, DeWitt, *Civic Environmentalism: Alternatives to Regulation in States and Communities,* Congressional Quarterly Press, Washington, DC, 1994.

Lamb, Berton L., Accommodating, Balancing, and Bargaining in Hydropower Licensing, *Resource Law Notes,* No. 25, Spring 1992.

McCaffrey, Stephen C., Water, Politics, and International Law, in Peter H. Gleick, ed., *Water in Crisis: A Guide to the World's Fresh Water Resources,* Oxford University Press, New York, 1993.

Miller, Ken, Endangered Species Act Turns 20, *Denver Post,* December 17, 1993.

National Regulatory Research Institute, Commission on Regulation of Small Water Utilities: Some Issues and Solutions, Columbus, Ohio, May 1983.

Pontius, F. W., Complying with the New Drinking Water Regulations, *Journal of the American Water Works Association,* Vol. 82, No. 2, February 1990, p. 32.

Pontius, F. W., A Current Look at the Federal Drinking Water Regulations, *Journal of the American Water Works Association,* Vol. 84, No. 3, March 1992, p. 36.

Pontius, F. W., *The SDWA Advisor,* American Water Works Association, Denver, 1994.

Rice, Leonard, and Michael D. White, *Engineering Aspects of Water Law,* John Wiley, New York, 1987.

Robinson, Bert, and Scott Thurm, Environmental Act May Not Survive, *Denver Post,* May 10, 1992.

Rubin, Debra K., US Faces a Draining Experience, *ENR,* September 21, 1992, pp. 34–38.

Savage, Roberta J., Groundwater Protection: Working without a Statute, *Journal of the Water Pollution Control Federation,* Vol. 58, No. 5, May 1986, pp. 340–342.

Schneider, Keith, US to Preserve More Species, *Denver Post,* December 16, 1992.

Tucker, L. Scott, Tucker-Talk, *Flood Hazard News,* December 1991.

U.S. Fish and Wildlife Service, *Endangered Species Consultation Handbook, Procedures for Conducting Section 7 Consultations and Conferences,* U.S. Fish and Wildlife Service, Washington, DC, November 1994.

Utton, Albert E., The Emerging Need to Focus on Transboundary Resources, Transboundary Resources Report, International Transboundary Resources Center, CIRT, University of New Mexico, Albuquerque, Spring 1987.

Vlachos, Evan, The Water Resources Development Act of 1986: Thoughts on Water Resources Today, Colorado State University, December 1988.

Vlachos, Evan C., Water, Peace and Conflict Management, *Water International,* December 1990.

Financial Planning and Management

Introduction

Finance is often the limiting factor in water management. Money may be the arena for water conflicts, and the need for financial accountability is high in our contentious and litigious society. The goal of this chapter is to impart an overall view of financial management skills needed by the water manager.

Costs for water services are rising due to health, safety, and environmental regulation, as well as population growth and decay of infrastructure. This brings public and political scrutiny, and it is no surprise that, when asked what subjects they would like to understand more about, water managers (and public works or utility managers in general) reply "financing," along with "politics" and "communications."

By 1972, the National Water Commission (1973) concluded that the grand total of all water resources investments, including state and local water and wastewater, was about $340 billion in 1972 dollars. Of this, the federal investment was about $88 billion, mostly in water projects. Most of this investment required little or no local cost sharing, but this changed in the 1980s. Updating that figure to the 1990s and adding in newer investments in the water quality field would probably yield a total installed investment in current dollars between $500 million and $1.0 trillion. See Reuss and Walker (1983) for the history of U.S. investment in water resources.

In the 1980s and 1990s, local investment became more significant than federal, especially to meet environmental and growth requirements and due to the phase-out of wastewater treatment subsidies. Investments are made in pipelines, urban water and sewer networks, watershed and agricultural improvements, dam construction and improvement, treatment plants, hydroelectric facilities, and systems control systems, including SCADA (supervisory control and data acquisition).

Issues facing water managers in other countries are similar, but all managers must work within their own cultures. Findings of a World Bank study drew attention to these issues: the need to pay attention to finances, to avoid bad investments, and to rely on user charges wherever possible (Bahl and Linn, 1984).

Inadequate budgets are the cause of many infrastructure problems, and an important cause of budget problems is inadequate planning and management. Budget plans indicate what managers intend to do and how that will be accomplished. The budgets display an agency's programs: the operating budget shows plans for operating and maintaining the agency, and the capital budget shows plans for expansion and renewal.

Public works management trends are toward a self-supporting basis, following the *enterprise principle,* meaning that services should be self-supporting and charged according to the benefits to users. The use of pricing through user charges is fundamental to several aspects of water resources management. If a service is self-supporting, revenue decisions are mostly under the control of managers rather than the political process. There is still a political component, however, and citizen preferences must be considered; thus a balance between agency and political control is needed.

Elements of Financial Management

The water manager must apply the elements of financial management to problems encountered (Grigg, 1988):

- Financial analysis and planning to enable the manager to develop viable plans and understand the financial situation
- Budgeting as a tool to plan and control the specific aspects of financing systems and organizational work
- Financial control and reporting mechanisms to be sure that the details of financial management are under control
- A revenue management program to obtain the funds needed for the programs
- Cost control measures to make sure that waste and losses are minimized

Analytical aids needed are accounting tools such as the cash flow analysis, the income–expense statement, and the balance sheet.

Although some training in accounting is necessary to fully understand these tools, the essential components of them are easy to understand.

Financial management takes place in several functional offices within or in support of water management offices: budgeting, accounting, auditing, assessments, purchasing, and the treasury function. The manager should get to know the directors of these offices so as to be able to integrate the financial management.

Financial Planning, Analysis, and Budgeting

Financial arrangements for both facilities and programs must be planned. According to Raftelis (1989), writing about water and sewer facilities planning, the steps are:

1. Evaluate the economic factors.
2. Prepare a comprehensive facility master plan.
3. Schedule capital financing requirements.
4. Evaluate alternative financing methods.
5. Determine annual operational and capital revenue requirements.
6. Calculate fees and charges.
7. Evaluate impact on customers.

Another list of steps in financial analysis and planning includes (Government Finance Research Center, 1981):

- Revenue analysis
- Cost analysis
- Institutional analysis
- Ability-to-pay analysis
- Secondary-impacts analysis
- Sensitivity analysis

The facility master plan and implementation plans will result in a cost analysis which should consider planning and economic factors, design and construction costs, operating and maintenance costs, and the cost of regulatory programs. This aspect of financial planning should also result in the scheduling of revenue requirements, both capital and operating funds.

Revenue analysis determines the feasibility of sources of money, from both an economic and a political basis. It is a projection of potential funds available. All potential sources should be considered, including fees and charges, debt financing, grants, intergovernmental transfers, and so on.

Institutional analysis is to evaluate the capability of institutions to manage the program. Ability-to-pay analysis is closely related, in that it determines the capability of customers to bear the cost of the program.

Sensitivity analysis examines the changes in outcomes of the analysis that might result from changes in assumptions. It is a final check on the reality of the assumptions and the analysis.

Financial planning and projections can be presented using projected financial statements, the income statement, the cash budget, and the balance sheet (Block and Hirt, 1981). The income statement provides estimates of revenues and expenditures over a time increment—say, one year. The balance sheet illustrates changes in assets and liabilities over the accounting period. These reports require accounting values such as depreciation and book value, and may be quite complex. The income statement displays receipts and disbursements, and the balance sheet displays debt structure and accounts payable. In a public enterprise, the cash budget shows the recovery of costs and self-sufficient financial operation.

Feasibility analysis is a critical part of project planning, and financial feasibility is an important aspect of it. Feasibility analysis was discussed in Chap. 4. Financial feasibility differs from economic feasibility in that it determines capacity to pay, whereas economic feasibility deals with whether a project should be built in general. Benefit–cost analysis (BCA), the best-known tool of economic feasibility analysis, deals with aggregate economic efficiency and does not focus directly on who pays the bills. That being said, it is only fair to say that BCA can be adapted to deal with who gets the benefits and who pays ("incidence analysis"). Economic tools are closely related to financial analysis tools; the difference is in the use of the information.

In most local plans, the main financial issues are what is the best way to achieve the objective, what it will cost, how it can be paid for, and how public approval can be gained for the plan. These were evident, for example, in a recent master plan for wastewater treatment expansion improvements for the City of Ft. Collins, Colorado (1990).

The planning staff used several background studies as a starting point. These included growth projections and service area studies; plans of adjacent sanitation districts; studies of expansion alternatives and future plan locations; water quality studies; wastewater rate studies; cost allocation studies; and studies of funding sources. Findings were that the city would need additional capacity within three years; that the potential service areas were not fully determined because of ongoing growth; that the development of a regional wastewater facility was likely; that there existed several alternatives for expanding the capacity; that the best alternative was to expand the existing plant first, then develop a larger regional facility; that construction costs should be allocated to both new customers for growth and to existing customers for enhancing treatment; that the plan would increase wastewater rates by around 6 percent per year for six years; and that the best options for financing the first part of the improvements were to issue revenue bonds or to apply for a low-interest loan from the state's Pollution Control Revolving Fund Program. The outcome of the planning process was satisfactory, and the elements of the plan were generally followed.

The Budget Process as a Tool in Financial Management

The budget process is the manager's most important tool for financial planning. The budget provides the links among planning, operating, and controlling. For example, in the Ft. Collins wastewater expansion just cited, the expenditures must be programmed into both the capital and operating budgets.

Figure 7.1 illustrates a typical budget process, which is valid for a large or small organization. For example, in the U.S. federal government, arguably the most complex organization in the world, the budget preparation phase takes place in agencies during the year preceding the President's budget message, presented in February of the year preceding the budget year. After the President's message, the legislative phase occurs. Last year (1995), the Republican Congress declared the President's budget "dead on arrival." That is a common tactic in the budget wars in the United States.

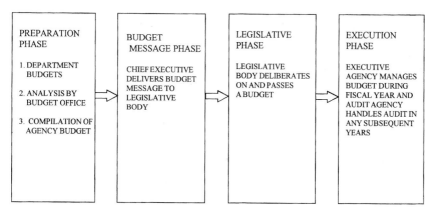

Figure 7.1   Phases in the budget cycle.

For a small organization—say, a water district—the preparation phase might be for a couple of months early in the year, the director might present the budget to the board in, say, October, and the board would act on it before January 1, when it would go into effect.

The budget is much more than just a tool for allocating money. It is an adopted plan for expenditures and revenues structured to follow the programs and divisions of an organization. After the budget is authorized, it becomes the official plan for the operation of the program for the fiscal year of concern. The budget process is the overall procedure of planning for, negotiating, presenting, adopting, following, and auditing the budget for the organization or program. The budget document is important as a communications vehicle. Budgeting is one of the most sensitive acts of the governing body, and an important management tool.

Budgeting involves many decisions about the policies and directions of the organization. The budget also states how the revenue will be made available, whether from debt, user charges, or other sources.

The budget goes far beyond fiscal accountability. New techniques, such as "planning, programming, budgeting systems" (PPBS) and "zero-based budgeting" (ZBB) have been tried, but due to practical difficulties, the focus always returns to the basic process.

Capital budgeting, the procedure for budgeting for capital items having lifetimes longer than about one year, is tied to capital planning and programming through the comprehensive plan, the capital investment program, and the capital budget. The links between planning and capital budgeting were presented by the U.S. General Accounting Office (1981). The most critical elements in success have to do with how well planning is linked with budgeting.

Budget politics and game playing affect water organizations just as they do other public agencies. Wildavsky (1984) presents some concerns in budget politics: for the agency, roles and expectations, deciding how much to ask for, deciding how much to spend, department versus bureau, the role of the budget office, and deciding how much to recommend; for the appropriations committees, roles and perspectives, deciding how much to give, and client groups.

Capital and Operating Budgets

Figures 7.2 and 7.3, from a document of the City of Ft. Collins (Colorado) Water and Wastewater Utility (1988), illustrate the separation of budgeting and financial management into operating and capital budgets. Figure 7.2 shows the system expansion area of operations to include purchase of water rights, treatment plants, and other infrastructure. As Ft. Collins has a philosophy of "growth pays its

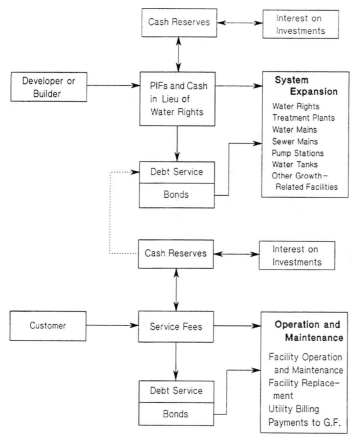

Figure 7.2 Capital and O&M cycles of utility. (*Source: Ft. Collins Water Utility.*)

own way," these items are generally paid for by the developer or builder, who then passes the cost on to the homebuyer or business. Figure 7.3 shows how this works in the "system expansion cycle" and the "O&M cycle." The different impacts of service fees versus plant investment fees and debt are shown.

The only additional aspect of capital budgeting that should be mentioned is that in Fig. 7.2, facility replacement is shown as an operation and maintenance item. That is true in distinguishing the system expansion cycle from the O&M cycle, but in budgeting, facility replacement is a capital item rather than an O&M item. However, it might be paid for by debt that is serviced by service fees, not contributions of developers or builders. This is shown in Fig. 7.2.

System Expansion Cycle

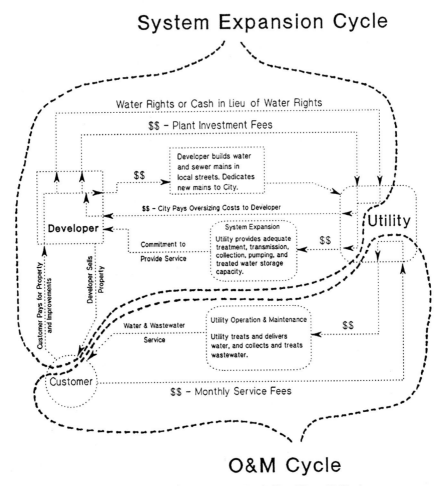

Figure 7.3 Funds flow in water utility. (*Source: Ft. Collins Water Utility.*)

Financial Control, Accounting, and Reporting

Financial control reports furnish the management information needed by boards of directors, customers, and regulatory agencies. Accounting provides information for management decisions and reporting and involves analysis and record keeping. Auditing checks on the validity of financial reports, and is generally carried out by accountants other than those who keep the regular accounts.

Utilities and enterprise funds should use the accrual method of accounting (Gitajn, 1984). Expenses and revenues are credited when they occur, rather than when the cash is received or disbursed; then financial reports reflect the actual financial health at all times.

Accounting uses "funds" with separate statements. The National Council on Governmental Accounting suggests eight categories of funds: the general fund; special revenue funds; debt service funds; capital project funds; enterprise funds; trust and agency funds; intergovernmental service funds; and special assessment funds.

For water supply utilities, the American Water Works Association suggests that the annual report should contain the balance sheet, the income statement, the statement of retained earnings, the statement of changes in financial position, and a five-year summary of operations, accompanied by management's analysis of performance.

The financial audit can be used to aid in broader "performance audits," including three elements: financial, economic, and programmatic. The definition of these three elements by the U.S. General Accounting Office is as follows:

> Financial and compliance—determines (a) whether financial operations are properly conducted, (b) whether the financial reports of an audited entity are presented fairly, and (c) whether the entity has complied with applicable laws and regulations.
>
> Economic and efficiency—determines whether the entity is managing or utilizing its resources (personnel, property, space, and so forth) in an economical and efficient manner and the causes of any inefficient or uneconomical practices, including inadequacies in management information systems, administrative procedures, or organizational structures.
>
> Program results—determines whether the desired results or benefits are being achieved, whether the objectives established by the legislature or other authorizing body are being met, and whether the agency has considered alternatives which might yield desired results at a lower cost.

Revenues for Operating Budgets

The operating budget has to be renewed each year and should be financed from recurring revenues linked to system benefits such as rates and user charges. There should be minimum reliance on subsidies, and close connection between the services rendered and charges imposed. Almost all of the revenues for water supply, wastewater, and stormwater systems should be from user charges and not from tax subsidies. Subsidies, such as the wastewater construction grants program, are diminishing. Sometimes it is appropriate that funds be provided from general taxes, but the primary source of operating revenue should be rates and user charges.

Rates and user charges allow the implementation of the "user pays" principle. User fees should lead to economic efficiency and equity in services. "Efficiency" means that there is no waste, and "equity" deals with fairness in providing and charging for services. Theories for how user fees should be assessed become complex because the specific aspects of efficiency and equity involve many parameters. For example, the Ft. Collins metering case and the water pricing case in Chap. 21 illustrate political complexities involved in setting charges and fees.

Opposition to user fees involves arguments such as: services bring social benefits that cannot be measured and charged for; tax payments for services provide for the redistribution of income to those who cannot afford vital services; public facilities and services attract economic development and the resulting tax revenues help pay for the services (this is actually the argument used to justify "tax increment financing"; dedicating tax revenues to specific services (a form of user charges) reduces flexibility to budget and manage public services in a time of changing priorities; a separate approach to managing specific services with dedicated user charges inhibits the coordination of different public services (Vaughan, 1983).

Subsidies cannot always be avoided, such as in the case of providing vitally needed services when they cannot pay for themselves. Examples of subsidies are the construction grants program for wastewater (capital grants) and subsidies for irrigation systems, such as allowance for deferred maintenance, with "catch-up" grants and loans for "rehabilitation."

Setting user fees high enough is important in developing countries that are trying to improve water and sanitation services. The Pan American Health Organization has been working to improve this situation for 25 years. Morse (1985) describes the obstacles—public mistrust of the government, the assumption that water should be free, and the assumption that increasing the cost of water will fuel inflation—and he concludes that lending agencies need to emphasize this issue.

Water Pricing: Determining Fees and Charges

Setting the prices for water use has become a prominent topic in water policy debates. In this section, several aspects of the issue will be discussed.

To begin, there is a healthy debate about water pricing, and as a result, one can see that pricing, like law, can be used as a tool to further societal objectives in managing water. Thus, all of the complexities of "social objectives" can be introduced into debates over water pricing. This is, however, a complex debate.

The Wall Street Journal (1991) captured some of the debate in an editorial about using pricing and regulatory policy to restore endangered species. The Journal stated that: "if a judge and his trusty biologist think salmon-species salvation is worth any price..." and "utilities meanwhile warn that protecting the salmon could deprive the region of several times the annual megawatts it takes to keep Seattle lighted..." and if the "broader public cared enough to protect the fish ahead of some other perceived need, legislators could appropriate funds to buy water flow at a marginal price higher than the power companies and the farmers are willing to pay." The *Journal* also opined that: "Approximating a free market in natural resources isn't going to be easy—especially when so many parties have careers and causes at stake. But it's hard to think of any other mechanism capable of arbitrating the myriad demands of millions of people in an economy....Without a pricing system to mediate the process, the swings of human nature can be violent."

Charles W. Howe (1993), a prominent water economist, set the stage for the discussion in an article entitled "Water Pricing: An Overview." Howe stated that, although determining the correct price for water is complex, the correct price ought to be "the amount paid per unit of water withdrawn from the supply system, for the next (or marginal) unit withdrawn." He chose this definition because it is the cost that "a rational user will compare with marginal benefits in deciding how much water to apply" to different uses such as domestic, industrial, irrigation, and other water uses. Howe concluded that "water prices, appropriately set and applied at different points of the water supply and use cycle, perform many valuable functions, namely to confront water users with the costs of providing water, to help signal water suppliers when supply augmentation is needed, and to help shape a rational approach to a healthy water environment."

Procedures for setting fees and user charges vary among the water services. Most attention has been given to water supply utilities which have well-developed models for setting rates and user charges. Procedures for wastewater are not as well established, and have been driven by the EPA's requirements related to federal grants. A 1973 report by a joint committee of three professional associations recommended a split between property taxes and user fees (APWA, 1973). Stormwater systems have initiated a stormwater utility concept, and user charges are based on parameters such as lot size and runoff coefficient. These charges are related to the property value, so it is still somewhat like a property tax. The future of stormwater user charges is still to be worked out. Procedures for setting charges to irrigation users are complex because some irrigation water is furnished through subsidized government projects. A discussion of this issue is included in Chap. 21.

After analyzing the general aspects of user fees, Vaughan (1983) presented five principles for their use: They should be levied on the beneficiaries of the services; prices or fees should be set at the marginal or incremental cost of providing the service, not the average cost; peak load pricing should be used to manage demand; special provisions should be made to ensure adequate access to services for low-income residents where burdens will result from marginal-cost pricing; and user fees should be responsive to inflation and to economic growth. These are good principles in general, but they must be applied to specific cases.

Another writer about water charges, John Boland (1993), outlined the objectives of water pricing as applied to urban water supply:

Economic efficiency—to promote patterns and levels of water use that minimize the total cost of meeting the service area's water needs

Fairness—to be perceived as fair by water users and the public

Equity—treating equals equally, or that all who purchase water with the same cost pay the same price

Revenue sufficiency—the charges support all costs including maintenance, pay-as-you-go capital outlays, and debt service

Net revenue stability—to provide a cushion against fluctuations in demand and natural conditions

Simplicity and understandability—avoid unneeded complexity and be understandable to water users and decision makers

Resource conservation—to promote conservation of scarce resources

In addition to these, Boland stated that, to provide for implementation, the tariff should avoid rate shock, provide for a smooth transition for easy implementation, and support good bond ratings.

In the water supply field, the general rate procedures are contained in a manual of practice of the American Water Works Association (AWWA) (1983). According to the AWWA, the rate-setting process consists of the determination of revenue requirements, the determination of the cost of service by customer classes, and the design of the rate structure itself. The AWWA manual specifies two basic approaches: the commodity-demand method and the base-extra capacity method. The difference between these two methods is essentially in the way they classify the costs.

These AWWA procedures are consistent with Vaughan's principles. By determining the "cost of service according to customer classes," the utility can be sure that the charges are levied on the beneficiaries of the service. Ideally, the procedure will also lead to charges being set at the marginal cost of providing service and to a valid consideration of added costs due to meeting peak loads. The other two principles, serving low-income residents and responding to inflation and growth, may require decisions outside the direct process.

The "cost of service" approach to setting rates and charges is a valid way from the point of view of business managers and for public enterprise. The main problem with it is that it is impossible to merge all the "social costs" into the "cost of service." For example, if a city can provide water to residents for $3.00 per thousand gallons, but in so doing it robs environmental uses, how will the "environmental costs" be considered in the rates? The answer is complex and relates to the discussion of the "public trust doctrine" in the chapter on water laws, but in a practical sense, the city might simply decide to override the "cost of service" and charge higher rates for conservation purposes.

In 1993 and 1994, the City of Ft. Collins undertook a "cost of service" study to allocate costs to customer classes for the water and wastewater utility. In general, for water supply the AWWA manual was followed, and for wastewater, cost allocation considered the cost impact of discharges from various industries and classes of customers. As a result of the study, five issues were identified and debated:

1. Whether to expand the utility's rate making to include cost of service and demand management policy

2. Whether to revise residential wastewater rates and link them to winter-quarter water consumption rather than average annual water consumption

3. Whether to introduce for nonresidential customers a "conservation block rate" to penalize excessive use

4. Whether to eliminate a standing 50 percent surcharge on outside-city customers

5. Whether to phase in cost-of-service rate adjustments

At the time of this writing, these revisions are for the most part being implemented, although some of the more controversial ones, such as charges to businesses for "excessive water use" are being sent back to staff for further evaluation. While a number of points of controversy remain, clearly the trend in this case is to move toward valid principles of rate making.

"Cost of service" is a way to assess costs in a fair way in proportion to how different parties benefit from a project or a service. At a larger scale, water projects have general, or *joint costs,* and costs that are identifiable with specific beneficiaries, called *separable costs.* Recognizing this, a method for multipurpose water resources projects has been developed called the *separable costs–remaining benefits method,* and it is sometimes used for cost allocation. A multipurpose water project might have, for example, three purposes: water supply, flood control, and hydropower. The hydropower might be produced by the government and sold wholesale to private utilities, so the separable costs for hydropower would be financed by user fees. Flood control might be negotiated and financed jointly by the federal and state governments. Water supplies might be financed through sale of the water to local governments through long-term contracts.

Another example of cost allocation might be a drainage and flood control project. Land developers benefit through improvements to their property, and some benefit accrues to the public at large. The city might therefore allocate costs to developers and pay for the general public benefits from tax revenues.

Water pricing and cost allocation are complex subjects, and it would be nice if they could be done through simple formulas and without negotiations, but there are too many actors and policy issues involved.

Capital Budgets

Capital financing is at least as complex as operations financing. Generally, the problem is how to come up with the funds needed to build, replace, modernize, or renovate a facility, such as, for example, the wastewater treatment plant expansion mentioned earlier.

Think of capital financing the same way as you would think of buying a new car. You could either save until you had the funds to buy it, or you could borrow the money to buy it. In the former case, you would be paying before you use it; in the latter, you would pay as you use the car. In the former case, you give up the interest you could be earning on your savings, so for your car you pay your savings and the forgone interest; and in the latter case you pay the principle of the loan and the interest charge.

It is mostly the same in water facility financing, except that it is, in general, illegal to charge today's customers for water project benefits enjoyed by tomorrow's residents. In the case of the car, the saver and the user would be the same, but it is different with customers of a utility. For this reason, and for utility financing in general, debt financing is mostly used for large capital needs. Normally, revenue bonds are used, but some variations are possible.

The basic reason for borrowing and debt financing is that future beneficiaries of a project are supposed to pay for it. If existing customers contribute capital to a fund (sinking fund) and if a project is constructed from the accumulated capital, then future customers would be financed by past customers, a situation that is basically illegal. The only remedy to this is if the revenue from future customers is used to repay former customers, but in this case the former customers would be charged excessive fees, almost amounting to a "forced savings plan."

How can borrowing best be carried out in the public sector? What should be the rules and procedures? *Capital budgeting* is the parallel topic in private-sector financial analysis to public-sector borrowing (Young, 1993).

In the private sector, the capital budgeting problem is: "Given the availability of X capital, what is the optimum way to invest it; or, given the opportunity to borrow capital, what is the optimum plan? In the case of borrowing, the risk is magnified.

In the public sector, the parallel problem is simpler. Given a project where capital is needed, what is the best way to finance the plan, to borrow the capital? The decision as to what project to build has been made in the public marketplace using social objectives that are determined politically (health, jobs, social welfare, safety, environment, preservation). *In effect, the political decision as to which objectives to pursue replaces the private-interest decision of profitability.*

What interest rate to use in studies is an important topic in public-sector decision making. For local studies, such as to finance a water treatment master plan, the cost of capital should be used.

Dealing with inflation is an important aspect of investment timing. The problem can be framed this way. Given increasing demand which causes a need for more capacity, when should investments be made and at what rate? To solve this problem, picture the variables that measure the rate of growth of demand, the changes in construction cost, the relationship between the cost of capital and the rate of inflation, and formulate an optimization problem that minimizes the real cost to consumers. As you can see, that makes a complex mathematical problem, and the data will not all be clear, as, for example, the future inflation.

Recurring capital improvements may be financed from current revenues. In this case, revenues will be placed into reserve accounts to be used when needed. Current revenues are a simple and efficient way to generate capital, but they are easy to divert when a crisis occurs.

Debt financing can come from the bond market or from loans. Water facilities are mostly in the public sector, so tax-exempt bonds are used. Debt financing may be attractive due to inflation and interest rate uncertainty. When interest rates are low, it pays to use debt to build. If interest rates are high, current revenues may be a better choice, with the option of borrowing at more favorable terms later. When the time period for repayment is the same as the life of the facility, debt financing is "pay as you use." The facility is fully paid for just as it needs replacing.

The emphasis is toward revenue-secured debt to obtain capital financing, and away from general-obligation (GO) bonds. The trend away from GO bonds can be explained by the requirement in some places that voters approve the bonds, and by the popularity of the "user pays" principle.

As revenue bonds are paid by dedicated revenues, the repayment scheme should make sure that all citizens have fair access to needed services. Also, if charges are too high or not well distributed, it may adversely affect economic development.

Debt Financing

Capital financing from debt can come from bonds or loans, though mostly it comes from bonds. Bond financing is primarily from general-obligation or revenue bonds, or from combinations. With difficulties in raising funds, attention is sometimes given to "creative bond financing," using techniques such as deep discount/zero-coupon bonds, variable-rate bonds, put option bonds, bonds with warrants, mini-bonds, and mini-notes, but regardless of "creative financing," the emphasis is on revenue-secured debt as the vehicle to obtain capital financing, and away from general-obligation bonds.

Vaughan (1983) reported that revenue debt was nearly three times general-obligation debt. This seems consistent with the emphasis on enterprise management. However, GO bonds may be more important to infrastructure than that ratio suggests, since revenue bonds are often used for home mortgage lending, industrial development, and quasi-private investment, leaving the GO bonds primarily to infrastructure-related purposes (Valente, 1986).

General-obligation bonds are backed by the full faith and credit of the organization issuing the debt. The bonds are usually paid off with some source of revenue, but the guarantee is with the taxing power of the entity. Of course, an organization must have taxing power to issue GO bonds. It makes sense to issue GO bonds when the project involved has community-wide benefits, such as municipal buildings, public schools, streets and bridges, and economic development programs.

The trend away from GO bonds can be explained by a number of factors. One reason is the requirement in some places that voters approve the bonds, in some cases with two-thirds approval. This kind of approval is expensive and difficult to secure. Another reason is the tendency to try to assign payment responsibility to the beneficiaries of projects, the "user pays" principle.

Interestingly, the Northeast is more likely to have local governments using GO bonds, with large, metropolitan areas being more likely to use them than small communities located outside metropolitan areas. The West seems to have stronger traditions of voter approval than the Northeast, with less use of GO bonds (Valente, 1986).

Revenue bonds are used when the dedicated revenues of a self-supporting project can be used to pay off the bonds. Revenue bonds can be issued by more entities than are able to issue GO bonds, and are usually viewed as being riskier, with correspondingly higher interest rates. Infrastructure services such as water, power, buildings, solid waste, parking garages, airports, and other facilities that can be used for a fee are candidates for revenue bond financing.

Since revenue bonds are paid by dedicated revenues, the same concerns that accrue to user charges for services need to be considered. First there is the question of equity. The repayment scheme needs to consider the need to make sure that all citizens, regardless of their income level, have a fair access to needed services. Then there is the question of economic development. If charges are too high, or not well distributed, it may adversely affect the community's ability to compete.

The approach to debt financing for managers is to determine how much money is needed and when, and to find the best deal for it. This will require professional advice, of course. There are many firms today vying for the right and business of providing this advice. The repayment of the debt will require funds to be allocated from revenues, usually from the operating budget, to retire the bonds. Preparing for a bond issue is complex and expensive. This is one of the disadvantages of going into the bond market.

The process of issuing bonds illustrates the roles of different parties in a bonding arrangement. Bonds are sold to finance a project which provides services to a governmental organization that deals with users of the service, such as water users. The bonds are issued by the "issuer," who goes through the trustee to sell them to the bondholders. The revenues then flow back from the users to the issuer and eventually to the bondholders. There are many variations of this basic plan, of course. Investment banking firms are active in the infrastructure financing area. Social events are organized by investment bankers to attract municipal officials and others who are empowered to issue bond debt. These events are often held at the conventions of groups such as the Government Finance Officer's Association, the National Association of State Treasurers, the Airport Operator's Council, the International City Manager's Association, and the International Bridge, Tunnel and Turnpike Association, as well as other infrastructure professional associations.

Touche Ross & Company (1985) completed a survey of infrastructure financing needs and plans. They received 19 percent response to 5000 questionnaires sent. The conclusion was that GO bonds and federal grants seemed the best ways to finance facilities, with revenue bonds and special assessments coming in next. Less than 30 percent favored privatization, tax increment financing, infrastructure banks, or other financing means. Again the regional difference is there, with the East favoring GO bonds and federal grants, and the West favoring revenue bonds more than GO bonds. While respondents did not favor tax hikes, they seemed to indicate that they may be inevitable. Also, even though the use of privatization and infrastructure banks was not favored by respondents, Touche Ross believes that they will be used more in the future.

The subject of debt financing is tied directly to capital budgeting and decision making. This is one of the most important subjects of financial management in all organizations, even the U.S. government. In discussing the topic, it is necessary to consider additional topics such as investments, inflation, interest rates, and currency exchange rates, especially when projects are financed in other than the home country's currency.

Of course, there is no such thing as a "free lunch," and there are limits to the use of debt financing. The statutory limits mentioned earlier apply for a reason, with usual limits of about 10 percent of assessed value. The default of the "Whoops" bonds made many headlines in the 1980s. "Whoops" is the Washington Public Power Supply System (WPPSS), which defaulted on $2.5 billion in municipal bonds in 1983. These had been used to finance nuclear power. With such a large default, there are many questions being asked by regulators and the public about the bonding process. Some brought forward by Leigland (1986), author of a book on the default, are:

Why were the bonds characterized as "obligations of the United States"?

Why did the bonds say "hydro-backed" when they were for a nuclear facility?

What is the responsibility of the underwriter?

Why were the bonds rated highly?

Why was the WPPSS pressured to build the facilities?

If there were staff difficulties, why were they not discovered before all the trouble?

These questions and the default of WPPSS show the limits to which debt financing can go, and offer a caution to both the issuer and the purchaser of high levels of debt instruments.

System Development Charges

System development charges have become an increasingly important part of overall capital financing strategies, since they provide a way to isolate the cost to serve a particular segment of a system, and to levy charges for it. In effect, they allow new users to "buy into" an existing system by paying their fair share of it. As a simple example, consider a community that has a water supply system already built, complete with adequate capacity for new developments. When a new development comes in, the fair cost of serving it is calculated and it is charged a system development fee to pay for its share of the system. This fee, of course, is passed onto the purchasers of developed property in the form of higher costs for the land, or they would be required to pay the fee themselves. Ultimately it passes the cost of the infrastructure on to the property owner.

In discussing the approach Ft. Collins uses to levy user charges for operating costs, some examples were given, including the transportation and drainage utility fees. Ft. Collins also uses the principle of "growth paying its own way" to justify the levying of user charges for new developments. These user charges or system development fees are used mostly for capital expenses. The ones in use are:

- Water plant investment fee
- Water rights acquisition charge
- Sewer plant investment charge
- Storm drainage fee
- Street oversizing fee
- Offsite street improvements
- Electric offsite and onsite fees
- Parkland fees

The total of these fees in 1982 for a $75,000 house on a 7200-square-foot lot was $7025, a figure that is about average for the region. Fees were slightly higher in the Denver metropolitan region, with a high of $9694 being registered by Lakewood, Colorado, a suburban city incorporated only in the late 1960s. The use of fees such as these to finance growth is common in the western United States, and becoming more accepted in other areas as well. Obviously, charges of the magnitude shown are of great concern to home buyers and home builders.

Opposition to system development fees, a form of user charges, will certainly appear when they begin to increase. "Impact fees," a particular type of system development charge which is levied on new development, attracts opposition from developers.

Grants and Subsidies

Grants have been an important part of the overall strategy of paying for infrastructure systems, but grants have dropped in popularity because of various trends, mostly the financial difficulties of central governments and the lack of accountability of the groups receiving the grants. Grants are essentially gifts to the entity receiving them, and they don't do much to encourage the "user pays" principle. Grants became popular in the United States in the period beginning with the Johnson administration and ending about 1980.

In the United States, the federal wastewater construction grants program, for example, financed some $40 billion in treatment facilities in the 10 years after the Clean Water Act passed. This was considered necessary by the government because there was at that time only a weak acceptance of the "user pays" principle for wastewater services.

Grant programs are scarcer now, and it appears that the emphasis will be on the enterprise concept and self-finance. Also, the federal government has devolved many responsibilities to state and local governments. Grants remain important sources of financing for local infrastructure; however, they are not as plentiful as they once were.

In terms of water system financing, the term *subsidy* means financial assistance granted by a government, and is essentially the same as a grant.

Development Banks

Development banks may be a principal source of funds for water program financing. Basically, a *development bank* is an institution created for the purpose of making loans to assist in economic development projects such as water projects. For example, in 1984, of the combined total of some $15.5 billion lent by the World Bank, $2.6 billion was for transportation projects, $0.6 billion was for water supply and sewerage, $3.5 billion was for energy projects, and $0.5 billion was for urban development. Another $3.5 billion was for agriculture and rural development, much of which was for water management through infrastructure systems.

In many countries the development bank is a principal source of funds for infrastructure development and financing. Many engineers and planners have worked with the World Bank, but there are many other development banks in operation as well. *ENR* magazine published a list of international development banks in 1985 (*ENR*, 1985): World Bank, African Development Bank, African Development Fund, Asian Development Bank, Inter-American Development Bank, Bank Ouest-Africaine de Development, Caribbean Development Bank, Central American Bank for Economic Integration, East African Development Bank, Abu Dhabi Fund for Economic Development, Arab Bank for Economic Development in Africa, Arab Fund for Economic and Social Development, Iraqi Fund for External

Development, Islamic Development Bank, Kuwait Fund for Arab Development, OPEC Fund for International Development, Saudi Fund for Development, European Development Fund, European Investment Bank, Overseas Economic Cooperation Fund, Japan, and the International Fund for Agricultural Development. With political shifts, development banking changes. For example, since the breakup of the Soviet Union, a new development bank has been created for European development, and after the North American Free Trade Agreement (NAFTA), a development bank was created for environmental cleanup along the U.S.–Mexico border.

In general, both regular and subsidized loans may be made by development banks. Also, deferred payment plans may be available to help with financing schemes. A regular loan would be repaid at market interest rates, but a subsidized loan would be repaid at less than market rates, perhaps even with no interest, as, for example, in a long-term rural development project. Of course, in times of inflation, a long-term, no-interest loan is essentially the same as a grant where the borrower never has to repay any significant sum.

The development bank is an attractive way to solve infrastructure problems since it offers a combination of self-finance and subsidy. This is illustrated by the World Bank with its hard window, IBRD, and soft window, IDA. When the United States was studying ways to improve its infrastructure systems in the early 1980s, the concept of an "infrastructure bank" surfaced many times. Such a bank would be, in effect, a development bank.

Although the U.S. infrastructure bank has not been implemented, some of the principles endure. The fund would provide predictable amounts of funds on a multiyear basis; in this way local and state governments could prepare realistic capital budgets. Compliance would be monitored by the General Accounting Office or another auditing bureau. Funds would be managed by the states, but the operation of the overall fund would be reported in a national capital budget.

Since the development bank may lose money, depending on the degree of subsidy, it is necessary to have make-up funds from the supporting governments. The bank will also be free to borrow additional funds from the bond market, these being repaid according to usual practices of bond financing.

Local, regional, and national development banks are in operation around the world. Any loan fund can be considered a development bank. Some of the major banks are the International Bank for Reconstruction and Development (World Bank), Inter-American Development Bank, and Asian Development Bank. In 1992, a European Development Bank was organized.

Privatization

Privatization means that public entities turn over the financing and/or operation of all or some part of an enterprise to private companies. Raftelis (1989) defines it as "private sector involvement in the design (if appropriate), financing, construction, ownership, and/or operation of a facility which will provide services to the public sector." According to Raftelis, privatization has attracted so much attention in recent years because of rising costs, growing demand for services, citizen initiatives for tax and fee limits, and disappearance of grants and tax incentives.

The potential roles for private companies focus on the delivery of services and on industry support such as design services. If effective regulation and coordination are provided, there is no reason that any water service cannot be provided by the private sector, at least in a healthy and functioning economy. Almost any service will be a monopoly, however, so if the private sector undertakes a service such as providing local water supply, then regulation of business activity as well as health and environment is required.

The basic philosophical question behind privatization is whether the government or the private sector should provide the needed service. The most visible examples are privatization of national industries, and the pendulum has swung back and forth between government and market-based industrial policies. After World War II, for example, the British government's Labor Party went on a campaign to nationalize certain industries, and in the 1980s the conservative government of Margaret Thatcher went on a campaign to reverse the earlier service and privatize the industries. Although the British water industry had been mostly in the public sector before, it was privatized along with the rest. In the 1990s, with the dismantlement of the Soviet Union and with the generalized discrediting of state-sponsored socialism, a large campaign was underway to privatize industries in Eastern Europe. Almost every country and international organization is promoting privatization as a valid alternative to provide water services.

Even in the poorest countries there are excellent opportunities for privatization of water services. In the World Bank's view, private-sector involvement and user associations are considered as two related ways to avoid control by central governments and large state companies and to devolve service closer to the level of the user. Avenues to provide services include concessionaire contracts, management contracts, private ownership, and participation by users and communities in managing water resources. According to the World Bank (1993), these can provide a sense of responsibility for water systems, improve accountability and concern for users' needs, constrain political interference, increase efficiency, and lower the financial burden on governments.

In the United States, privatization in the water industry became a popular topic in the 1980s with the general swing away from federal government involvement in the water industry's operations (not regulation). Whereas state involvement and centralization had been popular government programs in the 1960s, by the end of the 1970s democratization and privatization had become centerpieces of government policy, not only in the United States but in foreign policy as well.

The initial debates over privatization focused on advantages and disadvantages. The reasons why privatization works, according to Raftelis (1989), are construction cost savings because some government red tape can be avoided; procurement and scheduling efficiencies; risk reduction because the private firm will guarantee performance; operational savings because of private-sector advantages in hiring and training personnel; tax benefits such as accelerated depreciation and investment tax credits; debt capacity benefits because government debt limits are not imposed; and greater access to capital in the private markets.

The above benefits are hypothetical in some cases, and the jury is still out on some of them. They may or may not occur on a case-by-case basis. In any event, some of the disadvantages of privatization, again according to Raftelis (1989), are: loss of local control; and negative aspects of long-term contracts, such as getting out of a contract or correcting a contract if the deal goes sour.

Private water supply companies have operated for many years in the United States. Some public water utilities—Denver's, for example—got their start as private companies. Like electric companies, private water companies are regulated public utilities. The National Association of Water Companies coordinates their activities, and reports 86 private water companies with revenues in excess of $1 million. Many smaller water companies provide services in suburban and rural areas. The smallest ones may struggle to meet increasingly stringent regulations, and are candidates for merger or being taken over by larger, central city water utilities.

It has been reported that during the Great Depression the stocks of water companies held up better than most and were regarded as among the safest investments of that time. In 1983 water companies were rating higher price/earnings ratios on the stock exchange than electric companies. The reasons, in a time of financial turbulence, are reported to be stability and the exemption from taxation of earnings, if the company is bought out by the public sector (Blyskal, 1983). If such premiums occur, they come in spite of reported problems of private water companies: low profitability, regulatory hassles, resistance to rate increases, and local politics.

In Europe and Japan, privatization in the water field has gone, in many ways, even further than in the United States. In the United Kingdom, the largest water authorities have been privatized. According to Thackray (1990), the U.K. water authorities were facing large needs for repair, renewal, and upgrading of services and infrastructure for water supply, wastewater, and stormwater systems. The privatization program involved legislation, timed for the summer of 1989, sale of water authority assets to the shareholders, restructuring of rate structures, and new ways of doing business in a regulated environment. Proceeds of the sales went to the central government, with the share prices being set at appropriate levels to reflect risk and return. Two effects of the privatization program were a large reduction in the number of employees in the water industry and some near-term profits for shareholders due to rises in share prices. The big unknown in the privatization campaign was the level of deferred maintenance that had to be capitalized in the sale.

In France, private companies play important roles in water treatment and distribution. According to Drouet (1990), the arrangements in France can be considered "a tête à tête between 36,000 municipalities and three large private corporations, with the central administration providing the basic rules." Most of the some 50 private companies operating in the French market are concentrated in three groups: the Compagnie Générale des Eaux, the Société Lyonnaise des Eaux, and the Société d'Aménagement Urbain et Rural (SAUR).

Japan has more centralization and cooperation between the public and private sectors than the United States. The River Bureau of the Ministry of Construction promotes comprehensive water resources development, and the Water Resources Development Public Corporation (WRDPC) pursues water development projects. The WRDPC also relates to the Ministry of Health and Welfare for domestic water; the Ministry of Agriculture, Forestry and Fisheries for irrigation water; the Ministry of International Trade and Industry for industrial water; and the Ministry of Construction for river administration.

During the 1980s, much of the discussion of privatization has been about wastewater treatment applications, a service that has become a more important and expected part of municipal responsibilities in recent years with the passage of environmental legislation. By 1986 some 30 firms offered full contract operations and maintenance services at over 200 facilities.

Since most of the press is about success stories, we have to be cautious about privatization, however. One utility director, writing in response to "success stories," stated that performance has been less than expected with the operations contractor (Wallner, 1986).

There will be other opportunities for private-sector involvement in the future, but the pendulum between public and private-sector operations will continue to swing.

Questions

1. Wastewater utilities collect and take away wastewater, and protect the water quality in the environment. Which of these has the greater private purpose, and which has the public purpose, and who should pay for each?

2. The "enterprise principle" is an important principle of financial management. Explain it and how it influences the effectiveness of management. State when using it is a good idea and not a good idea in the financing of water enterprises. How does the principle relate to the debt problem in countries or at levels of government?

3. Planning, programming, budgeting systems (PPBS) came in with systems analysis as an attempt to make government work better. Explain what PPBS is and how it relates to water resources planning.

4. Would you use fees or debt or both to finance a capital budget? Explain why.

5. What factors explain why wastewater rates around the United States are being increased?

6. What is the main feature of the Water Resources Development Act of 1986? Why is it considered so important?

7. You borrow $100,000 from the bank at an annual interest rate of 12 percent for five years. Payments are to begin at the end of year 1 and continue for five years, when the loan will be completely paid off. Make a table and show the amount of the annual payments and the interest and principal you are paying at the end of each year.

8. You have $5000 to invest for 10 years. The interest rate is 8 percent, but the inflation rate is 5 percent. Compute what you will have after 10 years in today's dollars. What is the "real" interest rate?

9. Explain why using the "cost of service" principle to set urban water rates is not favored by environmentalists.

10. Developing user charges for wastewater is more difficult than for drinking water. Why? Can you suggest how it should be done?

References

American Water Works Association, *Water Rates,* AWWA Manual M1, 3d ed., Denver, 1983.

APWA, ASCE, and WPCF, *Financing and Charges for Wastewater Systems,* Chicago, 1973.

Bahl, Roy W., and Johannes F. Linn, Urban Finances in Developing Countries: Research Findings and Issues, *Research News,* World Bank, Washington, DC, Spring 1984.

Block, Stanley B., and Geoffrey A. Hirt, *Foundations of Financial Management,* Richard D. Irwin, Homewood, IL, 1981.

Blyskal, Jeff, Water Money, *Forbes,* February 14, 1983.

Boland, John J., Pricing Urban Water: Principles and Compromises, *Water Resources Update,* Universities Council on Water Resources, Carbondale, IL, Issue 92, Summer 1993.

City of Ft. Collins, Colorado, Water and Wastewater Utility, Michael B. Smith, Director, Cash Flow Cycle, April 11, 1988.

City of Ft. Collins, Colorado, Wastewater Master Plan Committee, Report to the Ft. Collins City Council on Wastewater Treatment Expansion, May 8, 1990.

Drouet, Dominique, The French Water Industry in a Changing European Context, in Kyle A. Schilling and Eric Porter, eds., *Urban Water Infrastructure,* NATO ASI Series, Vol. 180, Kluwer Academic Publishers, Dordrecht, 1990.

ENR, Funds for Development Growing, May 2, 1985.

Gitajn, Arthur, *Creating and Financing Public Enterprises,* Government Finance Research Center, Washington, DC, 1984.

Government Finance Research Center, Financial Management Assistance Program, *Planning for Clean Water Programs: The Role of Financial Analysis,* U.S. Government Printing Office, Washington, DC, 1981.

Grigg, Neil S., *Infrastructure Engineering and Management,* John Wiley, New York, 1988.

Howe, Charles W., Water Pricing: An Overview, *Water Resources Update,* Universities Council on Water Resources, Carbondale, IL, Issue 92, Summer 1993.

Leigland, James, Questions That Need Answers before We Go "Whoops" Again, *The Wall Street Journal,* July 10, 1986.

Morse, Charles, In Quest of Higher Water and Sewage Rates, *Newsletter, U.S. Section, Inter-American Association of Sanitary Engineering,* June 1985.

National Water Commission, *Water Policies for the Future,* U.S. Government Printing Office, Washington, DC, 1973.

Raftelis, George A., *Water and Wastewater Finance and Pricing,* Lewis Publishers, Chelsea, MI, 1989.

Reuss, Martin, and Paul K. Walker, Financing Water Resources Investment: A Brief History, U.S. Army Corps of Engineers, EP 870-1-13, July 1983.

Thackray, John E., Privatization of Water Services in the United Kingdom, in Kyle E. Schilling and Eric Porter, eds., *Urban Water Infrastructure,* NATO ASI Series, Vol. 180, Kluwer Academic Publishers, Dordrecht, 1990.

Touche Ross & Company, *Financing Infrastructure in America,* Chicago 1985.

U.S. General Accounting Office, Comptroller General of the United States, *Standards for Audit of Governmental Organizations, Programs, Activities and Functions,* Washington, DC, 1972.

U.S. General Accounting Office, *Federal Capital Budgeting: A Collection of Haphazard Practices,* Washington, DC, February 26, 1981.

Vaughan, Roger J., *Rebuilding America: Vol. 2, Financing Public Works in the 1980's,* Council of State Planning Agencies, Washington, DC, 1983.

Valente, Maureen F., Local Government Capital Spending: Options and Decisions, *Municipal Yearbook,* International City Management Association, Washington, DC, 1986.

Wall Street Journal, "Lox Horizons," editorial, May 29, 1991.

Wallner, Michael J., Utility Director of Fort Dodge, Iowa, letter to editor, *Waterworld News,* May/June 1986.

Wildavsky, Aaron, *The Politics of the Budgetary Process,* 4th ed., Little, Brown, Boston, 1984.

World Bank, Privatization and User Participation in Water Resources Management (Appendix C), in *Water Resources Management: A World Bank Policy Paper,* Washington, DC, 1993.

Young, Donovan, *Modern Engineering Economy,* John Wiley, New York, 1993.

8

Water Industry Structure

Introduction

Policy studies show that institutional and organizational issues of water management are critical in making systems work better and achieving sustainable development. The structure of the water industry sets the institutional framework for the responsibilities, authorities, tasks, and administrative functions required of water managers. The water manager must understand how water industries are organized so that he or she can get decisions made, succeed in business endeavors, or coordinate activities with related organizations.

The water industry is the collection of utilities, regulatory agencies, firms, interest groups, and support organizations that deliver water services, protect the water environment, or generally provide any function that focuses on water management. It is complex, multifaceted, intergovernmental, and, to some extent, a public–private enterprise. Figure 1.2 showed the general structure of the water industry in the United States.

This chapter outlines the structure of water industry organization and identifies management and service functions, as well as the needs for coordination. Related chapters are Chap. 4 (planning and decision making), Chap. 6 (water laws), and Chap. 10 (organizational case studies of water agencies). The chapter presents models of organization, overviews of the industry's components, and examples from the United States and a few other countries. The details of organization vary widely among nations. This is the reason for the "country focus" of the World Bank's comprehensive framework (see Chap. 9).

General Industry Frameworks

To see the range of public–private organizational possibilities, let us examine three simple cases of water industry structure: a country with a relatively simple economic system and an authoritarian political system without well-developed democratic traditions; a country with a more complex economy, well-developed democratic traditions, and a tendency toward high levels of government involvement in public services; and a similar country that favors the private-sector approach to providing public services. These countries might be, for example, a stable developing country, a European nation with a socialistic government, and another European nation that has privatized many services, such as, for example, England in the 1980s.

In the first case above, the water industry is likely to be what I would call "statist." Referring to Fig. 8.1, you see that there will normally be a "Ministry of Water Resources" and some form of state companies or regional offices of the ministry that deliver services. At this extreme, this is definitely a "command and control" approach to the water industry, and it is likely to have a number of problems, mostly for the reasons outlined in Chap. 7 under the discussion of privatization, and see Chap. 25 for a case study of how this can affect the water industry.

In the second case (Fig. 8.2), the structure is more ambiguous, but it will be a mixed water industry with some activities in the public sector, some in the private sector, and some shared. The U.S. water industry fits this model, as will be explained in this chapter.

In the third case, services would be mostly private. Figure 8.3 shows a hypothetical "integrated private-sector water company" that covers all functions of water management. England, after the privatization of the water industry, fit this model to some extent. England's approach will be described later in the chapter. Notice in Fig. 8.3 that regulators are separate from the company.

Regardless of the case, the functions will be similar. Many of these functions, such as service delivery, can be furnished by either public or private entities, but others, such as regulation, are suitable only for government. This illustrates the two dimensions of the water industry:

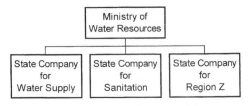

Figure 8.1 Statist water industry model.

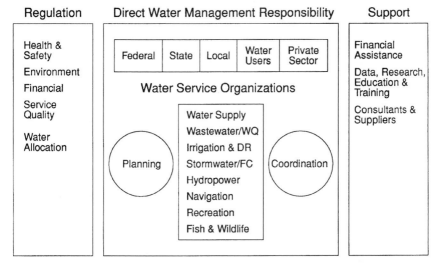

Figure 8.2 Integrated public–private water industry model.

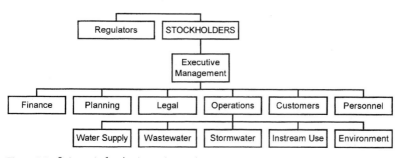

Figure 8.3 Integrated private-sector water company.

provision of services and regulatory structure. The statist versus private models refer to the two extremes of providing services, government or private. The other dimension, regulation, would have "command and control" as one extreme and "laissez faire" as the other.

If services are delivered by public entities, the enterprise principle should apply. As discussed in Chap. 7, this is the principle that all costs are recovered. Unless the enterprise principle is followed, and if subsidies are provided, you might end up with the bloated state companies that plague some countries.

In any event, most water services are monopolies; if services are provided by a private company, they will require regulation. See Chap. 6 for a discussion of the legal aspects of regulation.

Concept of a "Water Industry"

While water resources management includes important functions and activities, the concept of a "water industry" has not yet caught on. Where it has, it refers mostly to commercial aspects of water supply and wastewater activities, not the full range of an integrated industry. Government economic statistics are not kept for a "water resources industry," and the fragmentation shows up in policy studies. For example, the National Council on Public Works Improvement (1988) divided public works infrastructure into nine categories, and it took four of the nine to report on water (water supply, wastewater, water resources, hazardous waste).

In this author's view, the water industry involves four types of groups: service providers (those with direct responsibility for water services); regulators (those responsible for regulating rates, water quality, health issues, or service levels); planners (those responsible for planning and coordination functions other than in the course of providing services or regulating); and support organizations (those providing support in data management, consulting services, construction, legal advice, loans and financial support, research, equipment and supplies, and other support).

In Figs. 8.1, 8.2, and 8.3, the four groups are represented in different ways. In the "statist" model (Fig. 8.1), everything is provided by government. In the private-sector model (Fig. 8.3), everything but the regulators could be in the private sector, although depending on the extent of franchises, coordination between different groups would be necessary. In the U.K. model, private companies hold franchises for complete river basins, which reduces the need for coordination. In the French model, the river basin organizations are the coordination mechanisms. In the mixed U.S. model (Fig. 8.3), there are separate roles for the four groups, and some coordination is by the market, some by government, and some by informal mechanisms.

This author's estimate, adding industrial water and wastewater expenditures, government operations, irrigation expenditures, environmental water management, hydropower, and navigation, is a U.S. water industry total of about 2 percent of Gross National Product, a large part of the economy.

Service Providers in the Water Industry

Service providers are the principal management agencies, or the line organizations, of the water industry. They deliver the services and meet the fundamental purposes of the water industry; they are the "front-line troops." Services can be placed into five general categories: water supply; wastewater and water quality control; stormwater and flood control; in-stream uses management; and environmental water management.

Services are provided to four categories of "customers": people/cities (referred to as domestic, public, and urban water services); industries [when combined with domestic and public services, the term used is "municipal and industrial" (M&I)]; farms (referred to as irrigation and drainage services; if rain-fed agriculture, then only drainage may be involved); and the natural environment (this is a neglected customer needing more attention to achieve sustainable development).

Water supply utilities

Water supply is the primary urban water service, the other two being wastewater management and stormwater. A water supply utility provides water to the public for household and business purposes and is regulated by the government. A *public utility* is "a business that supplies an essential commodity to the public, as water, electricity, telephones, or transportation, and is regulated by the government, esp. as to rates" (*Scribner-Bantam,* 1979). The management goal in a water supply utility is to meet the needs of customers, within a political framework, while holding rates down and meeting regulatory standards.

In the United States, water supply utilities are for the most part divisions of city governments, private water companies, or special districts. Publicly owned companies, according to the American Water Works Association (AWWA), are usually found in one of the following four forms: part of a city department, with mayoral control; part of a city department, under the council–manager plan; a separate city department under a water board or water commission; or a separate utility district. Investor ownership can involve different forms, but the largest of the investor-owned water companies are corporations.

The Environmental Protection Agency's Office of Drinking Water distinguishes between "community water systems (CWS)" and "noncommunity water systems." As of 1983, there were some 58,700 CWS, with 11,000 being served by surface water sources and 47,700 being served by groundwater sources. The 158,100 noncommunity systems serve transients and customers and are regulated for water quality by EPA programs (U.S. Environmental Protection Agency, 1984).

Organizational structures of water supply utilities vary, but the Ft. Collins, Colorado, water and wastewater structure illustrates the main functions as they appear in an integrated organization. As Fig. 8.4 shows, supply is planned by the water resources and planning manager. Facility design and construction are managed by the engineering department, treatment is handled by the water production manager, and distribution by the distribution and collection manager.

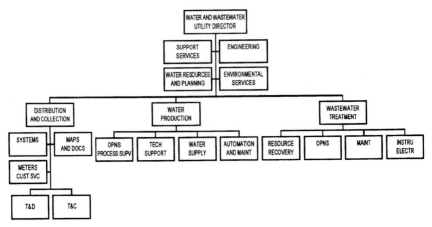

Figure 8.4 Utility organization. (*Source: Adapted from City of Ft. Collins, 1991.*)

The director of the water supply utility must deal with a tight regulatory climate and the political aspects of rate setting and other issues of concern to the public, including environmental issues. To succeed, he or she must combine technical knowledge with management instincts and communication skills. In small to medium-sized utilities, the manager must possess all of these skills. In large utilities, the emphasis shifts from technical to public and political issues.

The combined role of manager, policy specialist, and communicator in large utilities is described by Bill Miller, who was general manager of the Denver Water Department for 12 years. Miller had a background as a reporter, an attorney, a public relations specialist, and a director of local government agencies. While this background suggests an emphasis on management and communications, Bill also had access to an outstanding technical staff. His book, *Miller On Managing,* illustrates the management challenges faced in the industry today: employee relations, regulation, legal issues, communications, finance, relations with boards and commissions, and political issues (Miller, 1992). One of the most challenging political issues faced by Miller was to steer the Denver water utility through the Two Forks Reservoir case, described in Chap. 18.

The area where performance is under the most scrutiny is the quality or safety of the water supply. The Safe Drinking Water Act is the principal vehicle to regulate water supply but, as discussed in the chapter on water law, the cost of the program and public concerns about water are leading to alternative approaches, such as increased point-of-use treatment.

Interest groups are involved in engineering, health, environmental, or business aspects of water supply. The main association, the American Water Works Association (AWWA), located in Denver, Colorado, has the affiliated AWWA Research Foundation (AWWARF), a semiautonomous organization that is quite active in projects and publications. Other associations with programs in water supply include the National Association of Water Companies (NAWC) for private water companies, the American Public Works Association (APWA), the International Water Supply Association (IWSA), the American Society of Civil Engineers (ASCE), the Inter-American Association of Sanitary Engineers (AIDIS), and the American Society of Plumbing Engineers (ASPE). Also, utility contractors are an important industry group. State governments work through the Association of State Drinking Water Administrators (ASDWA).

Wastewater management utilities

The term "wastewater" generally includes dry-weather wastewater (sanitary sewage), wet-weather wastewater (combined sewage), and stormwater (storm sewage). In contrast to water supply utilities, wastewater utilities are newer and face potentially greater challenges. The service is a mirror image of water supply: Instead of source, treatment, distribution, the sequence of operations is collection, treatment, disposal.

The service provided by wastewater management, as viewed by the public, is getting rid of used water, but from society's viewpoint, the service is to keep the supply clean. Thus, wastewater management must balance individual and societal objectives.

Management of wastewater is mostly by city departments and special-purpose districts. While the water supply industry has relatively good inventories of utilities, and the AWWA has a water industry database, no similar data collection efforts exist for the wastewater industry. The best database on the wastewater industry is the needs assessment issued semiannually by EPA.

The distribution of size of wastewater utilities is similar to that of water supply: a few very large ones and many small ones. The large ones have the money to manage sophisticated programs; management of small systems is a challenge because of the complexity of wastewater treatment.

There are about 1000 communities in the United States with populations over 50,000, and most have a wastewater utility. There are another 3000 communities with populations over 10,000, and most of these also have wastewater programs. Apogee Research, Inc., estimated from the EPA's 1986 needs assessment that there were 15,438 wastewater treatment facilities in the United States, distributed in capacity as shown in this table (Apogee, 1987):

Flow range, mgd	No. of plants
0.01–0.10	4,960
0.11–1.00	7,003
1.01–10.0	2,898
10.1–	577
Total	15,438

In 1982 the EPA listed 32,511 "facilities" (U.S. Environmental Protection Agency, 1983). By their 1986 report, the number of facilities was 35,042 (Apogee, 1987). This included 15,438 treatment plants and 19,604 collection systems (some plants receive wastewater from several collection systems). The number of collection systems, about 20,000, is probably a better indicator of number of utilities than the number of treatment plants, although the numbers are not much different. The difference in number of wastewater collection systems and number of community water systems might be explained by more consolidation in wastewater systems, by accounting practices, and by the use of on-site treatment in a number of small communities that have community water systems.

The main wastewater industry association is the Water Environment Federation (WEF), which represents interests of operators, wastewater managers, design engineers, government officials, financiers, and equipment manufacturers. The Association of State and Interstate Water Pollution Control Administrators (ASIWPCA) represents state water regulators and interstate management officials. The Association of Metropolitan Sewerage Agencies (AMSA) represents some of the largest wastewater agencies. The American Public Works Association, with headquarters in Kansas City and a Washington, DC, office, also numbers wastewater managers among its members.

The WEF was formerly called the Water Pollution Control Federation (WPCF). The WPCF's board made a decision to change the name at its 64th annual conference in 1991, showing a maturation of the wastewater management industry. The federation's background illustrates interesting aspects of the nation's development of a wastewater industry. The federation has now had four names: Federation of Sewage Works Associations (1928–1950); Federation of Sewage and Industrial Wastes Associations (1950–1960); Water Pollution Control Federation (1960–1991); and Water Environment Federation (1991–).

Stormwater and flood control

Compared to other water services, stormwater management and flood control (SWMFC) have an "identity crisis." They do not provide a commodity, like water supply; and they do not provide a convenience, like wastewater. As a result, stormwater and flood control are often hidden from view and do not gain much public support.

Stormwater received little attention until recent years, but the quality of the living environment suffered. Now, with urbanization and nonpoint sources of water pollution, stormwater management is receiving more attention. Prior to the 1960s, the terms used were "drainage" and then "urban drainage." It was the 1970s before "stormwater management" started to stick. Some still use the term to refer to drainage only, and others use it to refer to water quality issues.

The wet-weather design problems of combined sewer systems are similar to those of stormwater; the main difference is that the combined sewer system is subject to the "overflow" problem, which means that the stormwaters reaching the receiving waters are contaminated with full-strength wastewater as well as with stormwater runoff. Stormwater runoff, however, is sometimes higher in the level of contaminants than some full-strength wastewater.

Stormwater systems drain minor stormwaters and control some flood flows. These two objectives are viewed somewhat separately because control of minor flows can be just for the convenience of the public, whereas controlling flood flows may prevent greater damage. In other words, the public is more concerned with flooding than with minor stormwater flows.

Stormwater programs can also provide benefits from erosion and sedimentation control, control of the quality of urban runoff through management practices and treatment of wet weather flows, and the beautification of urban stream environments. The cleaning function of stormwater is described by the caption on a 1970s EPA film on stormwater quality: "When a city takes a bath, what do you do with the dirty water?" When the city takes a bath, we have the cleansing service, and taking care of the dirty water is the job of the water quality protection service.

Although stormwater benefits are difficult to measure, stormwater programs can introduce multiple-purpose water/land improvement programs into development. There is an opportunity for "synergism" in development when stormwater programs serve the integrative function of bringing groups together for joint planning.

The concept of the stormwater "utility" has arisen in recent years as a method of financing the service. This is an attempt to overcome the "out of sight, out of mind" problem that plagues stormwater managers. Although there are successful utilities, there have also been failures.

The American Society of Civil Engineers has given attention to stormwater in the last 30 years through its Urban Water Resources

Research Council. The American Public Works Association has also undertaken investigations concerning stormwater. The flood management agencies have formed the National Association of Urban Flood Management Agencies (NAUFMA).

By the same token, flood-control interest groups do not have a unified focal point, but quite a few groups have an interest in flood control. These are the groups with an interest in dams, waterways, insurance companies, and disaster control.

Coordination of flood control programs is essential. Schilling et al. (1987) list 26 different federal agencies involved, and groups such as the Association of State Floodplain Managers and the Association of State Dam Safety Officials work along with the Inter-Agency Committee on Dam Safety to coordinate.

Irrigation and drainage services

Irrigation evolved as an agricultural technology to support farming in dry regions of the world, and it enabled people to settle arid and semi-arid lands. In effect, it enabled farmers to overcome the fluctuations of climate that limit food production (Lea, 1985).

Irrigation and drainage have made tremendous contributions to food production for centuries, but the practices now face greater competition for water supplies and changing societal attitudes. Today they are causing unresolved environmental problems of great concern. Although irrigation is receiving close scrutiny in the United States, we must remember that not all nations have adequate food supplies, and irrigation systems are even more essential in other countries.

The basic concept of irrigation is to supply either all of the water needed for crop production or to supplement rainfall when it is inadequate. Irrigation increases soil moisture content, which is then available for plant growth.

The five-year census of agriculture reports published by the U.S. Department of Agriculture (USDA) are the primary sources of data about irrigated agriculture in the United States. The *Irrigation Journal,* an irrigation industry publication, publishes data about irrigated acreage by state. Every five years, the U.S. Geological Survey (USGS) publishes reports on estimated use of water in the United States, and these are the primary sources of data about irrigation water use.

The first major irrigation program in the United States was instituted by the Mormon settlers in Utah, about 1847. The second one was at the Union Colony, near Greeley, Colorado. Irrigation spread from there across the West, and by the early 1900s it was extensive in the entire western United States. In the eastern United States, irrigation grew more slowly, with a focus on rice farming in the Mississippi Delta region and on citrus and vegetables in Florida. By the 1980s it was increasing along the eastern seaboard, with a number of other crops being brought under supplemental irrigation, mostly by sprinkler irrigation.

In 1974 only 12 percent of the nation's total harvested cropland was irrigated, but it yielded 26.9 percent of the total value. This amounted to 36.6 million harvested acres of irrigated farmland, yielding $15.1 billion of gross revenue, or $413.13 per acre. To compare, nonirrigated farmland yielded a gross revenue of $153.39 per acre.

The organizations involved in irrigation include federal agencies such as the Bureau of Reclamation, Department of Agriculture, the USGS, and increasingly, the EPA. State governments are not involved much except for regulatory programs, and local irrigation management districts in the West carry much of the management responsibility there.

State programs, all in the 17 Western states, included planning, technical assistance, research, education and demonstration, construction or cost sharing, loans, and water administration and regulation. For the most part, this has meant that the states have the usual state agency water programs, and they operate land-grant university cooperative extension and agricultural research programs. Actual technical and financial assistance from state governments is quite limited in the irrigation sector.

Local districts include both water delivery irrigation districts and soil and water conservation districts. There are nearly 8000 water system delivery organizations and about 1200 conservation districts in the 17 Western states.

Irrigation associations are intertwined with agricultural interest groups and water associations. For example, in Colorado the Four States Irrigation Council is an active force in coordinating irrigation water policy. Farm groups are involved, but they have so many issues to deal with that they do not concentrate on water policy. Water interest groups such as the Colorado Water Congress and the California Association of Water Districts (CAWD) provide a focal point for water policy discussions. The private sector also installs and manages extensive irrigation systems.

Irrigation and drainage systems have been under attack from environmental and urban interests for quite a few years. Cases pressed against them include charges of water waste, pollution, and harm to wildlife, wetlands, and sensitive waters, and also that they receive excessive subsidies. These charges are described in many books and articles from the environmental community, such as *Cadillac Desert* by

Marc Reisner (1986), and *Water for Agriculture: Facing the Limits,* published by the Worldwatch Institute (Postel, 1989). The latter, for example, attributes several environmental disasters, such as the falling levels of the Aral Sea in Russia, to irrigation schemes.

Some of the points made by environmentalists certainly are valid. The picture is much more complex than just the "bad hat" image, of course. In this author's opinion, however, even agricultural interests acknowledge that there are problems with irrigated agriculture.

Jan van Schilfgaarde, an irrigation scientist with the Agricultural Research Service, has asked in several articles if America's irrigation systems can last (van Schilfgaarde, 1990). Basic concerns with irrigation include both management problems and social attitudes about irrigated agriculture. Van Schilfgaarde describes irrigation problems that include the salinization of ancient Mesopotamia, drainage problems due to British systems in the Indus Basin, and selenium impacts on wildlife in Kesterson Reservoir in California. He acknowledges that irrigation always degrades water quality because drainage water carries salt that requires disposal, and without proper management, land becomes waterlogged or salinized.

Cultural attitudes probed by van Schilfgaarde focus on reducing preferential treatment for irrigators and doing more to protect the environment. Although van Schilfgaarde believes that irrigation will survive and flourish, he believes that radical reforms will be required. These reforms will require irrigation to deal with concerns about toxics in drainage water as well as salts; avoiding salinity issues by adequate drainage; finding alternative paths to comprehensive water management; and dealing with the political issues of subsidies.

In-stream flows and uses management

Surface water used for municipal, industrial, or irrigation supply is diverted from streams. However, there are important uses for water remaining in the streams, including navigation, hydropower, recreation, and environmental purposes. These "in-stream uses" are part of the water industry, but sometimes work at the edges of it. If we are to succeed in coordinating water policy, they must be given recognized roles.

Water-based navigation supports the transportation of water-borne commerce. A diverse navigation industry has an interest in maintaining a system of navigable streams to enable operators to transport goods efficiently under all weather conditions. Figure 8.5 shows the organization of waterways in the United States.

The hydropower industry captures the available energy from flowing water and converts it to electrical energy that goes into systems for distribution to consumers. To power advocates it is "clean energy,"

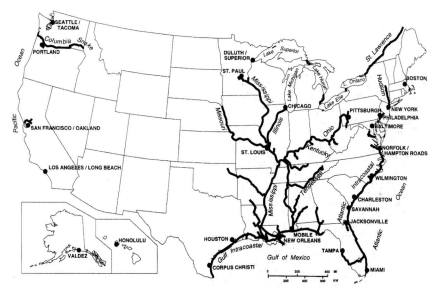

Figure 8.5   U.S. waterways organization. (*Source: U.S. Army Corps of Engineers.*)

but to some environmentalists it is one of the principal environmental culprits, responsible for extensive damage to streams and habitat.

The U.S. Army Corps of Engineers is the principal flood control agency in the United States, having responsibility for major flood control projects. In addition, owners of any dam have the responsibility to operate it in a manner that will not aggravate flood damage. There are relatively few nonfederal agencies with a primary mission in flood control, but many agencies have some responsibility for it. The Miami (Ohio) Conservancy District is an example of a regional water management district that was created primarily for flood control. Denver's Urban Drainage and Flood Control District has flood and stormwater responsibilities. The Los Angeles County Flood Control District has similar responsibilities. Emergency control agencies, ranging from the Federal Emergency Management Agency (FEMA) to state emergency control agencies, also become involved in setting flood control policy, especially emergency response.

From the standpoint of in-stream flows, the water quality management issues are to maintain stream standards to enable desirable species of fish and wildlife to prosper. The industry subsector dealing with water quality management includes the dischargers (wastewater organizations, farms, cities, industries, etc.), the regulators, and the planner/coordinators.

Preserving habitat for fish and wildlife is a water management purpose that for the most part lacks agencies with direct responsibilities to provide it. However, water management agencies with other direct missions are increasingly aware of the need to consider fish and wildlife habitat requirements. Agencies of the federal government such as the U.S. Fish and Wildlife Service are becoming more focused in their application of statutory authorities to protect habitat, especially for endangered or threatened species. State fish and game agencies are also increasing their work with water management organizations to improve habitat.

The recreation industry takes advantage of streamflow, lakes, and reservoirs to provide recreational opportunities such as boating, swimming, rafting, fishing, water skiing, picnicking, sightseeing, and general recreation. Recreation is advocated by a very diverse and uncoordinated group of agencies and private interests. Mostly, park management agencies, such as the National Park Service and state parks departments, are involved, along with the same agencies of government that deal with fish and wildlife.

Regulators in the Water Industry

Regulation in the water industry deals with health and safety, water quality, fish and wildlife, quantity allocation, finance, and service quality. The legal aspects of regulation are described in Chap. 6.

Water industry regulation has developed piecemeal through principles, rules, or laws that are intended to control behavior in the public interest, although defining the public interest is an elusive goal. It began with water allocation systems in the West, then with public health related to drinking water. Now, it has been extended to environmental issues such as endangered species. Finance and service quality are not regulated much in the water industry, but they could be in the future.

Who are the regulators? Regulators work in government agencies with missions to protect the public interest in health and safety, water quality, fish and wildlife, quantity allocation, finance, and service quality. Let us discuss a few examples.

Anyone working for the U.S. Environmental Protection Agency is a regulator, although some more so than others. By the same token, the state EPAs or water quality agencies are regulators. The Corps of Engineers is a little schizophrenic about being regulators because they formerly developed projects, but now they have been delegated the authority to implement Section 404 of the Clean Water Act, a regulatory program. The U.S. Fish and Wildlife Service and the U.S. Forest Service have entered the regulatory arenas with the Endangered Species Act and enforcement of federal water-related rules in National Forests.

In the West, state engineer offices are regulators in the sense that they control the diversion of water from streams and wells. Eastern states are increasing their activity in this area. State natural resources departments with dam safety missions regulate some aspects of safety.

State public service commissions regulate costs of water service for some utilities. These commissions, where they are concerned with water at all, regulate only private water companies. The public is largely ignorant of whether they are receiving the most cost-effective water supply service possible. Rate decisions for electric, gas, and telecommunications utilities are made public, and comparisons of costs are easier for the public to make.

Environmental organizations are key players in regulation. They are deeply interested in laws such as the National Environmental Policy Act (NEPA), the Endangered Species Act, and various authorities given to federal and state environmental and resource agencies. The 20 or so major environmental organizations have perhaps 15 million members and budgets of about $600 million total (Brimelow and Spencer, 1992).

Water Planners and Coordinators

The need for planning and coordination to solve water problems, and the various roles and responsibilities, are often unclear and neglected. There are powerful reasons why this is so. You might hear: "We got so burned out going to meetings that we decided to go it on our own." This attitude explains in a nutshell the difficulties involved in planning and coordination. Another reason is the difficulty in collaborative decision making.

The disincentives for coordination are formidable, and include, in addition to meetings and the need to collaborate, conflicting missions between line agencies; political issues; competing services such as data collection, education, and research; conflict between economic development and regulatory programs; the need for role clarification amid bureaucratic infighting; the need for leadership to focus coordination; the complexity of the issues; financial implications; political posturing related to water; and conflicts between disciplines and interest groups. With all of these disincentives, it is amazing that anyone perseveres with planning and coordinating.

Planning and coordination within the water industry are not fully provided by the service providers and regulators. Actually, the support organizations of the water industry carry out much of the coordination: the consultants, lawyers, suppliers, researchers, associations, and data management agencies.

Water Industry Support Organizations

To describe the water industry, it is necessary to include its support organizations. These organizations, from both the public and private sectors, provide policy analysis, design, construction, equipment, legal services, research, publications, education, and data management, and they help in providing coordination by networking through water industry associations.

Data management

Data management for the water industry is a very dispersed support service, difficult to organize and coordinate. The main federal water data agency is the U.S. Geological Survey (USGS). The single largest USGS water-related program, over 40 percent of its Water Resources Division activity, is the Federal-State Cooperative Program. Through this program, the USGS works with over 1000 state and local agencies, which finance 50 percent of the activities.

Consultants

Consultants make up a large part of the support force for the water industry. While consulting engineers are the largest component, there are also management consultants, geologists, financial consultants, researchers, policy analysts, and attorneys.

Lawyers

Attorneys are necessary to the water industry because conflicts often are resolved in the legal arena. One cause is the regulatory issues that arise, for example, when a regulatory agency issues an order against a water management agency for violating an environmental statute.

Construction industry

The construction industry in the United States is on the order of $500 billion, dwarfing the water industry in expenditures. The activity and role of the construction industry in water projects and programs is substantial, especially utility contractors as a specialty group.

Equipment suppliers

Suppliers of equipment and services are critical to the water industry. Many of them participate in the Water and Wastewater Manufacturer's Equipment Association (WWEMA).

Research organizations

Research for the water industry includes a broad range of activities carried out by government agencies, universities, and private organizations such as institutes and think tanks. Although much of the innovation in the industry comes from formal research offices such as these, there are also many innovators in support firms, in agencies, and in consultant offices.

Education and training providers

Like research, education in the water industry includes many providers. They provide degree programs and training through nondegree programs and continuing education. Today, two trends characterize the supply and demand for education: the increasing role of technology in supplying educational wares, and the increasing performance demands and pressures on individuals at all levels.

Industry associations

Histories and roles of several key associations have already been described. Of their many functions, knowledge development, networking, education, and training are key functions of associations. Associations representing state government water managers include the ASIWPCA, the ICWP, the National Governor's Association, the Western States Water Council, the Conference of State Sanitary Engineers, the Council of State Governments, and others.

Interest groups and environmental organizations

Interest groups, mainly environmental organizations, are influential in the water industry. Since water touches so many aspects of the economy, interest groups are naturally active. A few of them are in the categories of trade associations that represent particular industries such as, for example, pulp and paper; environmental organizations ranging from focused groups, such as the National Wildlife Federation and the Audubon Society, to more extreme, across-the-board environmental zealots; consumer groups; and public-interest organizations such as the League of Women Voters.

Publishers

Publishers in water resources play an important role. Although there are a few publishers that specialize in water publications, many mainline publishing houses have a broad range of titles. Journals are an important product of publishing houses, and those available include the main journals of the water associations as well as a few stand-alone journals. There are also quite a few magazines, newsletters, and newspapers serving the water industry.

Governmental and Nongovernmental
Players in the Water Industry

In addition to a functional view of the water industry, we can look at it from the perspective of the players. All three levels of government and the private sector, both profit and nonprofit parts, are players in the water industry. In the government, all three branches, legislative, executive, and judicial, have key roles. In the nongovernmental world, players include industry, private water businesses, support organizations, and environmental organizations.

Local government

The main local government organizations involved are city or county departments and special-purpose districts or authorities organized either under local charters or under state law. According to the U.S. Census of Governments, there were in 1982 a total of 82,290 units of local government. Of these, 3041 were county governments, 19,076 were municipal governments, and 16,734 were township governments. Special districts totaled 28,588, including natural resources (6232), fire protection (4560), and housing and community development (3296) (U.S. Department of Commerce, 1985).

State government

State governments play important and increasing roles in water management. The main organizations are state environmental agencies, health departments, and water management agencies, usually within departments of natural resources. Most state water agencies operate regulatory programs. Some state-level water management agencies may be engaged in planning, designing, financing, constructing, and operation of projects, or in only some of these activities. Some state-level organizations have regulatory responsibilities mixed with project development missions, but this usually creates problems and the functions are often separated (see Chap. 10). Substate special districts operate with some of the authority of state government, but with regional subdivisions.

Federal government

The federal government has several primary water management agencies, and other agencies with more limited roles. The primary agencies are the Army's Corps of Engineers, the Interior Department's U.S. Geological Survey and Bureau of Reclamation, the Environmental Protection Agency, and the Department of Agriculture's Soil

Conservation Service. The Corps of Engineers and the Bureau of Reclamation have developed the larger projects, and the Soil Conservation Service (SCS) helps farmers with water management through small projects and programs. The SCS has technical employees in most counties to work with the nearly 3000 soil and water conservation districts. The USGS is the primary data collection agency, and the EPA is the primary regulatory agency. A number of congressional committees have jurisdiction over one aspect or another of water management. The judicial branch is responsible for court decisions, such as interstate water allocation decrees (see Chaps. 11 and 19) and interpretation of statutes.

Industry and the private sector

Private-sector groups involved in the water industry include water users, consultants, equipment manufacturers and suppliers, contractors, water well drillers, lawyers, and the "knowledge industry," including universities, think tanks, publishers, associations, and various nonprofit entities.

International organizations

International organizations involved in water policy include the UN family of agencies: the United Nations Development Programme, UNESCO, the Food and Agriculture Organization, the UN Environment Programme, the World Bank, the World Health Organization, and the World Meteorological Organization. The United Nations has had much influence on water policy through its international conferences, such as the UN Conference on Environment (Stockholm, 1972), the World Water Conference (Mar del Plata, 1977), and the UN Conference on Environment and Development (Rio, 1992, UNCED). Bilateral aid organizations such as the U.S. Agency for International Development, and many others, work in water management. Regional development banks exert a lot of influence. These include the Asian Development Bank, the Inter-American Development Bank, the Arab Development Fund, and the European Development Bank. International associations are an important force for the water industry. These include, for example, the International Water Resources Association and the International Association for Hydraulic Research.

International Models

How have other countries approached water industry structure and coordination?

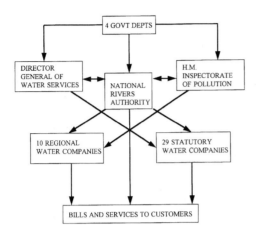

Figure 8.6   British water organization. (*Source: Adapted from Thackray, 1990.*)

United Kingdom model

The U.K. model received worldwide attention in the 1970s and 1980s because the nation gave much attention to its water industry. First, Britain regionalized water services in the 1970s and then it privatized them in the 1980s. The British seem to believe they now have management arrangements about right.

The general structure of the water industry as it exists in England and Wales is as shown in Fig. 8.6. The 10 regional water companies and the 29 statutory water companies provide the services, and they are generally overseen by the National Rivers Authority for environmental regulation. There remain regulatory roles for H.M. Inspectorate of Pollution and the Director General of Water Services (regulation of price and service levels). Four government departments provide additional oversight (Thackray, 1990).

French model

According to Drouet (1990), the French model is a tête à tête between 36,000 municipalities and three large corporations, with the central administration providing the basic rules. In France, the country has been divided into six river basin regions for the purpose of coordinating management (World Bank, 1993, Ministere de l'Environnement, 1992). This means that water resources are managed at the level of the river basin, an important advance in integration. Institutionally, there are six river basin committees and six river basin financial agencies (Agences Financières de Bassin). The six basin groups are

Seine-Normandie, Loire-Bretagne, Rhin-Meuse, Rhône-Méditerranée Corse, Adour-Garonne, and Artois-Picardie. According to the World Bank, they have performed efficiently for 25 years. The river basin committees facilitate coordination and are at the center of the negotiations and policy making at the river basin level. They are composed of 60 to 110 persons representing interest groups and agencies. The committees approve long-term master plans and action plans to improve water quality. Figure 8.7 shows a scheme of how the basin management works. The Basin Committee or "Water Parliament" has considerable authority over charges and selection of projects. The financial agencies implement action plans, and aim at targets for water

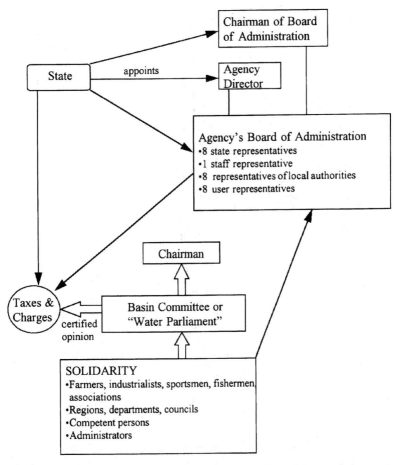

Figure 8.7   French water organization. (*Source: Adapted from Ministere de l'Environnement, 1992.*)

quantity and quality. They propose the long-term plans, the five-year plans, and the charges. Water users pay two fees: one for the amount of water used and the other for pollution. The financial agencies collect the fees, make grants and loans, collect and process data, and carry out studies and research. With access to considerable talent, they have become the primary water planning organizations.

In France, the urban water supply and wastewater systems are owned by various levels of municipal government (Whitely, 1994). France is almost unique in that it has a long tradition of having these systems operated by private companies. Large private water companies dominate the water services sector. The three main water companies are the Société d'Aménagement Urbain et Rural (SAUR), a division of Bouygues; Société Lyonnaise des Eaux—Dumez; and SOGEA, which is affiliated with Compagnie Générale des Eaux. These companies treat and distribute water under franchise agreements.

Japanese model

With Japan's small scale and tight centralization, water resources management seems less daunting than in the United States. At a national level, the River Bureau of the Ministry of Construction administers comprehensive programs for major flood control, urban flood control, preventing damage due to landslides, improving waterfront environments, and promoting comprehensive water resources development. There is a Water Resources Development Council under the National Land Agency, which provides "fundamental affairs coordination." The organization chart of the Water Resources Development Public Corporation (WRDPC) shows it as relating to several ministries: the Ministry of Health and Welfare for domestic water; the Ministry of Agriculture, Forestry and Fisheries for irrigation water; the Ministry of International Trade and Industry for industrial water; and the Ministry of Construction for river administration. Rivers in Japan are rigidly classified (class A, class B, and class C) with the largest (class A) being the responsibility of the national government, the next largest (class B) being the responsibility of prefectures, and the smallest (class C) the responsibility of the local governments. The river classification system goes back to the Meiji era, about 1870. The River Act was first enforced about 1896.

Questions

1. Why is coordination in the water industry a difficult problem?

2. There are four basic functions in water management: to provide direct services, to regulate, to plan and coordinate, and to provide support services.

Discuss which level of government should undertake each of these different functions in the case of municipal water supply.

3. What is the reason for the "small system" problem among U.S. water supply organizations? How does it compare to a similar problem in developing countries?

4. From the point of view of the state governors in the United States, what are the proper roles of the federal and state governments in water resources management?

5. Is a public organization or a private organization better equipped to provide urban water supply? In either case, explain what kinds of regulation are required. Should urban water supply be a monopoly, and if it is, what influence does this have on the preference for a public or private undertaking?

6. Do you believe that either the British model or the French model of water management is better than the U.S. model? Discuss the pros and cons.

7. Which form(s) of privatization do you believe are most appropriate for the U.S. water industry? Why?

References

Apogee Research, Inc., The Nation's Public Works: Report on Wastewater Management, National Council on Public Works Improvement, Washington, DC, 1987.

Brimelow, Peter, and Leslie Spencer, "You Can't Get There from Here," *Forbes,* July 6, 1992.

City of Ft. Collins, Colorado, Water and Wastewater Utility, Water Board Annual Report, 1991.

Drouet, Dominique, The French Water Industry in a Changing European Context, in Kyle E. Schilling and Eric Porter, eds., *Urban Water Infrastructure,* Kluwer Academic Publishers, Dordrecht, 1990.

Lea, Dallas M., Irrigation in the United States, U.S. Department of Agriculture, Economic Research Service, Staff Report No. AGES840815, Washington, DC, April 1985.

Miller, William H., *Miller On Managing,* American Water Works Association, Denver, 1992.

Ministere de l'Environnement, *Water, A Common Heritage: Integrated Development and Management of River Basins, the French Approach,* Paris, 1992.

National Council on Public Works Improvement, Final Report, Washington, DC, 1987.

Postel, Sandra, *Water for Agriculture: Facing the Limits,* Worldwatch Paper 93, Washington, DC, December 1989.

Reisner, Marc, *Cadillac Desert,* Penguin Books, New York, 1986.

Schilling, Kyle, Claudia Copeland, Joseph Dixon, James Smythe, Mary Vincent, and Jan Peterson, The Nation's Public Works: Report on Water Resources, National Council on Public Works Improvement, Washington, DC, May 1987.

Scribner-Bantam English Dictionary, New York, 1979.

Thackray, John E., Privatization of Water Services in the United Kingdom, in Kyle E. Schilling and Eric Porter, eds., *Urban Water Infrastructure,* Kluwer Academic Publishers, Dordrecht, 1990.

U.S. Department of Commerce, *Statistical Abstract of the United States,* 105th ed., 1985.

U.S. Environmental Protection Agency, The 1982 Needs Survey, EPA/43019-83-002, June 15, 1983.

U.S. Environmental Protection Agency, Office of Drinking Water, FY 1983 Status Report, The National Public Water System Program, Washington, DC, 1984.

van Schilfgaarde, Jan, America's Irrigation: Can It Last?, *Civil Engineering,* March 1990.

Whitely, Hugh, The Effectiveness of Regulatory Structure in the Promotion of Water Conservation in Ontario (Canada), the United Kingdom and France, Transatlantic Seminar on Water Management, Paris, July 1994.

World Bank, Water Resources Management: A World Bank Policy Paper, Washington, DC, 1993.

9

A Comprehensive Framework for Water Management

The previous chapters described management issues of the water industry: science, engineering, politics, systems, law, finance, and organization. This chapter weaves these together into a comprehensive framework of principles for water industry management. The issues, policies, scenarios, processes, principles, tasks, tools, roles, and players make up the elements of a comprehensive management framework for an integrated water industry. Emphasis is on large-scale water management actions, but the principles apply at smaller scales too.

Problems and Issues in the Water Industry

Water problems vary around the world. While problems in the United States are not as severe as in some nations, we still struggle with water policies. Numerous policy study groups have developed lists of water issues, and the following is this author's concept of their "top 10" issues.

1. *Water policy reform and coordination.* This overarching and comprehensive policy issue remains on the list because of the complexity and difficulty of water policy. It is approached through specific issues such as the Clean Water Act, the Endangered Species Act, specific projects, etc.

2. *Geographic coordination of water management.* This deals with the need for a watershed focus, river basin commissions, and regional authorities.

3. *Pricing and water allocation.*   This very broad issue has two main aspects: pricing as a water allocation tool, and pricing to provide operating revenue and capital.

4. *Water supply security.*   The security of water supply quantity and quality involves safe drinking water, drought water supply, and related issues.

5. *Water quality policy.*   This deals with problems related to ambient water quality, nonpoint sources, industrial wastewater, and making wastewater treatment effective and affordable.

6. *Environmental systems and policy.*   This issue deals with sustainable development and maintaining biodiversity and with large-scale problems such as the Everglades or the Ogallala regional aquifer system.

7. *Small systems.*   This deals with the limited ability of small system operators in rural areas to pay for and operate water supply and wastewater systems that meet regulations. Many problems of water supply and sanitation in developing countries are in this category, which is closely linked to the issue of capacity of the local areas to solve their own problems.

8. *Water management conflicts.*   This highly political category includes area-wide water supply controversies; state-wide water supply issues such as California's Bay-Delta region; large-scale agricultural–urban deals; interstate river conflicts; relicensing of controversial hydroelectric projects; and large-scale water transfer proposals.

9. *Issues of efficiency, including public–private issues.*   Service organizations work relatively well, but regulation is too expensive, the command-and-control system is too adversarial, and the industry needs better coordination. Privatization may be one answer to efficiency, but it will not help solve the adversary problem.

10. *Information services.*   This deals with alleged inadequate research, education, training, and data support for the water industry. With the complexity and links among parts of the industry, and with opportunities for the private sector, it cannot be resolved by a single program.

In its policy paper on water resources management, the World Bank (1993) described a vicious cycle in developing countries of poor water services, leading to consumer unwillingness to pay, inadequate operating funds, and a further deterioration of services. These problems, which affect health, welfare, and safety around the world, also exacerbate environmental problems at a time when water is becoming increasingly scarce as a result of rapid population growth and urbanization. In developing countries, solutions have been elusive because of misallocation and waste of water, institutional weakness, market failure, distorted policies, fragmented public investments, and excessive reliance on government agencies.

As examples of specific problems, Frederiksen (1992) describes how cities in India with populations greater than one million do not have 24-hour water service, how groundwater that is overdrafted in Beijing will force population relocations or large-scale water transfers, and how poorly understood are Mideast shortages, including their links to regional conflicts. Frederiksen (1996) takes exception to some of the current policy studies, and has published a paper to clarify some of the misconceptions about water needs in the developing world. He sees four basic contraints that are inadequately addressed: little time to act, limited measures available, competing demands for funds, and minimal ability to manage droughts.

Players and Roles in the Water Industry

Improvements in the water industry will come about when the industry's players assume proper roles. Ultimately this is the main issue in coordination: having the players play the right parts. As the main players, one might think of elected and appointed officials, the utilities, the development interests, the environmentalists, the regulators, and the support groups.

At all three levels of government, both elected and appointed officials are involved in water because it is a central concern of government. Their main roles are policy, oversight, and problem solving. They must get together on large, multijurisdictional problems. If they do not, solutions will be impossible.

The utilities and the development interests have the most direct interest in exploiting water resources. They are the water providers, wastewater agencies, irrigators, industries, navigation interests, and power companies. These groups play strong special-interest and advocate roles in the water industry. Comprehensive water management agencies, such as the water management districts and regional authorities, are involved, but in a different way than direct provider groups. Hydropower, navigation, and flood control interests are involved in in-stream issues, not so much in water quality regulation. Industrial and urban development interests have strong interests in the water industry because they depend on water for development. Agriculture and resource development sectors are strongly involved in the water industry, both because they may depend on water for irrigation and because they may be regulated by water quality agencies.

Environmentalists include, of course, environmental organizations, but also a broad group of public interest organizations. In their role as environmental auditors and, to some extent, enforcers, they are among the strongest players in the water industry.

Regulators have tremendous authority and are important players. In addition to agency regulators, the court and legal systems, while operating behind the scenes, are in this category.

Support groups are too numerous and diverse to summarize here (see Chap. 8). Data collection and assessment agencies are support players. Consultants, suppliers, and contractors are among the largest private-sector players in the water industry. Also, universities and research institutes sometimes become involved in research and assessment. Often, they express themselves through professional, scientific, and trade associations. These groups can play significant roles in raising capability and in coordinating in the water industry.

Requirements of the Management Paradigm

Large-scale water problems are characterized by complex linkages and interdependencies among many stakeholders and geographic areas, as well as between the water environment and related ecological systems. To characterize the conflicts inherent in large-scale systems, some writers refer to the institutional issues that arise (see, for example, Viessman and Welty, 1985). Engineers who try to quantify the problems refer to multiple, noncommensurate objectives; multiple decision makers; large numbers of variables, parameters, and coupled subsystems; complex input–output relationships; and high variability (Haimes et al., 1987).

Words used to describe the needed approaches include comprehensive, coordinated, integrated, total, holistic, collaborative, and systemic. A number of writers have suggested management principles and practices that must be included in a comprehensive framework for water management.

Regardless of the descriptors, an effective management paradigm for large-scale water resources problems calls for combinations of regulatory actions, capital investments, citizen involvement, public education, and behavior modification. These are what Gilbert White (1969) called "multiple means" in his book *Strategies for American Water Management.*

Conceptual Frameworks

Water resources managers are bombarded with new terms, perhaps as a spinoff from jargon in the business and management fields. They describe goals, processes and procedures, organizing frameworks, philosophies, and leadership styles. Taken together, they form a powerful group of concepts which, if they can be focused, promise to improve management of large-scale water resources systems. Some of the terms are reviewed in this section.

Integrated water management is a popular conceptual framework, but the term is hard to explain because it includes so many concepts. Mitchell (1990) edited a volume of national case studies about integration, and although he did not present a definition, his approach gives clues to the concept: "problems that cut across elements of the hydrological cycle, that transcend the boundaries among water, land and environment, and that interrelate water with broader policy questions associated with regional economic development and environmental management." "At a conceptual level, there is often agreement that it is necessary to define and tackle problems characterized by their interrelationships and linkages...these problems are frequently referred to as metaproblems, wicked problems, or messes." Mitchell's classification system is that integrated water management can be interpreted in at least three ways: integration of components of water (what I call purposes and functions), integration of water with land and environment (ecologic and intersectoral integration), and integration with social and economic development (intersectoral integration).

Some use *comprehensive water management* to mean the same thing as integration (Wagner, 1995), but there is an important difference: Integration implies linkage, whereas comprehensive implies broad coverage. Alternative terms are *total water management* (American Water Works Association, 1993), "an attempt by the water supply industry to assure that water resources are managed for the greatest good of people and the environment and that all segments of society have a voice in this process," and *holistic water management* (Kirpich, 1993), an approach for the irrigation sector that emphasizes interagency coordination, performance standards for water users and staff, use of indigenous knowledge, local participation for corollary activities, top-down and bottom-up coordination, and the linkage between water and agriculture policy.

Some might consider *sustainable development* as a framework for integration, but it is an overarching goal, a rather new term that has become popular quickly.

The World Bank's (1993) *comprehensive framework* is a policy framework meant to apply in developing countries: "An analytic framework for water resources that views water as a single resource with many uses and interlinkages with the ecological and socioeconomic system."

Integrated resource planning (IRP), developed in the electric utility industry, has been translated into a procedure for the water industry

(AWWARF, 1995). It involves "defining the overall goals and objectives and establishing milestones; identifying all of the stakeholders and their concerns and involving them throughout the process; determining the problems, critical planning issues, and potential conflicts to be addressed during the process; identifying and managing risks and uncertainties; implementing the IRP; and evaluating the effectiveness of the process and making appropriate adjustments."

Also in the category of actual processes are the National Environment Policy Act (NEPA) or *environmental impact statement* (EIS) process, the *recovery planning process* under the Endangered Species Act (ESA), and the EPA's *management planning process* under the National Estuary Program. All three stem from regulatory authority, but with differences. The EIS process requires that a proposed action be studied from all viewpoints, with a full statement prepared. A federal agency then decides if the proposed action can be approved. The recovery plan process stems from Section 7 of the ESA (U.S. Fish and Wildlife Service, 1994), "the section of the Endangered Species Act of 1973, as amended, outlining procedures for interagency coordination to conserve Federally listed species and designated critical habitat." A Section 7 consultation involves "the various section 7 processes, including both consultation and conference if proposed species are involved." If, under the ESA, the Fish and Wildlife Service issues a "jeopardy opinion," that a species is in jeopardy, then a "recovery plan" is necessary, with elements as required by the federal agencies.

The EPA's management planning process has a different character. The EPA introduced the term under the National Estuary Program (USEPA, 1988), with an approach based on "collaborative, problem-solving approaches to balance conflicting uses while restoring or maintaining the estuary's environmental quality." It seeks to combine the straightforward planning process with a politically realistic one. It has four phases: building a management framework; characterization and problem definition; creation of a Comprehensive Conservation and Management Plan; and implementation. The overall process includes the governor's nomination; the convening of a "Management Conference" by the EPA administrator; estuary characterization; a Comprehensive Conservation and Management Plan (CCMP); implementation plans; and continuous monitoring. It is summarized briefly by the EPA as follows: "the program is woven together by two themes: progressive phases for identifying and solving problems and collaborative decision making."

Generalized concepts from other fields which also apply are *systems thinking, civic environmentalism,* and *collaborative leadership.* These offer useful frameworks and ideas for principles and processes, but do not give specific guidance for water managers. The business tool called systems thinking (Senge, 1990) can be a valid framework, but its elements have not been widely applied to water resources other than in systems analysis applied to operations. John's (1994) "civic environmentalism" is a search for alternatives to political confrontation and a new role for the federal government as a participant in decisions made at the state or local level on problems such as nonpoint problems, pollution prevention, and protecting ecosystems. It would make extensive use of nonregulatory tools and interagency and intergovernmental cooperation. Chrislip and Larson's (1994) "collaborative leadership" provides insight into complex issues, engaging frustrating and angry citizens, and generating civic will to break gridlocks. How well a community pulls together to solve its problems is a measure of its "social capital" or "civic infrastructure." Chrislip and Larson cite the Clark Fork River in Montana, where 11 federal agencies are involved along with a host of local and state agencies. The issue—no one in charge—is a critical problem in water management. As Chrislip and Larson state, "collaboration...goes beyond communication, cooperation, and coordination...to create a shared vision and joint strategies to address concerns that go beyond the purview of any particular party."

Finally, the field of *total quality management,* and related concepts such as *management principles and practices,* offer promise. At first glance, total quality management would not seem to apply to large-scale problems, but parts of it might through the linkage with management practices. It might be worth exploring, for example, whether the ISO 9000 process could be a venue to implement best management practices for water. ISO, the International Organization for Standardization, is a "worldwide federation of national standards bodies representing 90 countries," with the purpose to facilitate the international exchange of goods and services and to develop intellectual, scientific, technological, and economic cooperation. "The ISO 9000 concept is that certain generic characteristics of management practice could be usefully standardized, mutually benefiting producers and users alike" (Voehl and Ashton, 1994).

This is quite a mixture of concepts, but the question is how they apply to water resources management.

A Process for Coordinated Action—The Greatest Challenge

Although many tasks are involved, the greatest challenge is succeeding with coordinated management actions for large scale systems. As noted in Chap. 19, which covers river basin management, two wise men of water management, Jacques Costeau and Abel Wolman, described the problems with coordinated, cooperative, collective actions: on the one hand, badly needed; and on the other, extremely difficult.

The challenge includes determining what to do (analysis, planning, and decision), involving stakeholders and gaining their commitment (negotiation and dispute resolution), and implementing individual and group actions (implementation and monitoring). Although the process is complex, the basic model of problem identification, study of alternatives, decision, and implementation, still applies. The complexities are caused by multiple agendas of stakeholders and institutional constraints faced by water managers.

Of the conceptual frameworks, the ones that offer process models are integrated resource planning, the environmental impact statement process, the management planning process under the National Estuary Program, and recovery plans under the Endangered Species Act. Integrated resource planning is still focused on supply projects and the environmental impact statement and recovery plans have built-in adversarial qualities, so the best of these models for management planning seems to be the EPA process.

Coordinated action is, of course, a challenge on many fronts. One approach to it is "partnering," which has become popular as a way to avoid litigation and disputes in business. In water management, the concept can be applied as a way to develop collaborative solutions. One place where these approaches is essential is in estuary management, and an example of partnering for estuaries is Coastal America. Coastal America is a "collaborative partnership process for action" that joins the forces of federal, state, and local agencies with private interests to address environmental problems along the nation's shorelines (Coastal America, 1993).

The Tennessee Valley Authority (TVA) has undertaken another experiment in partnering through what it calls "river action teams." This initiative is part of the TVA's clean water action plan, meant to make the Tennessee River the "cleanest and most productive commercial river system in the United States by the year 2000" (Tennessee Valley Authority, 1993). The TVA sees the "integrator role" as critical in problems like this.

Multiple Agendas

Including multiple players, regions, agendas, and purposes is the challenge of comprehensive water management. A variation of an old saying is: "If you have something important to do, don't waste time talking with others." That is just the opposite of comprehensive water management—it is imperative to collaborate.

Institutional Issues

It is in the institutional issues that the uniqueness of water problems arises. Each locality, each type of problem, will feature different institutional issues. Institutional issues are not well understood, and engineers need more education in them.

Management Practices in a Coordinated Framework

Engineers who study total quality management have identified "benchmark" management practices, and provided lists of attributes of quality organizations. The management practices are attempts to transfer the concept of "standards" to the fields of management. Generally, management practices are presented with detail so that they can be applied at the individual, team, or organizational levels. They fit within an overall management framework, but they are not the framework. Broad management practices provide guidance for the attributes of the framework, however, and more detailed practices are embedded into it.

Over the past few years, I have reviewed suggestions by writers about "lessons learned" and principles for good water management. Examples of suggestions include those of Wagner (1995), the "lessons learned" publication of the Water and Sanitation for Health Project (U.S. Agency for International Development, 1990), the keynote speech at the VIII World Congress on Water Resources in Cairo in 1994 by Ismail Serageldin (1995) of the World Bank, recommendations about water quality by Professor Dan Okun (1977), the management practices project of the American Public Works Association (1991), and suggestions by drought water managers (Grigg and Vlachos, 1993). Although they come from widely differing sources and problem areas, the suggested principles fall into patterns and can be classified in groups: comprehensive approaches; watershed focus; coordination mechanisms and stakeholder involvement; voluntary, cooperative, regional action; public involvement; conflicts and disputes; local responsibility and accountability; organizational management and role setting; conservation approaches and environmental ethics; training, education, and capacity building; market focus, pricing, and incentives; risk management; decision support; finance; and regulation. In addition, many platitudes and general concepts, as well as good management practices, such as effective supervision, will apply in any situation.

Each of these groups contains a number of specific principles. An earlier draft of this chapter attempted to delineate some of them, but the list was too long. A couple of examples might serve as well. For ex-

ample, the enterprise principle (under finance and under local accountability) is a key requirement for organizations to be self-supporting, and is widely endorsed in the water industry. Capacity building is
another consensus principle, with applications ranging all the way
from the need for wastewater treatment plant operators to be fully capable to the ability of villagers to take matters into their own hands.

For purposes of coordinated frameworks, it is helpful to observe
that some of the practices apply to large-scale problems and some
more to single organizations. In the next section those applying mostly to large-scale situations are delineated.

Model

From the broad groups of management practices and principles, 15 attributes are probably essential for a comprehensive approach to large-
scale problems. The attributes can be placed into five main groups:
one that requires a coordinated framework, one that deals with inclusion, one that sets up control requirements, one that outlines process
principles, and one that contains requirements for the framework.

> *Coordinated framework principle.*   This requirement sets out the
> concept of a coordinated framework for problem solving, an organi-
> zational structure to coordinate the efforts, a name for the program,
> something for the players to identify with.

> *Sustainable development.*   This is an overarching goal of the
> process.

> *Process-based.*   This principle is that the framework has within it
> a decision and implementation process that can be identified and is
> repeatable, not arbitrary.

> *Comprehensive.*   This is the inclusion principle, which requires
> that the planning and management be broad in concept, include the
> stakeholders, purposes of water management, geographic regions,
> and related planning sectors.

> *Integrated.*   This is the linkage principle, which requires that not
> only are the planning and management comprehensive, their parts
> are linked together to optimize the whole and to avoid optimizing
> just the pieces. Some integrating force is needed, going beyond just
> getting the players together.

> *Collaborative.*   This is the voluntary, collaborative requirement,
> which includes incentives for the players to cooperate with each other.

> *Stakeholder involvement.*   While the stakeholder involvement prin-
> ciple is addressed to some extent in the comprehensiveness princi-
> ple, it is so important that it is given a separate category of its own.

Action-oriented.   This principle ensures that the planning and management exercise will, in fact, lead to results.

Adaptive.   This principle ensures that the planning and management process is dynamic, leading to periodic reevaluation of goals, needs, and actions, and not to a static plan.

Effective management practices.   This principle requires that effective management practices in the sense of total quality control are identified and implemented.

Science-based.   This requirement is that the impairment to be addressed is identified and defined by scientific means.

Risk-based.   This principle incorporates uncertainty and risk into decision making and ensures that decisions are taken with a proper perspective on risk, cost, and measures to reduce uncertainty.

National policy framework.   The national policy framework principle requires the federal government to identify and set goals and standards to guide local officials.

Local control.   This principle asserts that authority and decision making ought to be devolved to the lowest levels possible, to ensure maximum incentive to identify problems that are real and to implement cost-effective solutions.

Capacity building.   Local authority requires the capacity to decide, implement, and manage. Planning and management framework may require capacity building in different forms to form effective water managers and public officials.

Discussion of Cases and Model Attributes

In a related paper, I presented six brief cases to illustrate the model. Five of the cases are described in the book, and a sixth has been added (Wagner, 1995). Not all attributes of the model are discussed in each case, but each attribute is covered somewhere in the six cases. The cases illustrate a range of situations: two illustrate water quality problems in Eastern coastal waters, one deals with Western coastal waters and includes water rights issues, another deals with large-scale water transfer in the East, another deals with endangered species recovery in the West, and the sixth deals with a regional water supply project in the Mountain West.

Framework principle.   In each case a framework evolved to put boundaries on the possibilities and the action. In Jamaica Bay (Wagner, 1995), it was the bay plan. In Albemarle-Pamlico (Chap. 22), it was initially the governor's plan, later the national estuary study. The Bay-Delta issue (Chap. 22) is so large that it might be impossible to have a single framework, but the Governor's Oversight Council provides one. In Virginia Beach (Chap. 15), it is the need for the water supply, driven by the city. In the Platte (Chap. 19), it is the endangered species issue, and in Two Forks (Chap. 10), it was the proposed water supply reservoir.

Sustainable development.   Sustainable development is, of course, an underlying issue in all of the cases.

Process-based.   Each case has one or more processes. Take Virginia Beach, for example, There was an interstate committee negotiation process, later a 404 process, then a settlement process, and so on. One way to conceptualize these is to think of the framework as the umbrella for the processes.

Comprehensive.   It is difficult to say how comprehensive the cases are, other than to observe that each involves many issues.

Integrated.   Unfortunately, integration in each case seems to be worked out in the heat of conflict, with few examples of groups working out linkages voluntarily. One exception was the regional integration of Two Forks, but it became defunct with the permit veto. Should the Virginia Beach issue be resolved with an interstate compact, that will be a success, although it will be difficult to achieve.

Collaborative.   The cases show little voluntary collaboration. The attempts that were made seem to have been blown apart by stronger forces. Examples are the Two Forks partners and the initial attempts by Virginia and North Carolina to negotiate.

Stakeholder involvement.   All cases show a high level of stakeholder involvement, even if it is in conflict situations. The Bay-Delta process, for example, shows an enormous range of stakeholders: citizens from all of California, federal agencies, state agencies, local agencies, and water users.

Action-oriented.   Wagner's Jamaica Bay plan is an excellent example of action orientation. Governor Hunt of North Carolina stressed the need for an action plan in 1979. Politicians know by instinct that they cannot just call for "studies," as the public wants action.

Adaptive.   Again, Wagner's Jamaica Bay plan exemplifies the adaptive principle. It is difficult to see it so clearly in the other cases, if for no other reason than that they typically involve larger numbers of decision makers.

Effective management practices.   The need for or use of effective management practices can be seen all through the cases. In the case of Albemarle-Pamlico, for example, the use of "best management practices" for agricultural runoff has been stressed from the beginning. In Two Forks, the call by some for water conservation and smaller, low-impact projects has been steady.

Science-based.   In the heat of conflict, the use of science as a basis for decisions sometimes gets lost. Exceptions are in the Jamaica Bay plan and in Albemarle-Pamlico, where computerized estuary models played prominent parts. Of course, in the Platte case, both sides utilized numerous models, ecological arguments, and other testimony, often with conflicting conclusions.

Risk-based.   The Jamaica Bay plan illustrates risk assessment. Another place it arises is in Two Forks, when the issue of water supply security is confronted. In the Platte the question of risk in hydrology is part of the plans for furnishing flows for the species recovery.

National policy framework.   One might say that each case unfolds within a national policy framework. Jamaica Bay and Albemarle-Pamlico are driven mainly by the Clean Water Act. Virginia Beach was played out by 404 process, the Federal Energy Regulatory Commission (FERC), and in federal court. The Platte was also played out through the FERC process, but also with Endangered Species Act issues driving the process. Two Forks became a 404 issue, and Bay-Delta has a strong component of federal agency involvement.

Local control.   Jamaica Bay illustrates local control, but none of the other cases really feature this principle. In the case of Two Forks, critics of the veto suggested that it was inappropriate for a Washington official to make the decision, but that is how it happened. There is a built-in conflict between the local control principle and the national policy framework objective.

Capacity building.   The capacity-building objective is not too visible in these cases. Perhaps we can observe that citizen education, and building local environmental values and better leadership potential in local areas as a result of the conflicts, is an example of capacity building.

Conclusions

What should be the paradigm for management actions to solve America's serious large-scale water problems, which are characterized by highly complex linkages and interdependencies and are difficult to analyze, much less solve? Existing management and problem-solving concepts offer powerful tools to apply to the problems. The main challenges are determining what to do, involving stakeholders and gaining their commitment, and implementing individual and group actions. Although incentives and human nature work against such joint action, in comprehensive water management it is imperative to collaborate.

The uniqueness of water problems can be seen in the institutional issues. Each locality, each type of problem, will feature different institutional issues. Regardless, a comprehensive, integrated framework applies to all of the problems and is needed to penetrate the confusion that arises due to the complex structure and lack of focus found in large-scale problems.

A model framework is offered with 15 elements. The main element is that a coordinated framework for action exists. Three process elements require that the framework be comprehensive, collaborative, and with stakeholder involvement. Two control elements require local control within a national policy framework. To promote action, three process requirements include having identifiable processes, an action orientation, and the quality of being adaptive. Six further requirements are that the framework promote sustainable development, be integrated, promote good management practices, be science-based and risk-based, and build capacity.

The cases (Grigg, 1995) show that management frameworks must provide incentives, continuity, and a coordinated structure for the management actions required in actual situations. Sustainable development is an underlying fundamental principle that must be included in a national framework of environmental objectives.

Large-scale problems involve multiple decision processes that unfold under the umbrella of the management framework. For this reason the word "process" is not broad enough, and the term "framework" is preferred to describe the overall approach.

Problems are comprehensive, not by design, but by necessity. They show a high level of stakeholder involvement, again more by necessity than by design. Unfortunately, integration seems to be worked out only in the heat of conflict, with little working out of linkages voluntarily. The cases showed little voluntary collaboration, with attempts being blown apart by stronger forces. This is an important reason that regionalization is so difficult.

The integrator role is very important, but who will take it on? Consider the incentive structures of water agencies, regulators, planner/coordinators, and support organizations. Who has the incentive to converge on multifaceted solutions to large-scale and cross-cutting problems? In a democracy, when confronted with this issue, people might conclude that a "benevolent dictator" is needed, but the benevolent dictator should be the integration and coordination mechanism.

Neither "command and control" regulatory models nor "invisible hand" market models are likely to solve large-scale problems.

Politicians know by instinct that an "action orientation" is required, and not just "studies." Wagner's (1995) Jamaica Bay plan illustrates the adaptive principle, but other cases illustrate it only because they are drawn-out and inconclusive.

The need for effective management practices can be seen all through the cases.

In the heat of conflict, the use of science as a basis for decisions sometimes gets lost. Risk assessment enters into the cases by necessity more than design and as a result of established procedures such as hydrologic analysis.

Each case unfolded under a national policy framework, but there is a built-in conflict between the local control principle and the national policy framework objective. Although the capacity-building objective is not too visible in the cases, citizen education, and building local values and leadership potential, are examples of capacity building. The capacity-building concept appears more in studies of developing-country problems than it does in studies of U.S. problems, but we must remember that capacity is needed to govern and work out civic problems as well as to understand and manage technical systems.

Five principles stand out: coordination, inclusion, harmonization, integration/linkage, and the need for science and politics. The important lessons from the study are that the framework must be put into place and understood by participants. Long-term personal, institutional, and financial commitments are necessary to make them work; otherwise they will unravel and/or become static plans. More citizenship is needed along with science.

Educators face several dilemmas. Do we try to teach management of large-scale systems? Should it be at the undergraduate, graduate, or professional level? Do we emphasize management actions, which apply to the United States, or do we emphasize project development, which is an urgent requirement in developing countries? Of course, the answer is that we must teach both, but they are not mutually exclusive. Management actions involve management principles, and these principles should be embedded in project design.

The case study approach is necessary for engineers to learn about such complex issues and processes. This should be introduced into school, but it is difficult at the B.S. level. Management actions involve management principles, and these principles should be embedded in project design education with cases.

The words, comprehensive, management planning, framework, are not new, certainly to engineers. They will continue to be fuzzy because they are so broad and general. What is new is recognizing water management of large-scale systems as a comprehensive, coordinated, integrated process, in something more than an abstract way.

If water service providers and regulators can take on, in addition to their principal roles, extended "citizenship" and "statesmanship" roles for the greater good, the water industry will be better coordinated. This will have to take place, of course, in local areas, in watersheds, and with help from the nongovernmental sector. There will also be abundant opportunities for citizen leadership in bringing opposing forces together to work out problems and issues.

The shift away from government solutions is explained by management thinker Peter Drucker (1994), who concluded that organizations are the necessary integrating mechanisms for social problems of all kinds, but in addition to their intended *social functions* (such as providing a water supply), organizations must also take *social responsibility* (such as promoting cooperation in the water sector.) One of the biggest contributions of these social sector organizations, according to Drucker, is that they create "citizenship," a badly needed quality in the water industry.

Looking for mechanisms to increase coordination, cooperation, and integration will be a continuing site-specific challenge. In some cases there may be a role for the federal government when more than local interest is involved. Voluntary forums are to be desired, if they can be made to work. Management conferences can be convened by different authorities, including nongovernmental groups such as scientific and professional associations. Reporting at such conferences and in technical papers can encourage integration.

While it is inevitable that an adversarial climate will remain in the water industry, by moving toward a comprehensive framework such as this, a great deal of progress is possible. I personally am optimistic that the industry can solve them and even improve its management of all aspects of water resources. In the developing countries, the World Bank's assessment seems bleaker.

Two wise men of water management, Jacques Costeau and Abel Wolman, described the problems with coordinated, cooperative, collective actions: on the one hand, badly needed; and on the other, extremely difficult. These actions will not get easier, but they will remain badly needed.

Questions

1. This book includes around 50 case studies. Select one and discuss: what is the problem or issue as it would be seen by policy makers at the local, state, or national level; what lessons from planning and management could be applied to improve the situation; and what would be your advice to those in charge to improve the situation?

2. *A legal situation.* North Korea planned to build the Kumgangsan Dam on the Han River, thereby diverting the water before it reaches South Korea. From what you know about water law, both national and international, discuss the legal aspects of this matter.

3. *A political situation.* One example of an interbasin transfer situation is the movement of water from Northern California to Southern California. Another is the problems created in Colorado by moving water from one region to another. Consider that you are a water resources manager in Colorado and the governor asks you to brief him on the political aspects of such interbasin transfers. Write the main points of your briefing.

4. *A financial situation.* Assume that you are the manager of a large water organization. You require a financial plan for a new water supply project. You will retain a consultant to prepare the plan. Write instructions to the consultant so that the final report submitted by the consultant will provide you with everything you need to know about the financial feasibility of the project.

5. *An environmental situation.* The environment of South Florida, like that of other similar regions, is quite fragile. Describe the environmental difficulties in providing an adequate and balanced water supply to the region as it experiences population growth; and describe the water management approaches that will be necessary to mitigate the environmental problems.

References

American Public Works Association, *Public Works Management Practices*, Kansas City, MO, August 1991.

American Waterworks Association, Principles of Total Water Management Outlined, *AWWA Mainstream,* November 1994.

Chrislip, David D., and Carl E. Larson, *Collaborative Leadership: How Citizens and Civic Leaders Can Make a Difference,* Jossey-Bass, San Francisco, 1994.

Coastal America, *Building Alliances to Restore Coastal Environments,* Washington, DC, January 1993.

Drucker, Peter F., The Age of Social Transformation, *Atlantic Monthly,* Vol. 274, No. 5, November 1994, pp. 53–80.

Frederiksen, Harald D., Water Resources Institutions: Some Principles and Practices, *World Bank Technical Paper No. 191,* Washington, DC, 1992.

Frederiksen, Harald D., Water Crisis in the Developing World: Misconceptions about Solutions, Resources Institutions, *Journal of Water Resources Planning and Management* (American Society of Civil Engineers), January/February 1996.

Grigg, Neil S., and Evan C. Vlachos, Drought and Water-Supply Management: Roles and Responsibilities, *Journal of Water Resources Planning and Management* (American Society of Civil Engineers), September/October, 1993.

Haimes, Yacov, J. Kindler, and E. J. Plate, eds., *The Process of Water Resources Project Planning: A Systems Approach,* UNESCO, Paris, 1987.

John, DeWitt, *Civic Environmentalism: Alternatives to Regulation in States and Communities,* Congressional Quarterly Press, Washington, DC, 1994.

Kirpich, Phillip Z., Holistic Approach to Irrigation Management in Developing Countries, *Journal of Irrigation and Drainage, American Society of Civil Engineers,* March/April 1993.

Mitchell, Bruce, ed., *Integrated Water Management: International Experiences and Perspectives,* Belhaven Press, London, 1990, p. xiii.

Okun, Daniel A., Principles for Water Quality Management, *Journal of the Environmental Engineering Division, American Society of Civil Engineers,* December 1977.

Senge, Peter M., *The Fifth Discipline: The Art and Practice of the Learning Organization,* Doubleday Currency, New York, 1990.

Serageldin, Ismail, Water Resources Management: A New Policy for a Sustainable Future, *Water International,* vol. 20, pp. 15–21, 1995.

Tennessee Valley Authority, Announcement of TVA River Action Teams, Communication Plan, Board of Directors, January 20, 1993.

U.S. Agency for International Development, Water and Sanitation for Health Project, *Lessons Learned from the WASH Project,* Washington, DC, 1990.

Viessman, Warren, Jr., and Claire Welty, *Water Management: Technology and Institutions,* Harper & Row, New York, 1985, pp. 51–53.

Voehl, Frank, Peter Jackson, and David Ashton, *ISO 9000: An Implementation Guide for Small to Medium Sized Businesses,* St Lucie Press, Delray Beach, FL, 1994.

Wagner, Edward O., Integrated Water Resources Planning Approaches the 21st Century, Presented at the 22nd Annual Conference of the Water Resources Planning and Management Division, American Society of Civil Engineers, Cambridge, MA, May 8, 1995.

White, Gilbert F., *Strategies of American Water Management,* University of Michigan Press, Ann Arbor, 1969.

World Bank, *Water Resources Management: A World Bank Policy Paper,* Washington, DC, 1993.

2

Problemsheds of Water Management: Case Studies

Cases are the "problemsheds" of water management because they are where the principles, actors, and situations of water management converge. They host integrated situations similar to the way a watershed hosts integrated hydrologic and geologic conditions. Case studies offer promising vehicles for learning about complex management issues and introduce experience and wisdom into different decision-making situations.

Cases, as problemsheds, also introduce new principles into the study of water resources management. Part 1 of the book presents a set of "core" principles (hydrology, systems, planning, finance, law, organization), but the application areas also introduce principles such as drought water management, river basin planning, flood control, water supply, and quite a few others.

The case study chapters represent cross-cutting areas that illustrate the principles covered in the book—both core principles and those from application areas. The cases have two main goals: to present the state-of-the-practice in cross-cutting areas; and to summarize actual ongoing water management problems and processes. They do not present complete expositions about problem areas, but introduce the reader to them.

The case study chapters include about 50 different situations, but the list is far from exhaustive. I considered including a case for each role played by the water resources manager, in each industry sector, but decided that this was not necessary. For example, there is a case for water supply management but not one for wastewater utility management. However, the chapter on water quality covers many of the essential features of the wastewater game. Appendix D, a list of cases cross-referenced to water industry roles, illustrates this.

Among business educators, the case study method is widely used. Leenders and Erskine (1973) say that a case is "a description of an administrative decision or problem. It is normally written from the point of view of the decision maker involved." The John F. Kennedy School of Government (1992) uses the case method to present "accounts of decisionmaking in public policy and public administration and, as such, represent frontline research about the nature of public sector operations both in the US and abroad." Through case instruction, an interactive method of learning, students experience problems as seen through the eyes of the participating government officials. Cases fall into two types, action forcing (what would you do?) or retrospective (telling the whole story) (Kennedy and Scott, 1985). A good case has pedagogic utility, is conflict-provoking and decision-forcing, has generality, and is brief (Robyn, 1986).

Of the roughly 1000 cases in its catalog, the Kennedy School has only a few that deal directly with water issues. These include Tocks Island Dam, Managing the EPA, Federal/State Relations: The 1972 Federal Water Permit Program, Groundwater Regulation in Arizona, Massachusetts Water Resources Authority, Millonzi Commission: Preference for New York Hydro Power, Replacement of Locks and Dam 26, Saving the Tuolumne, and Wastewater Wars. These help students to understand getting a dam approved, implementing the Clean Water Act, organization and management of an environmental agency, groundwater regulatory programs, development of a water supply policy and plan, rate issues related to hydropower, a wild and scenic river issue, wastewater plant conflicts, and a lock-and-dam issue. The topics illustrate the overlap of interests of students of public administration and of water resources management. See Grigg (1995) for a discussion of cases in water resources education.

About 1970, Colorado State University created a Water Resources Planning and Management (WRPM) program to teach graduate students about water resources systems and interdisciplinary problem solving. The gateway course in the program is Water Resources Planning, which was created to teach project planning principles. By the mid-1980s, the course was emphasizing broader problem solving and decision making, as well as project planning based on sustainable development principles. Since its inception, the course has emphasized actual projects and examples from practical experience, and by the late 1980s it began to include cases in a formal way. One goal in the course is to explain cooperative solutions with water management partners (cooperation), involving the public and other governmental agencies in the planning process (public involvement), and resolving disputes and conflicts after they arise (conflict resolution). Decision support systems are recognized as critical in each of these skills.

Nearly 50 cases are included in the following chapters. In addition, the earlier chapters contain numerous examples and partial cases. Still, these do not describe the full range of situations encountered by water resources managers. Nevertheless, they are offered to provide a glimpse, from the perspective of a participant, of the application of the principles of water resources management.

The cases are difficult to classify because they involve different principles and management elements. In the following chapters, they generally follow the sequence of water industry structure presented in Chap. 8: service delivery and management functions; regulation, administration, and environmental protection; planning and coordination; and organization and policy. Along the way, interesting crosscutting issues are encountered. These include the following.

Water supply industry. Unique problems facing water supply utilities in today's regulatory climate, such as the quest for new raw water supplies, and metering and demand management.

Storm- and floodwater industry. Stormwater utilities and flood control agencies. Examples include organization of a stormwater utility and evaluation of alternatives for flood control in cities.

Water as infrastructure. Water facilities and systems as capital investments. Examples include how to plan projects, financial planning, and maintenance management systems.

Regional environmental issues. Issues such as lake level control or remediating past environmental problems and restoring natural systems. These problems normally involve water quantity and quality planning in large natural systems, and require management and regulatory strategies.

Regional water quantity issues. Water quantity issues requiring coordination or regulation to balance uses within geographic regions such as river basins or metropolitan regions. Normally, issues of development, water law, and ecology are involved, and resolution requires coordinated work in multiple political jurisdictions.

Regionalization of investment. Regionalization of investment in water infrastructure. This involves coordination of investment strategies over a region for systems such as water supply or wastewater, and deals with the thorny political problem of regional cooperation.

Drought water management. Water management measures taken to prepare for and to respond to droughts. Topics include drought contingency planning, coordination, and drought response.

Water planning and management organizations. The category addresses the question: How should water agencies be structured? Aspects that deal with the unique characteristics of the water industry are stressed, especially the coordination of assessments and policy setting.

Groundwater coordination. Regulation of groundwater systems is a difficult problem requiring coordination of policies.

Financial issues. Financial issues are approached from the policy standpoint. Examples include pricing strategies for urban and agricultural water supplies.

References

Grigg, Neil S., Teaching Water Resources Management: Use of Case Studies, *Journal of Engineering Education and Practice,* American Society of Civil Engineers, January 1995.

John F. Kennedy School of Government, *The Kennedy School Case Catalog,* 3d ed., Harvard University, Cambridge, MA, 1992, p. i.

Kennedy, David M., and Ester Scott, Preparing Cases in Public Policy, John F. Kennedy School of Government, Harvard University, Cambridge, MA, 1985.

Leenders, Michiel R., and James A. Erskine, *Case Research: The Case Writing Process.* School of Business Administration, University of Western Ontario, London, Ontario, 1973, p. 11.

Robyn, Dorothy, What Makes a Good Case? John F. Kennedy School of Government, Harvard University, Cambridge, MA, 1986.

10

Water Supply and Environment: Denver Water's Two Forks Project

Introduction

The Two Forks Project is a landmark case study of conflicts between water utilities and environmentalists. It shows how securing new water supplies for growth in the face of environmental conflicts will be a tough challenge for U.S. communities in the future. It illustrates how this challenge was approached by one of the nation's leading water utilities: the Denver (Colorado) Water Department.

As described by Frederiksen in Chap. 9, the challenge will also be felt in other nations. Environmentalism is one factor, and, according to Frederiksen, in some places there just is not enough water to go around. Two Forks illustrates how this conflict was played out in one locality, but the conflicts, the players, and the process will be found in other places as well.

In a nutshell, a large water supply project that had been planned for decades by Denver was vetoed by the Environmental Protection Agency, throwing the water supply planning process into turmoil in spite of an environmental study costing over $40 million.

There is plenty of conflict—and complexity—in the Two Forks case. As the reader will see, the conflict lasted for several years, and the complexity required studies and processes that cost over $40 million, but still did not yield a permit for the project.

Water supply issues like Two Forks illustrate the dependence of different sectors of the water industry on each other. When an entity seeks new supplies, it affects the rights of others and usually generates environmental conflicts. Regional cooperation is needed, and the principles of coordination, outlined in Chap. 4, outline processes to consider the rights of others and the environmental issues. However, as shown by this case study, good intentions for coordination and attempts at regionalization may not carry the day. In other words, getting a new water supply project built is a political battle, filled with conflict.

The main lessons in this case study are that securing new water supplies is an extraordinarily complex exercise, and that projects must be planned with care and consider the needs of neighbors and the environment if they are to have a chance to succeed.

The case study also illustrates, perhaps as well as any of the case studies, the extraordinary demands placed on public sector managers today. Water utility managers handle internal organizational issues at the same time that they deal with the conflict of water supply projects.

A professor at the University of Colorado at Denver, Lynn Johnson, developed a Two Forks simulation game, and dozens of students learned about water management by playing the game on personal computers. At the University of Colorado at Boulder, I participated on an interdisciplinary team of the Center for the American West that developed a Chautauqua theatrical event portraying a conflict over developing a reservoir in the mythical state of Lincoln. The 1991 event was the Third Annual American West Symposium and was entitled "Western Rivers from Grand Wash to Coyote Flats: Conflict and Community." Historical characters that were portrayed included John Wesley Powell, famous explorer of the Colorado River Basin; John Muir, famed early environmentalist; William Mulholland, influential engineer of the Los Angeles water system; Mary Hallock Foote, a frontier writer; and several others. I played a hydrologist who testified at the hearing.

The Setting

The proposed Two Forks Dam above Denver was one of the most controversial water conflicts of the 1980s in the United States. The conflict erupted when Denver, along with a group of suburban water agencies, banded together to plan a joint water supply project. The joint project seemed to represent a triumph for regional cooperation and negotiation, but environmentalists opposed the project, and finally they were able to convince the Bush administration to veto it under the authority of Section 404 of the Clean Water Act. Getting to the point of proposing the dam required years of negotiation, a governor's water roundtable for conflict mitigation, and much personal political work.

Several aspects of the Two Forks case are of interest. One topic, planning for water supply in the urbanizing area along the Front Range of the Rockies, will be on the national stage for years to come as a result of the rapid urbanization, scarce water supplies, and high level of conflict over water in the area.

Another topic of interest is metropolitan cooperation to hold the water purveyors together in a regional venture. Conservation enters the picture as the main proponent, the Denver Water Department, was obligated to play the conservation card before building a new project. Environmentalists alleged that the proponents had not considered alternative, low-impact sources of water. Other topics of interest included financing of the plan; the role of the players, including state government; the lack of consensus among the public; the $40 million environmental impact statement; the national political agendas of the environmentalists; the national implications of using Section 404 to veto water projects; and the aftermath of the process.

The case illustrates a number of components of today's water planning and coordination process. It relates to a number of other cases: the CBT project (Chap. 12); water transfers (Chap. 15); conservation (Chap. 17); Platte River Basin management (Chap. 19); regional water management (Chap. 21); Colorado state government water management (Chap. 23), and Western water management (Chap. 24). To paint the picture of these interacting issues, let us begin with the context of the Denver water planning scene.

Denver's Water Supply System

Denver's growth and prosperity have historically depended on the availability of an adequate water supply. This is true in all localities, of course, but is especially apparent in water-short areas.

Note that I did not say that the water supply caused Denver's growth. There is a difference: Water supply is a necessary but not sufficient condition for growth to occur. Water regulations and opposition to projects such as Two Forks act as deterrents to growth. The growth-versus-antigrowth conflicts are an important component of water conflicts around the nation, East and West.

In the nineteenth century, Colorado's pioneers saw the need to develop water to provide for farms, industries, and cities. Their activities to secure the water make a colorful history (see James Michener's popular book *Centennial,* which became a television miniseries). Gordon Milliken (1989) has written an excellent summary of the historical development of the Denver water supply. To see how the system developed is to see how the Front Range water supplies came about; there are definite parallels of water development all along the Front Range, but Denver was the leader.

The history of the water supply for Denver begins about 1859, when the city was founded as a mining camp (Cox, 1967). From 1859 to 1872, residents relied on individual supplies from private wells or streams. The Auraria and Cherry Creek Water Company was formed in February 1859, but its plans were never carried out.

In 1872, the Denver City Water Company began to deliver water. From 1872 to 1878 it provided service from South Platte supplies, but by 1878 it needed new supplies. The Denver City Irrigation and Water Company was formed to build a lake on the South Platte, and it was consolidated with the Denver City Water Company in 1882 to form the Denver Water Company. From 1886 to 1890 several competing private water companies were formed. These were consolidated with the Denver Water Company in 1890 to form the Denver City Water Works Company, which became the American Water Works Company of New Jersey. This company went into receivership in 1892, and in 1894 it was purchased by the Citizen's Water Company, which had been formed by two former directors of the Denver Water Company, D. H. Moffat and W. S. Cheesman.

In October 1894 the Denver Union Water Company (DUWC) was incorporated, and by the articles of incorporation it took over the assets of the competing water companies and obtained a monopoly to serve domestic water in Denver. It also obtained a 20-year franchise from Denver sometime in that period.

The DUWC built Cheesman Dam in 1905, which is an important part of the Denver system today. In 1907, Denver chose to exercise its option to buy the DUWC, but the board of arbitration could not agree on the price. In 1910, Denver decided to build its own system, but was stopped by an injunction filed with the U.S. District Court by the DUWC. This dispute went to the U.S. Supreme Court. After further disputes, Denver finally got an option in 1916 from the DUWC to buy its assets, and in 1918 a bond issue was passed by the voters to buy the assets. In this same election, a charter amendment was passed that established the Denver Board of Water Commissioners, the governing board of the Denver Water Department.

Thus, by 1920 the Denver system had been converted from private to public ownership. Most U.S. water utilities are publicly owned, but it could have gone the other way and Denver's system could have remained private, as the water utility of Indianapolis, Indiana, did. Chapter 8, which describes the organization of the water industry, compares aspects of public and private ownership of water utilities.

The next 40 years saw tremendous growth in Denver's water system. In 1921–1923 filings were made on West Slope waters on the Blue, Frazer, and Williams Fork, all tributaries of the Colorado River. In 1936 the Moffat Tunnel was completed. This brought into reality the dream of bringing West Slope water to Denver.

In these years, Denver had the opportunity to create a truly regional water supply system, but it did not happen due to conflicts with the suburbs. The proliferation of suburban water agencies goes back to 1948, when the Denver Water Board (DWB) raised rates on Englewood customers. Englewood, a suburb of Denver, was angered by this action and sued Denver to halt the rate increase and to place the DWB under Public Utilities Commission jurisdiction. After losing in court, Englewood decided to build its own system, and was followed in subsequent years by Aurora, Boulder, Golden, Morrison, Northglenn, Westminster, and Thornton. Arvada buys its raw water from Denver at cost and treats it separately. Aurora's decision to establish its own system goes back to the early 1950s when Denver established the "Blue Line," outside of which there was a 50 percent rate differential, no limiting clause on rate increases, and only a year-by-year guarantee of water. Why there was no more cooperation than that is a mystery that can be understood by looking at today's city–suburb conflicts, not only in Denver, but around the nation.

The 1950s drought tested the Denver and suburban systems. By the early 1960s Denver had completed Dillon Reservoir, which holds 254,000 acre-feet. In the 1960s the Blue Line was abandoned, and Denver embarked on an annexation program. Aurora and Colorado Springs launched their Homestake Project in the 1960s, just in advance of the rapid rise of environmentalism.

By the 1970s the state legislature had passed the Poundstone Amendment, which halted Denver's aggressive annexation program. In 1973, Denver proposed to add dramatically to its treatment capacity with the Foothills Treatment Plant. This proposal unleashed considerable environmental opposition, and after a number of legal and political disputes, the issue was finally settled in a political compromise in which the Denver Water Department (DWD) agreed to a strict conservation program, to release in-stream flows, and to appoint a citizen's advisory committee.

Part of the concerns of groups opposing Foothills (antigrowth groups) was that Foothills was to be part of a larger, integrated water supply system that would give Denver the key to a new surge of growth. This concern led into the Two Forks controversy.

A map of the Denver water system is shown in Fig. 10.1. The features just described are shown on the map, including the location of the Two Forks site.

Evolution of the Two Forks Controversy

As an attractive dam site, Two Forks was being studied by the DWD's predecessors as early as the 1890s. Denver filed on water rights for the area in 1931. However, it was only in 1982 that 40-odd suburban

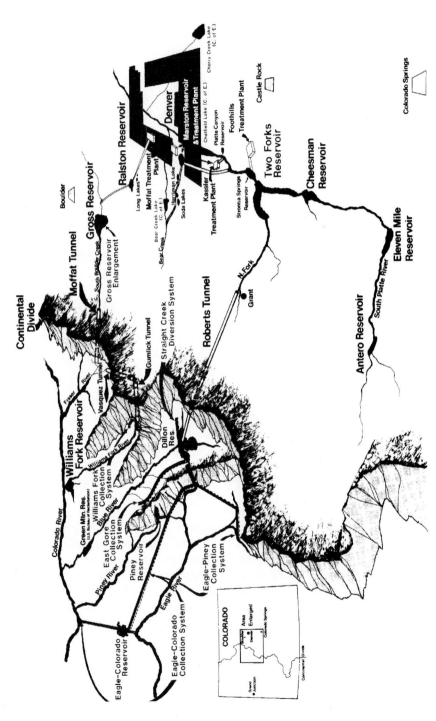

Figure 10.1 Denver water supply system. (*Source: Denver Water Department.*)

governments and water districts united to form the Metropolitan Water Providers and to join Denver in the Two Forks Project. In 1986 the DWD filed for a permit to build the Two Forks Dam. The main permit involved is under Section 404 of the Clean Water Act (see Chap. 14 for a description of the Clean Water Act).

A system-wide environmental impact statement (EIS) was already being prepared by the U.S. Army Corps of Engineers. In 1988 the Corps issued the EIS.

Many technical issues were raised and debated in the EIS process. The process illustrates how water resources planning and management in the United States unfolds today with a combination of technical and political facets.

The planners projected the growth of demand under different scenarios of conservation that range from little water conservation to a great deal. The projected deficit unfolds over time. When you add the controversy over the accuracy of population projections to that of water conservation, the uncertainty in demand forecasting is evident.

Figure 10.2 illustrates the staging of hypothetical additions of supply components to the system. In this scenario, one of several presented in the EIS, Two Forks would bring on line about 100,000 acre-feet of safe yield to the system.

In June 1988, after an extensive period of study, Colorado Governor Roy Romer recommended to the Corps to approve the permit, with a 25-year shelf life. In January 1989 the Corps announced its intent to permit the dam, but in March 1989 the EPA administrator, William Reilly, announced his intention to veto the permit, and it was officially vetoed a year later.

Much of this chronology and description of the Two Forks case is from a Western Governors' Association report (1991). This report is included as an appendix to this chapter because it represents a description of the issue that has been examined by both sides and is considered to be an accurate recount of a complex series of events. The Western Governors' Association identified the following issues for discussion:

- Conflict between traditional water development interests and the environmental community

- The existence of reasonable and practicable alternatives within the Section 404 process

- The forum for decision making, which involved municipal water providers with the means to build the project, federal regulatory agencies with the means to permit or deny it, and only a roundtable to coordinate the actions

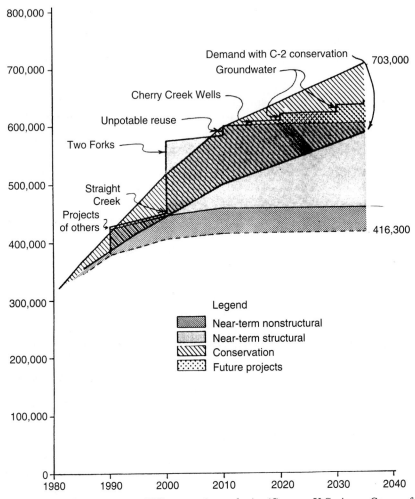

Figure 10.2 Denver water EIS: scenario analysis. (*Source: U.S. Army Corps of Engineers, Omaha District.*)

To these issues, I would add the following:

- What obligation does government have to conserve the taxpayers' money in a process like this?
- Is a $40 million EIS a financial disaster?
- What was Denver's obligation to play its conservation card; did it live up to its Foothills obligations?
- Was the alleged lack of consideration of low-impact alternatives a valid issue?

- Was the financing plan for Two Forks reasonable?
- What were the roles of players in the process?
- Was there a public consensus or a lack thereof about the need for Two Forks? If not, should that have been a factor?
- What were the national agendas of environmentalists? Did the veto turn on the personal connections of environmentalists with bureaucrats in EPA?
- What are the national implications of the use of Section 404 to veto projects?
- What was the aftermath, and what are the implications for water supply projects?

Aftermath of Two Forks

The Two Forks controversy has national implications, and was not just a local water skirmish. I became aware of two similar cases, and I am sure there will be others.

The issues identified in the previous section make up a complex bundle of questions. I do not have the answers to all of them, but we can look at some things that have happened in Colorado after Two Forks.

Most apparent is that the Denver Water Department announced that it could no longer serve as the main water supplier for all of the suburbs, and it refused to join a suit to overturn the Two Forks veto (Farbes, 1992; Obmascik, 1991). This means that Denver has withdrawn from its role as the leader of regional water efforts and will take care of its own needs with existing water supplies, which are adequate.

There have been several new organizational initiatives. A Front Range Water Authority and a Metropolitan Water Authority have been organized. There were meetings of a "Group of 10," a metro cooperation group that included water supply among its aims, and the governor organized several conferences and task forces for state-wide initiatives. At this writing (1996), none of these seems to be leading to any major projects or programs, but Colorado's growth is rapid at this time, and the consensus is that new water supplies will be needed.

The City of Thornton, a northern suburb of Denver, announced a "City-Farm Program" which involved buying up farms and water supplies from northern Colorado. As of this time, the program has been through most of its court hurdles and it appears that the project will become a reality. Private developers have also announced proposals for several new projects.

A state view of the situation was presented by Governor Roy Romer (1993), who noted that in the past few years the state had invested millions of dollars in planning for water projects that had not come about. He said we must "blow the whistle on what has become an unacceptable level of administrative gridlock, litigation, expense and delay whenever water development or transfers are proposed." He said that after Two Forks, water supply planning in the metro Front Range had proceeded piecemeal, with little direction or momentum. He stated that the water wars have focused attention on potential economic and environmental impacts of water transfers. He identified issues of statewide concern: waste of public and private funds; environmental consequences, extensive lead time required to produce new supplies; and impact on future development in other parts of the state. He suggested new directions and alternatives: a regional water coordinating organization led by the state; state incentives such as loans or grants to encourage cooperation; state water project; cooperation with agricultural water users; and enhanced information and decision support systems by state agencies.

Clearly, many things are still to develop in the Front Range of Colorado in the wake of Two Forks. The project was a harbinger of change in the water supply industry's quest for new supplies. As a result of Two Forks, and related developments, the rule in the future, at least in the West, will be to stretch out old supplies much further through innovations such as reuse, water banking, conjunctive use, conservation, transfers and exchanges, regional cooperation and sharing, and other management measures.

Other disputes will be similar. Of course, the EIS process will be costly and protracted in a number of places. Under the U.S. legal system, it is the forum for working out disputes between water developers and environmentalists. One similar case involving a veto of a Section 404 permit was for a project called Lake Alma, in southeastern Georgia. There, the Corps of Engineers decided to issue the permit, but it was vetoed by EPA. A lawsuit was filed to overturn the veto, but at this writing the outcome was still unknown. A more visible case was Ware Creek, a proposed reservoir in southeastern Virginia. In this case, James City County, Virginia, sued over an EPA veto of the Section 404 permit, and an appellate court upheld the EPA. In 1994 the U.S. Supreme Court refused to hear the case, so the appellate court ruling was allowed to stand, and the EPA apparently is free to veto water supply projects on the basis of environmental impacts (American Water Works Association, 1994).

References

American Water Works Association, Supreme Court Refuses to Hear Ware Creek Case, *AWWA Mainstream,* November 1994.

Cox, James L., *Metropolitan Water Supply: The Denver Experience,* Bureau of Governmental Research and Service, University of Colorado, Boulder, 1967.

Farbes, Hubert, Denver Can No Longer Promise to Supply Water to the Entire Metro Area, *Denver Post,* April 25, 1992.

Milliken, J. Gordon, Water Management Issues in the Denver, Colorado, Urban Area, in Mohamed T. El-Ashry and Diana C. Gibbons, eds., *Water and Arid Lands of the Western United States,* Cambridge University Press, Cambridge, 1989.

Obmascik, Mark, 8 Suburban Districts Sue over Two Forks Veto, *Denver Post,* November 23, 1991.

Romer, Roy, The Role for the State of Colorado on Front Range Water Challenges, 1993 Colorado Water Convention, January 4, 1993.

Western Governors' Association, The Two Forks Project, prepared for a 1991 conference, Denver.

Appendix   The Two Forks Project*

Characterization

Application of federal law, Section 404 of the Clean Water Act, resulting in a veto of a locally sponsored, state supported decision to authorize a water storage project, on the basis that less environmentally damaging practicable alternatives to the project existed.

Background

On November 23, 1990, the United States Environmental Protection Agency vetoed the proposed Two Forks dam and reservoir project, near Denver, Colorado, citing adverse environmental effects that would result from the project. The agency concluded that less environmentally damaging practicable alternatives exist to provide future water supplies for the Denver area.

The EPA veto was the culmination of a lengthy and expensive process. For eight years, the Denver Water Board and more than twenty co-sponsors sought to win public support for the $400 M project, arguing that it was necessary to meet the needs of a metropolitan area population projected to reach 2.25 million by 2010. The dam would supply some 90,000 acre-feet of water a year, but would also flood a 28 mile stretch of the South Platte River upstream of Denver.

Note: This summary of the project was recommended to the author as a balanced description, and was prepared for a 1991 conference by, and is published here with the permission of, the Western Governors' Association, 600 17th Street, Suite 1705, South Tower, Denver, CO 80202.

Physical setting.   The Two Forks Dam and Reservoir would capture and store water from river basins on both sides of the Rockies to which the Denver Water Board has rights but which it does not utilize for lack of adequate storage on the east side of the Continental Divide. This water includes spillover from Dillon Reservoir, occasional excesses on the South Platte to which the Denver Water Board has rights in times of flooding, and other miscellaneous rights on the river. The reservoir was planned to be a short distance upstream from Strontia Dam, the diversion facility which provides water to the Foothills Treatment Plant, and was intended to regulate the flow of the South Platte and its north fork as well as water deliveries through the Roberts Tunnel of the Blue River system. Located 24 miles southwest of Denver, the reservoir would be a short distance upstream from the Foothills Treatment Plant.

The conflict.   Denver's water system now serves a metropolitan area of about 1.8 million. In order to gain support from Denver's suburbs, Denver sold shares in a project to 42 other towns and water districts, which, together with Denver, constituted the Metropolitan Water Providers (MWP). MWP spent $5M for land acquisition for the project and $40M for the most expensive environmental impact statement in US history. The debate over the dam deeply divided Colorado's leaders and its citizens. The various views are summarized as follows:

Proponents.   The desire of the metropolitan Denver area to secure a permit to build the Two Forks project arose in a unique political context. A decade ago, concerns over the city of Denver's potential to annex surrounding lands and communities led to a constitutional amendment prohibiting Denver from annexing land without permission from the county that would lose it, leaving Denver with its excess water supplies as the only bargaining chip in negotiations over a number of issues meriting metropolitan cooperation i.e. schools, hospitals, cultural facilities, etc. However, in order for Denver to market its water to surrounding communities, the city sought more dependable future water supplies for its citizens. A permit to build Two Forks provided the necessary security in this regard. This was true, proponents argued, notwithstanding the fact that other alternatives were available to augment supplies and to reduce demand in the metropolitan area. Denver agreed to pursue conservation programs but argued that these measures if successfully implemented would fall well short of the future needs of the metropolitan area. Further, they argued, without the ability to build Two Forks, Denver would lose its advantage to negotiations with the suburbs. Furthermore, the proponents argued that a large storage system on the South Platte would be the most effective way of retaining water diverted from the western slope's Blue River through the Roberts Tunnel into the north fork.

With regard to the environmental impacts, while it was clear that the extraordinary trout fishery as it now exists would disappear, they pointed out that the mitigation plan recognized and compensated for the unique values represented by Cheesman Canyon and noted that the excellence of the fishery was due in part to the placement and operation of Cheesman Dam, built at the turn of the century. The plan aimed for replacement of 90% of the losses of instream biomass, modifications to reservoir releases throughout the South Platte and Blue River systems to improve stream values, improvements in recreational accesses to streams, and recreational development of the proposed Two Forks Reservoir. Furthermore, the proponents argued that without the cooperation among metropolitan entities which the Two Forks project would facilitate, competition among these entities for new water to attract suburban growth would result in continuation of wasteful and fragmented water development and distribution along the Front Range with its own potential for adverse environmental consequences.

Opponents. West slope concerns centered around water availability for their future growth, for meeting the needs of a recreational economy, and for environmental quality, as well as the potential need to build new sewage treatment plants because dilution capacity of streams would be lost.

A group of environmental organizations, known as the Environmental Caucus, were also active in opposing the project. They contended that it would remove the "last vestige of free-flowing water on the South Platte River." They also raised concerns about the potential impact of the project on endangered species such as the whooping crane, wetland vegetation, fish, and wildlife habitats, increased salinity on the Colorado River, and the anticipated draw-down of Dillon Reservoir with an attendant reduction in its recreational benefits. They also cited the continually increasing costs as a reason to reevaluate the project. If revenue from the project did not come from additional sales taxes, which must be approved by voters, they pointed out that the funds would come from water sales. They argued that such a revenue source would eliminate any incentive for the Denver Water Board or other metropolitan water purveyor to encourage conservation by their customers because of the need to increase revenues from water sales to pay for the project. They also pointed to other alternatives available to the Denver area. In addition to water conservation, and the purchase of water from existing projects, such as the Windy Gap project, they suggested that Denver could pursue construction of small projects, including a small dam on the west slope. Environmentalists also suggested that the metropolitan area could purchase water now used for irrigating crops and convert it to municipal use.

Processes for resolution

The state. The Governor's Metropolitan Roundtable, originally under the chairmanship of then Colorado Governor Richard Lamm, began meeting in 1982 to provide a means for interested or affected parties to take a comprehensive look at ways to satisfy Denver's future water demands with special attention to the Two Forks issue. The Roundtable was composed of thirty representatives of such groups as the metropolitan suburbs, western slope interests, Denver Water Board commissioners, environmentalists, public interest groups and agricultural interests. This was the first time that some of these interests were included at the table to discuss water decisionmaking. It had not decisionmaking authority, and there was no state permitting or review process into which its advice could be channeled. As such, its greatest value was in providing interaction among diverse interests. While the Roundtable was still active, however, Denver began its EIS preparation, drawing much of the energy away from the Roundtable process.

Following completion of the environmental impact statement, but prior to its decision on whether to grant the Section 404 permit for the Two Forks Project, the Corps of Engineers asked then Colorado Governor Roy Romer for his comments. The governor expressed serious concerns about the present need for the project and about the consequences for the environment, especially Cheesman Canyon. Nevertheless, he recommended approval of the permit, for the Two Forks Project, finding that it would serve Colorado in three ways. "First, it will allow the metropolitan area water providers to jointly develop and share interim supplies. Second, it will preserve the fragile institutional relationships among the water providers in the metropolitan area. Third, it will serve as the long term structural supply in the event that other alternatives cannot be utilized."

He recommended, however, that the permit be given a shelf life of 25 years so as to allow additional time to review whether the construction of Two Forks is truly necessary. He urged that full implementation of the mitigation plan negotiated by state agencies take place, that there be established a metropolitan-wide water authority by July 1990 pursuant to state law, that prospective recipients of Two Forks water supplies be required to undertake conservation programs sufficient to achieve reduction in demand of 42,000 acre-feet by 2010, and that prospective recipients of Two Forks water be required to develop and share interim supplies prior to the building of any South Platte structure.

He agreed that opponents of the project were absolutely correct that significant supplies of water currently exist for the Denver metropolitan area and that sensible plans for water conservation and the development of known interim supplies could add years to the region's water supplies. However, he emphasized that in order to obtain the kind of cooperation that would be necessary to develop the interim supplies, it would be necessary to hold an "insurance policy" guaranteeing additional South Platte storage. He said that Two Forks should be built only as a last resort. To accomplish the goal of meeting the metropolitan area's long-term water needs and leaving the South Platte a free flowing river between the Cheesman and Strontia Springs dams, he proposed a plan of action requiring Colorado legislative approval.

With regard to this plan of action, he suggested that it would bring the decision back to Colorado "where it belongs." "It says to the federal agencies," he continued, "you grant us the permits, but we will decide which to use." He did recognize, however, that neither he nor any state agency had authority to grant, deny or condition the manner, timing, or size of water development within the state if a proponent holds valid water rights.

The federal government.   In March of 1989, the Corps of Engineers announced its intention to issue a permit to build Two Forks Dam within certain conditions imposed, but without the long lead time requested by the Governor.

Days later, William Reilly, newly appointed EPA Administrator, announced that EPA would review the permit for possible violations of the Clean Water Act. Thus began a series of hearings and investigations which ultimately led to the EPA veto. Concerns within EPA had surfaced in May of 1988, when three staff members reviewing the Two Forks Project for compliance with Section 404, wrote a long memorandum to Regional Administrator James Scherer, arguing that Two Forks did not meet the law's guidelines and should not get a permit. They argued that "several less environmentally damaging practicable alternatives exist," citing several smaller dam proposals. The staff argued in a later memo that the Corps' EIS did not consider "all reasonable alternatives." The Corps alternative relied heavily on local ground water development to replace Two Forks, which the EPA staff described as a "straw man predicated on biased assumptions, such as no sharing and each water provider in the area building independent facilities." The result, these EPA representatives argued, was to make Two Forks appear less costly.

EPA and the Corps had an even more basic difference. According to the second EPA memo, the Corps' first choice was to seek mitigation of the project's environmental impacts. The EPA approach, on the other hand, was to avoid those impacts. EPA staff argued that destruction of Cheesman Canyon would be irreplaceable, while the Corps found that the impacts had been "mitigated essentially to zero" because of the promised wetland and habitat replacement. Regional Administrator Scherer chose to negotiate with Denver for additional mitigation measures and to recommend the permit. However, the Environmental Caucus utilized the internal memos to convince Reilly to intervene.

Major issues

The lengthy and complex process which culminated in the EPA veto of the permit to build the Two Forks Project raised several issues. To most people, it was a classic confrontation between traditional water development interests and the environmental community over the use of water in the West. "High Country News" concluded as follows: "Two Forks will help decide how water in the West will be used in the future: to benefit large, dense urban areas, or to maintain the natural environment in a smaller, diverse population. . . . If Two Forks is built, it will mark the continuing strength of the urbanizing vision that has marked the West for the last four decades. If Two Forks is rejected, it will mark the end of an era, and announce the need for the West to march in a different direction."

The focal point of the controversy was a Section 404 permit. The central question, therefore, focused on the issue of whether reasonably practical alternatives existed to the project. Virtually everyone agreed that such alternatives existed in the near term, but differed over the long term. Further, some thought the matter had to be considered in a much broader context. The only forums for resolving these issues, however, were the Metropolitan Water Roundtables, which brought parties to the table but could not do more than advise municipal water providers; and the federal agencies who had ample power to halt or condition development proposals but lacked political sensitivity or support for its alternative approaches.

Resolution on Two Forks

In justifying her final decision, EPA Assistant Administrator Wilcher referred to the Corps' EIS: "We agree with the EIS analysis that less environmentally damaging, practicable alternatives exist. Under the Clean Water Act, EPA cannot sanction construction of a project that would destroy valuable aquatic habitat and resources when such al-

ternatives exist. We look at many factors—including cost—to determine whether identified alternatives are, indeed, feasible."

Observations and analysis

While this decision appears to have put the final nail in Two Forks'
coffin, controversy and conflict continue over many of the issues that
were raised. Only time will tell whether the cooperation will occur
that everyone agrees is necessary between Denver and its surrounding
metropolitan area. Further, it is clear that fundamental questions regarding federal and state roles in water development and use were
also raised by the Two Forks process, but not resolved. In announcing
his decision, Governor Romer said: "I have been placed in this decision-making chain, but only with the option to recommend approval or
denial on the specific project which has been applied for. That process
does not give me the authority to change the recommended solution
and to see that it happens. Therefore, I am trying to devise a solution
that I can make. But I need the legislature to help get this done."

The Governor made recommendations based on consideration of all
of the major issues. In contrast, EPA, following what it saw as its mandate under Section 404, vetoed the project because of the existence of
practicable alternatives. While local proponents railed against EPA's
veto as an example of federal overreaching, project opponents were apparently able to fairly say that proponents of Two Forks were never
able to convince the majority of the people of Colorado of the need for
the project.

However, because EPA's decision necessarily turned on language of
the Clean Water Act, little attention was given to the major concern
of the Governor and Denver metropolitan interests regarding a permit for Two Forks as a catalyst for metropolitan cooperation. In addition, regardless of the merits of the project itself, some felt that the
Two Fork's project raised a fundamental question of whether the
Section 404 process can adequately address the broad range of issues
and considerations that may rightly be applied to proposed water
projects in the West.

On the other hand, environmentalists, generally pleased with the
results of the federal veto, criticized Colorado state law as inadequate
to safeguard environmental values and consider important economic
and social issues. They object to a system which declares private
property rights to free flowing waters and which is administered by
what they view as a cumbersome court system unique in the West.
They point with criticism to provisions that water rights can only be
awarded for "beneficial uses" and that a diversion is required to be
considered beneficial.

Chris Meyer, representing the National Wildlife Federation in the debate over Two Forks, made the following statement. "We must shift the way we think about water from the traditional way—as an absolute property right—to a shared resource that certainly has property rights to it but also is subject to the concerns and needs of the people of the entire state." Meyer warns that until this happens, other big water projects will fail as Two Forks did. "The federal government is going to continue to fill the void."

Questions

1. In the Two Forks planning process, what would have been different if the project proponents had "planned with care and considered the needs of neighbors and the environment"?

2. What influence did metropolitan cooperation have on the Two Forks process? What impact did the veto have on metropolitan cooperation for the future?

3. Who were the main players in the Two Forks process, and what were their roles?

4. The state's governor said that we must "blow the whistle on what has become an unacceptable level of administrative gridlock, litigation, expense and delay whenever water development or transfers are proposed." What did he mean by this statement?

5. Two initiatives suggested by the governor were a regional water coordinating organization led by the state, and enhanced information and decision support systems by state agencies. How might these make up for the loss of water in Two Forks?

6. Explain what the "404 program" is and how it relates to the siting of new reservoirs.

7. The Two Forks Project involved aspects of *regionalization, cooperation,* and *integration* in water management. Define these concepts and explain how the project illustrated both the theory and practical aspects of implementing them.

11

Flood Control, Floodplain Management, and Stormwater Management

Introduction

On an overall basis, floods cause more property damage than any other natural hazard faced by the United States. Although the timing of floods varies, most regions of the country face significant flood hazards.

Water resources managers must control floods and prevent damage from them, but this is a different type of mission than providing water supply or controlling water quality: It is a protective mission. This means that flood control—and its cousin, stormwater management—involve a mixture of land use and water management, and, as a result, involve political issues related to land use.

This case study chapter outlines the principles of flood control and floodplain management, both in large river systems and in urban stormwater situations. It does not deal with stormwater quality, which is covered in another chapter, nor does it deal with the infrastructure issues of urban stormwater management, which are discussed briefly in Chap. 3.

The chapter traces the historical development of U.S. policy of flood control and floodplain management. It also illustrates different kinds of flooding, ranging from the slow-building but long-lasting Mississippi River floods to devastating flash floods that strike without warning. It illustrates an organizational response—Denver's Urban Drainage and Flood Control District—and it underscores why a national organization such as the U.S. Army Corps of Engineers is necessary to oversee reservoir and dike systems.

While the case study briefly introduces quantitative tools of flood analysis, mainly the hydrograph, it does not present much detail on them. The reader should consult a text on hydrology for more information.

The terms used in the chapter title—flood control, floodplain management, and stormwater management—hint about the two approaches to the problem: controlling storm and flood waters, usually through reservoir storage and conveyance systems, and preventing potential damage by managing land use in the flood plain. These two general approaches, along with associated tools and measures, make up the "structural" and "nonstructural" approaches to the flood problem.

While engineers have debated about structural versus nonstructural approaches, nature favors the nonstructural approach. Regardless, there are situations where reliable structural features make the difference between life and death. In the winter of 1994–1995, for example, there was an emergency in Holland due to Rhine River flooding. Much of Holland's usable land is made available by diking and holding back the rivers and the sea, and high flood waters threatened the dikes. Fortunately, the emergency passed, but if any dikes had failed, the flooding would have been extremely damaging. One might say that "the dikes shouldn't have been put there in the first place," but that would not be a practical answer to a difficult problem.

The competition between structural and nonstructural approaches to flooding highlights, to some extent, the water project development era of 1920 to 1970. Reservoirs were the solution of choice to flood control early in the period, then emphasis swung to nonstructural solutions. In fact, the Report of the Interagency Floodplain Management Review Committee (1994), prepared after the great Mississippi River floods of 1993, stated:

> By controlling runoff, managing ecosystems for all their benefits, planning the use of the land and identifying those areas at risk, many hazards can be avoided. Where the risk cannot be avoided, damage minimization approaches, such as elevation and relocation of buildings or construction of reservoirs or flood protection structures, are used only when they can be integrated into a systems approach to flood damage reduction in the basin.

The Interagency Committee's report also underscored the importance of a coordinated approach: "The...Committee proposes a better way to manage the floodplains. It begins by establishing that all levels of government, all businesses and all citizens have a stake in properly managing the floodplain."

Flood Problems and Responses
in the United States

Flooding is the most destructive and costly natural disaster faced by the United States (Schilling et al., 1987). Floods account for 85 percent of disasters declared by the President annually, and about 160 million acres, or 7 percent of all U.S. land, are in floodplains.

During the Summer of 1993, the nation received a wakeup call in the form of the great Mississippi River flooding which is reviewed later in this chapter. The misery and damage caused by this flood was shown to the nation each night on television, and flood policy is certain to be affected by the attention it received. A landmark report (Interagency Floodplain Management Review Committee, 1994) was issued, and its recommendations will be reviewed later.

Historically, four federal agencies have had lead roles in flood control projects: the Corps of Engineers, the Soil Conservation Service, the Bureau of Reclamation, and the Tennessee Valley Authority. Other federal agencies, such as the Federal Highway Administration and the Federal Housing Administration, become involved as flooding concerns facilities that they manage. In addition, the Federal Emergency Management Agency has a key role in floodplain regulation and in flood response.

The history of U.S. flood programs begins with the "structural, federal era" according to the Natural Hazards Research and Applications Research Center 1992 report on floodplain management in the United States (1992). This era began with the settlement and development of the nation, with local efforts based on levee districts, conservancy districts, and individual landowners. About the time of the Civil War, Congress authorized stream gauging by federal agencies, but it was not until after serious flooding in the decades after 1900 that Congress authorized flood control works in the Flood Control Acts of 1917, 1928, 1936, and 1938. The emphasis in these acts was on works such as dams, levees, and channel modifications.

By the 1960s the federal government had completed a vast array of projects, with an authorization for the Corps of Engineers alone for 220 reservoirs, over 9000 miles of levees and floodwalls, and 7400 miles of channelization at a cost of $9 billion.

Losses mounted in spite of these projects, and in the 1960s the emphasis began to change to nonstructural measures such as zoning, flood forecasting, flood insurance, relocation, and alternative storage techniques such as land spreading. In 1960 the Flood Control Act authorized the Corps of Engineers to begin to provide floodplain information, and the Corps began to produce floodplain information reports as a service to local governments.

At this point in the discussion, I would like to pay tribute to Dr.

Gilbert White, a well-known geographer who made landmark contributions to the fields of water resources and environmental management. White, who remains active at this writing as an emeritus professor at the University of Colorado, continues to contribute with incisive publications. His own Ph.D. dissertation, "Human Adjustment to Floods," written at the University of Chicago in the 1940s, is widely recognized as one of the most influential statements of the need to use a nonstructural approach to floodplain management. White's contributions are also mentioned in Chap. 4.

Returning to the evolution of floodplain policy, in 1966, House Document 465, "A Unified National Program for Managing Flood Losses," was submitted to Congress by the Johnson administration. It stated the case for a unified approach, and made 16 recommendations for federal agency actions such as research and information.

Additional legislation was followed in 1976 by a report from the U.S. Water Resources Council presenting "A Unified National Program." Among other issues, it cited coordination as "the weakest component of current management efforts." As the reader knows by now, this author views the constant search for coordination mechanisms as a central problem of water resources management.

After additional legislation and executive orders during the Carter administration, the Federal Inter-Agency Floodplain Management Task Force submitted revisions to the Unified National Program in 1979. The task force expanded the strategies to include restoration and preservation of natural values and emphasized the need for additional information to make decision makers aware of alternative approaches.

In 1986 the Federal Emergency Management Agency assumed responsibility for the task force, and it submitted a revision to the 1979 program. The revisions included a federal disaster mitigation strategy, additional research, and attention to coastal zone flooding.

Schilling et al. (1987) state that the degree of fragmentation of flood authority in the federal government is a cause for concern. They display a table that illustrates how programs are diffused through 26 agencies and nine program purposes. The program purposes include flood insurance studies, floodplain management services, floodplain information reports, technical and planning services, flood-modifying construction, flood preparedness, emergency and recovery, warning and forecasting, research, and open-space activities.

The focus remains on nonstructural measures. I cited earlier the reports of the Interagency Floodplain Management Review Committee, prepared after the great Mississippi River floods of 1993, and to add to this, the President's budget of 1995 would make any investment in structural projects even more difficult. Clearly, trends are toward nonstructural, comprehensive approaches, and away from flood control reservoirs.

Flood Causes and Risk Factors

Hydrology as a basic science of water management was discussed in Chap. 2, but I would like to add some detail here that relates directly to causes of flooding.

Flooding is caused by excess rainfall or snowmelt—that is, runoff that is in excess of that needed to replenish aquifers or surface water features. All aspects of the hydrologic cycle are involved in flooding: precipitation, runoff, infiltration, and channel flow.

For a relatively large basin, flooding does not depend on the complexities of urban stormwater systems. For example, Fig. 11.1 illustrates a 1965 storm that caused devastating flooding on Fountain Creek in the city of Pueblo, Colorado. This flood was reviewed in my text, *Water Resources Planning* (Grigg, 1985).

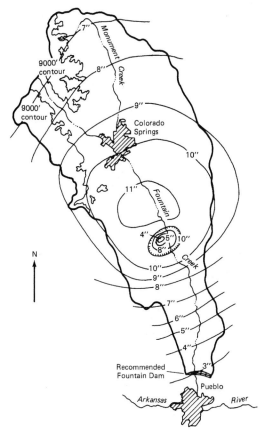

Figure 11.1 Storm hyetograph: Fountain Creek, Colorado. (*Source: Grigg, 1985, adapted from a Corps of Engineers report.*)

The first consideration in flood risk is the chance of extreme rainfall depths or intensities which cause flooding. They are described by the historical records of rainfall at a particular location. The basic tool to assess the risk is historical experience.

For example, if the flood risk to be evaluated is a 10-in rain within 24 hr, historical data will show the chance that such a rain will occur in a given year. Perhaps that chance will be 1 in 200, or 0.5 percent. Said another way, 10 in would be a 200-year 24-hr rainfall depth for that location. These rainfalls are in the range of several historical events described in this chapter.

Next, flood risk depends on runoff, which in turn depends on a number of hydrologic factors. A heavy rain can fall on a dry watershed and cause little flooding; or a moderate rain can fall on a saturated watershed and cause heavy flooding. Runoff is predicted hydrologically from watersheds. The rate of runoff as a fraction of rainfall is often difficult to predict, and is one of the main objectives of engineering hydrology. Flow in channels causes flooding when the rate of flow exceeds the capacity of the channel to carry the flow without overtopping the banks and covering the floodplain with water.

Large floods are caused by widespread rainfall and runoff. For example, as described later in the chapter, the rainfall that caused the great Mississippi River floods of 1993 was widespread in time and space. Also, widespread flooding in Bangladesh is caused by unique conditions of heavy runoff, a large population, and vulnerable lands.

The rain–runoff–channel flow sequence makes up the principal cause of flooding. Consequences of flooding are due to depth, velocity, and extent of flood waters, as well as the vulnerability of people and their property.

In general, the deeper the flood waters and the higher the velocities, the greater the damage will be. Damage can be predicted from depth–damage curves for different classes of residential properties, but for commercial and industrial properties the variation of damage types is too great to place into just a few categories. For commercial and industrial properties, a survey of each property is usually necessary.

Figure 11.2 illustrates conceptually the sequence of rainfall, runoff, stream flooding, and damage that results from floods. Damage can be predicted using methods of economics, and the result should be damage–frequency curves that enable the computation of benefit–cost information (Fig. 11.3).

Flood Quantification for Small and Large Basins

There are many methods for computing stormwater or flood runoff, but they are only as good as the input data. A full discussion of these meth-

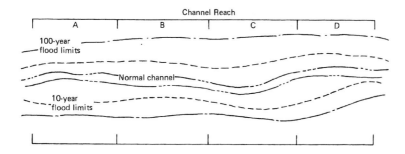

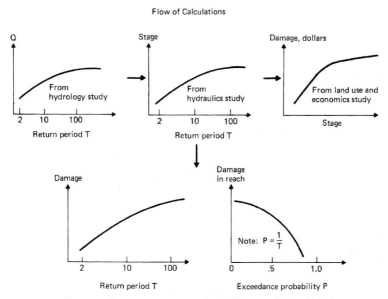

Figure 11.2 Damage–frequency computations. (*Source: Grigg, 1985.*)

ods is beyond the scope of this chapter, but two elementary concepts, the rational method and the hydrograph, will be presented briefly because they are at the heart of the methods. Think of these two methods as characterizing the way to compute runoff for small flooding situations, such as in a city ditch, and large situations, such as in river basins.

Small basins: The rational method

The rational method, used for stormwater calculations on small basins, is over 100 years old, and is based on a simple equation. Although it is not considered very accurate, it has passed the test of time and is widely used to plan and design urban stormwater and related small facilities such as road culverts.

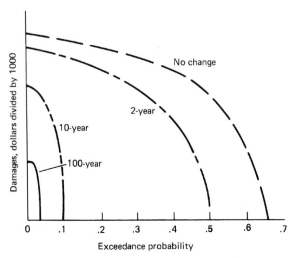

Figure 11.3 Damage–frequency curve. (*Source: Grigg, 1985.*)

To those who would criticize the method, I want to say that I am not recommending it as the method of choice; I am merely pointing out that it illustrates the basic principles of runoff computation for small basins.

The equation is:

$$Q = CiA$$

where C = the runoff coefficient, or the fraction of rainfall that appears as runoff
  i = the rainfall intensity at the basin time of concentration for the frequency of rainfall of interest
  A = basin area in acres

The method illustrates the correlation between rainfall, a runoff factor, and the resulting storm or flood runoff. All flood prediction methods require estimates of these parameters. The reader can turn to texts on hydrology or stormwater for a discussion of the parameters.

Stormwater management in urban areas has advanced quite a bit in the last 30 years. Now, the dominating issue is water quality, as discussed in Chap. 14. Planning for stormwater quantity issues requires a sophisticated process by itself.

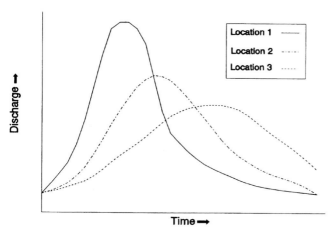

Figure 11.4 Flood wave movement. (*Source: National Weather Service, 1994.*)

Large basins: Hydrographs

For basins larger than about 200 acres, including those up to thousands of square miles in area, the hydrograph is the basic quantitative tool to describe flooding. Engineers and scientists have devised numerous ways to estimate hydrographs, and the reader is referred to textbooks on hydrology for discussions. To illustrate the elements of a hydrograph, Fig. 11.4 shows a hydrograph with reduced flood crests as the flood wave moves downstream, a common situation.

As shown on hydrographs, flood events can cover anywhere from a few hours to many days. From the hydrograph, one can see the rise, peak, and recession phases of the flood event.

Measures against Flooding

Guarding against floods has received much attention due to the damage that floods cause. The Natural Hazards Research and Applications Research Center (1992) lists four basic strategies for floodplain management:

- Modify susceptibility to flood damage and disruption (zone or regulate land use in the floodplain).
- Modify the flooding (use flood control reservoirs).

- Modify the impact of flooding on individuals and the community (use mitigation techniques such as insurance and floodproofing).
- Restore and preserve the natural and cultural resources of floodplains (recognize the values of floodplains and use them for recreation and other appropriate activities).

Tools for these measures are as follows.

Strategy A: Modify susceptibility to flood damage and disruption

1. Floodplain regulations
   a. State regulations for flood hazard areas
   b. Local regulations for flood hazard areas
      (1) Zoning
      (2) Subdivision regulations
      (3) Building codes
      (4) Housing codes
      (5) Sanitary and well codes
      (6) Other regulatory tools
2. Development and redevelopment policies
   a. Design and location of services and utilities
   b. Land rights, acquisition, and open space
   c. Redevelopment
   d. Permanent evacuation
3. Disaster preparedness
4. Disaster assistance
5. Floodproofing
6. Flood forecasting and warning systems and emergency plans

Strategy B: Modify flooding

1. Dams and reservoirs
2. Dikes, levees, and floodwalls
3. Channel alterations
4. High flow diversions
5. Land treatment measures
6. On-site detention

Strategy C: Modify the impact of flooding on individuals and the community

1. Information and education
2. Flood insurance
3. Tax adjustments
4. Flood emergency measures
5. Postflood recovery

Strategy D: Restore and preserve the natural and cultural resources of floodplains

1. Floodplain, wetland, coastal barrier resources regulations
   a. Federal regulations
   b. State regulations
   c. Local regulations
      (1) Zoning
      (2) Subdivision regulations
      (3) Building codes
      (4) Housing codes
      (5) Sanitary and well codes
      (6) Other regulatory tools
2. Development and redevelopment policies
   a. Design and location of services and utilities
   b. Land rights acquisition and open space
   c. Redevelopment
   d. Permanent evacuation
3. Information and education
4. Tax adjustments
5. Administrative measures

There is some overlap in these measures. A simpler classification might be between "structural measures" (strategy B) and "nonstructural measures" (strategies A, C, and D).

Structural measures use traditional engineering tools and are still being implemented in some countries. The most significant structural measures for controlling floods are reservoirs and levees (or dikes).

Chapter 13 describes how storage space can be reserved in reservoirs to control floods. If a flood with a certain volume can be stored in the available space, downstream areas will be spared from damage, and the stored floodwaters can be released gradually after the flood. Sometimes, a flood exceeds the storage space of the reservoir, and it is necessary to release waters during the flood via the service spillway. If a flood is very large, the emergency spillway may have to be used.

Levees are protective structures meant to shield property from floodwaters. In many places, levees and associated floodwalls are essential parts of comprehensive flood protection systems. They require considerable maintenance and are risky. During the great Mississippi River floods of 1993, to be described later, a number of levees failed. In the 1994–1995 Rhine River flooding that threatened Holland, dikes did not fail, but there was widespread concern.

In the United States, attention has turned to nonstructural measures, and in particular, to floodplain management. The Unified

National Program for Floodplain Management, promulgated by the Inter-Agency Task Force in 1976, lists three principles for floodplain management:

1. The federal government has a fundamental interest in how the nation's floodplains are managed, but the basic responsibility for regulating floodplains lies with state and local governments.
2. Floodplains must be considered in the context of total community, regional, and national planning and management.
3. Flood loss reduction should be viewed in the larger context of floodplain management, rather than as an objective in itself.

These principles remain valid, some 20 years later.

Colorado Flooding: Mountains and Plains

The first case study is flooding in Colorado. Although Colorado is generally known as a semiarid state, it suffers, as do other dry regions, from flooding. Neither dry nor humid regions are immune from flooding.

Three types of floods will be discussed briefly: flash floods from heavy rain in the mountains, generalized flooding from rainfall, and snowmelt flooding.

Flash floods: The Big Thompson disaster of 1976

Colorado, like other mountainous areas, has to be careful about rapidly developing flash floods. In August 1976 the state got a lesson about flash flooding when the Big Thompson River had a flash flood that killed 139 people in the space of a couple of hours (McCain, 1979).

The Big Thompson is a mountain stream with headwaters in Rocky Mountain National Park near Estes Park, Colorado (Fig. 11.5). The watershed is rocky and steep and has a drainage area of 305 mi^2 at the mouth of the Big Thompson Canyon above Loveland. By the time the river empties into the Platte River near Greeley, the watershed area has increased to 828 mi^2.

The storm that occurred on July 31–August 1, 1976, was extraordinary, to say the least, and poured 6–10 in of storm rainfall over a wide area of the basin. Figure 11.6 shows the rainfall buildup at two locations, and Fig. 11.7 shows the physics of the storm. The estimated peak discharge was more than four times the previous maximum during 88 years of record. Stream velocities were 20–25 ft/sec. The peak

Figure 11.5   Location of Big Thompson flood. (*Source: McCain, 1979.*)

discharge at one point in the canyon was 3.8 times the 100-year flood level. However, prior floods on several other streams in the foothills have approximately equaled the Big Thompson experience.

Regions like the Front Range of the Rockies are susceptible to flash floods. They do not occur often, but when they do, they can be deadly. Figure 11.8 is a photo of damage done by the Big Thompson flood.

The flooding of 1965: South Platte and Fountain Creek

The Big Thompson flood was by no means the only large flood to occur in Colorado; on the contrary, large floods seem to occur relatively frequently, at least in the context of decades. The June 1965 floods along

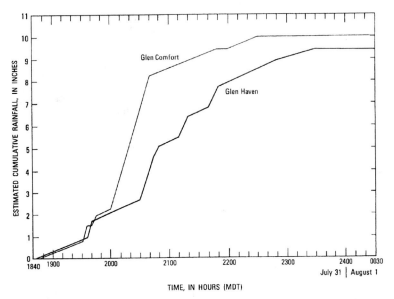

Figure 11.6   Rainfall during Big Thompson flood. (*Source: McCain, 1979.*)

the foothills greatly exceeded the Big Thompson flood levels, for example.

In the June 1965 flooding, peak discharges were caused by heavy, intense rain that fell over a 3-day period when the ground was already somewhat wet. Peak discharge of the South Platte River through Denver was 40,300 ft^3/sec, some 1.8 times the previously recorded peak of 22,000 ft^3/sec (record starting in 1889). Eight deaths were attributed to the floods, and some $508 million in property damage occurred, mostly in the Denver area (Matthai, 1969). Storm cells for this flood were extensive, similar to those of the 1976 Big Thompson flood. Rain cells with maximum depths of 14 in were widespread, and similar to the storm transposition reported for Fountain Creek, Colorado (see Fig. 11.1) (Grigg, 1985).

Snowmelt flooding

Snowmelt flooding is the other principal mode of flooding in the Rocky Mountains. The Poudre River flooding of 1983 near Ft. Collins, Colorado, was an example. Heavy snows, combined with rapid warming, create conditions for these floods. If rainfall occurs at the same time as snowmelt, flood conditions can worsen.

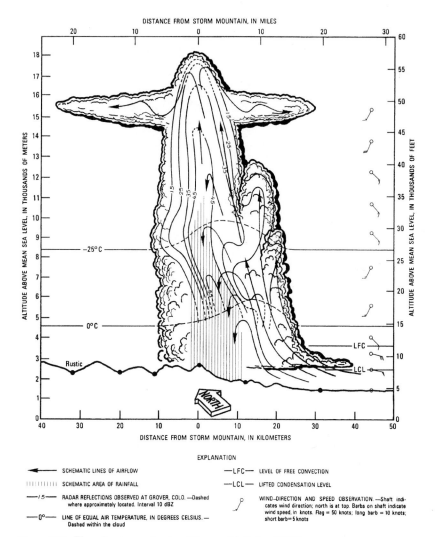

DISTANCE FROM STORM MOUNTAIN, IN MILES

DISTANCE FROM STORM MOUNTAIN, IN KILOMETERS

EXPLANATION

◄———— SCHEMATIC LINES OF AIRFLOW

||||||||||| SCHEMATIC AREA OF RAINFALL

——/5—— RADAR REFLECTIONS OBSERVED AT GROVER, COLO. —Dashed
where approximately located. Interval 10 dBZ

——0°—— LINE OF EQUAL AIR TEMPERATURE, IN DEGREES CELSIUS. —
Dashed within the cloud

—LFC— LEVEL OF FREE CONVECTION

—LCL— LIFTED CONDENSATION LEVEL

WIND–DIRECTION AND SPEED OBSERVATION. —Shaft indi-
cates wind direction; north is at top. Barbs on shaft indicate
wind speed, in knots. Flag = 50 knots; long barb = 10 knots;
short barb = 5 knots

Figure 11.7   Thunderstorm physics. (*Source: McCain, 1979.*)

One Colorado Response: The Urban Drainage and Flood Control District

After the 1965 flooding in Colorado, and with increasing recognition of the need for regional solutions to flood problems, Colorado decided to organize a regional flood control district. I shall describe the district because it represents over 25 years of experience with regional drainage and flood control. My source is Tucker (1994).

Figure 11.8 Damage caused by the Big Thompson flood.

The organizational meeting of the Urban Drainage and Flood Control District (UDFCD) was held on July 28, 1969. The district boundaries consisted of five counties containing some 30 cities and towns and the city and county of Denver, an area with a population of somewhat more than 1 million people. The population has about doubled in the last 25 years. While there are many flood control districts in the United States, most do not include more than one county. Figure 11.9 shows the UDFCD boundaries.

Early business of the district included accepting the offer of the Denver Regional Council of Governments' drainage technical committee to be the Urban Storm Drainage Advisory Committee of the UDFCD, and the Urban Storm Drainage Criteria Manual was adopted as a model for the Denver area. The district had authority to levy a 0.1-mill tax which generated $276,500 in 1971. By 1994, the budget had grown to $19.3 million.

Initially, due to the limited funds, the district concentrated on planning, and by now has planned some 800 miles of major drainageways.

Early results of planning showed the need for an extensive floodplain management program. About 1000 miles of floodplains have been delineated, and local governments can use the information for land use decisions.

The district's mill levy was increased to 0.5 mills in 1973 to provide funds for design and construction. The district's board decided on a

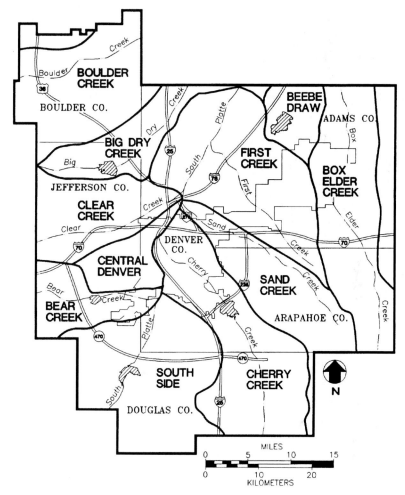

Figure 11.9   Denver Urban Drainage and Flood Control District. (*Source: Tucker, 1994.*)

50–50 matching program with local governments. District funds of $67,200,000 were spent by 1994, and local governments provided at least as much. Rough calculations show that to be about $90 per capita for the 20-year period.

Maintenance proved to be a problem because, after construction, when the drainageways were dedicated to the local governments, maintenance was a low priority. In 1979 the district was authorized by the legislature to levy an additional 0.4 mills for maintenance. The program worked well: Facilities were owned by local governments and the UDFCD managed the maintenance, done mostly by private firms. In the district's 1995 work program, some 250 maintenance projects are defined at a cost of $4.7 million. The program worked so well that the legislature made the maintenance levy permanent.

By 1983, the district had four active programs: master planning, floodplain management, capital improvement, and maintenance. The next program to add was a South Platte River maintenance program. A levy of 0.1 mills was authorized, and now an annual program of about $1 million is in place for planning, design, construction, and maintenance of projects along about 40 miles of river.

In 1989 the district's boundaries were expanded to accommodate growth of, among other things, the new Denver International Airport. Now the total area is 1608 mi^2.

The district reluctantly entered the water quality field when stormwater quality became an issue, beginning in the 1970s. The district prepared the National Pollutant Discharge Elimination System (NPDES) permit applications for Denver, Lakewood, and Aurora, and committed to continued technical support, as needed.

The district's staff is only 18 to manage an annual budget of about $19.3 million. Scott Tucker, the executive director for 23 of the district's years, sees more emphasis on the preservation of natural waterways, multiple use of urban drainageways, and water quality. The district's vision is to use the drainageways as open green corridors, and most projects have public access and multiple-use flows. Erosion control will become of more concern with the shift to natural, as opposed to lined, channels. The district hopes for a more cooperative working arrangement with the federal government, but feels the heavy hand of "command and control" regulation.

This brief history of the UDFCD shows what can be accomplished when local governments get together and, with the cooperation of the state legislature, organize a regional approach to stormwater management and flood control.

Southeast United States: Black Warrior River

In humid areas, flooding occurs more regularly than in dry areas, and often with great intensity. A major flood on the Black Warrior River is described in Chap. 13. The issue is control of run-of-river reservoirs during flooding.

Rainfall charts and flood hydrographs shown in Chap. 13 illustrate that, even in relatively flat country, peak discharges can develop quickly. The Black Warrior River case also shows a little of flood damage scenarios.

Great Mississippi River Floods of 1993

During the summer of 1993, the Upper Mississippi River Basin suffered the worst flooding in memory (*Water,* 1993). July rainfall in 10 basin cities ranged from 131 percent to 643 percent of the 30-year average, and new records were set at 42 gauging stations on 33 streams in seven states.

The images of the great Mississippi River floods of 1993 were recorded on television, in magazines, and in the memories of those who were involved. The images are too vast to capture here, but one small diagram may serve to illustrate. At one location, the Missouri River broke its levee on July 16, and floodwaters surged toward the Mississippi, where the waters joined 20 miles upstream from their normal confluence.

Figure 11.10 shows the general area affected by the flooding.

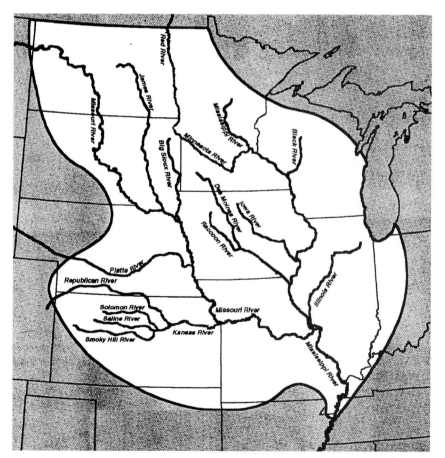

Figure 11.10 Area of great Mississippi River flood of 1993. (*Source: National Weather Service, 1994.*)

Cumulative rainfall amounts ranged between 45- and 300-year return periods for April–July, especially June–July 1993. For example, Iowa had 18.1 in of rain during June–July, a 260-year depth. Floodwaters remained above the previous record from July 11 until about August 10. This was truly a record-breaking flood, but it shows the extent to which flooding can occur over a widespread area.

As a result of the flood, the efforts to combat it, and the damage it caused, the interagency committee that reviewed the flooding and the response reported that nonstructural measures were clearly the best policy for events such as this.

Bangladesh Flooding

Lest we think that nonstructural solutions are the way to go everywhere, and to place the Mississippi River flooding in perspective, I would like to mention flooding in Bangladesh. There, the country sits atop a vast floodplain formed from one of the world's largest river deltas, where the Ganges, Brahmaputra (Jamuna), Padma, and Meghna Rivers converge. While the Mississippi River floods are impressive, Bangladesh flooding involves much more water. According to the U.S. Agency for International Development, the combined flow of the Brahmaputra, Ganges, and Meghna Rivers that meet in Bangladesh have an annual discharge five times the Mississippi's average (*Water*, 1993).

Bangladesh's population of 110 million, which is expected to double in 30 years, creates one of the highest population densities on Earth. Eighty percent of the population is rural, 60 percent is landless, and just 10 percent of the rural residents own 50 percent of the arable land. Bangladesh is one of the poorest nations in the world, with a per-capita income of only about $200 a year. Only 30 percent of men and 19 percent of women are literate. Sixty percent of children are malnourished, and infant mortality is 120 per 1000 live births. The nation sits at the head of the Bay of Bengal, right in the path of some of the world's most powerful tropical storms. High-intensity cyclones hit regularly, usually during spring or fall. In 1991, a cyclone and storm surge killed an estimated 139,000 people and left about 10 million homeless. The dimensions of the disasters are hard to grasp.

The 1988 flood was particularly devastating. At their peak, flood waters covered three-fourths of the country, some 40,000 mi^2, and left 28.5 million people homeless. The official death toll was 650, but diarrhea and other waterborne diseases claimed more lives. Much of the infrastructure was washed away, and crops and stored grain were ruined. The nation bounced back, however, perhaps because the capital city, Dhaka, was flooded and many important people experienced flood misery first hand, not just on television (Khondker, 1992).

The rivers constantly form and move small pancakes of land called chars. Most of these rise only a foot or so above the water during flood season, but many people have no choice but to live on them (Cobb, 1993).

Flood control dominates policy attention, but the problem seems unsolvable. The government, working with the World Bank, has fashioned the Flood Action Plan (FAP), which is controversial because changing the rivers could upset social and environmental balances. Some believe that the FAP is focused too much on the structural solutions, and that there is too much engineering in it.

The FAP was organized by the World Bank and a consortium of governments after the floods of 1987 and 1988. The plan has 26 components and a time frame of several decades (Wescoat, 1992). The centerpiece of the FAP is embankments, but they often fail, and they prevent the fertilization of floodplains with floodwaters. Controversial elements of the FAP include technological, social, political, environmental, and financial questions about the FAP.

Indeed, the solution in Bangladesh may not be structural in the long term. Oxfam America, a Boston-based relief and development agency, believes that investment in flood shelters, disaster-proof housing, community reforestation, small-scale earthworks, and other projects can mitigate problems better than massive and misplaced infrastructure investments (Charney, 1991).

The flood problem in Bangladesh may be best summarized by James and Pitman (1992), who wrote:

> To date, no country has fully integrated structural and nonstructural measures, and yet many would ask one of the world's poorest and most backward countries to do so. We talk of floodproofing, an information-intensive technology, in a country where few people read. We talk of protecting a delicate natural environment for people who scarcely have enough food. Our greatest challenge is to overcome ideologies and find a practical road to better lives for a poor and suffering people.

Into the Twenty-First Century

Clearly, flood control should be seen as part of integrated strategies of river basin management and not just as a single-purpose water project goal. The need for new directions was outlined by the Interagency Floodplain Management Review Committee (1994), and their recommendations will be outlined here to provide a perspective on the future situation in the United States. Keep in mind, however, that in other parts of the world, solutions may differ.

Three major problems in floodplain management

- People and property throughout the nation remain at risk from flooding, many do not understand the risk, and fiscal burdens of the risk are unevenly shared.
- Over the last two centuries, there has been tremendous loss of habitat in the Upper Mississippi River Basin.
- The division of responsibilities (role definition) among federal, state, tribal, and local governments needs better definition.

Recommendations

- The President should propose enactment of a Floodplain Management Act, issue a revised Executive Order outlining responsibilities and roles, and activate the Water Resources Council.
- The President should see that the Principles and Guidelines are revised to accommodate new objectives for environmental quality and economic development.
- The administrative should support collaborative efforts among governments to enhance coordination.
- The administration should provide for cost sharing in predisaster, recovery, response, and mitigation activities.
- The administration should take full advantage of federal programs which enhance the floodplain environment and provide for natural storage in bottomlands and uplands.
- The administration should enhance the efficiency and effectiveness of the National Flood Insurance Program.
- The administration should work to reduce the vulnerability to damages of those in the floodplains.
- The administration should arrange for periodic reviews of completed projects to ensure that they meet their goals.
- The administration should assign management responsibility for levees under federal programs to the Army Corps of Engineers.
- States and tribes should take responsibility for nonfederal levees.
- The administration should take steps to capitalize on the successes during the 1993 flood by using the National Flood Insurance Program rating system, providing funding for buyouts in the floodplain, provide block grants to states, assign Federal Emergency Management Agency responsibility for integrating federal disaster responses, and encourage federal agencies to use nondisaster funding to support hazard-mitigation activities.

- The administration should establish Upper Mississippi River Basin and Missouri River Basin Commissions to provide integrated hydrologic, hydraulic, and ecosystems management.
- The administration should provide timely flood data by establishing a clearinghouse at the U.S. Geological Survey and exploiting science and technologies such as geographic information systems.

Final Note on Flood Control, Stormwater Management, and Floodplain Management

From the material in this chapter, the reader can see that controlling floods, managing stormwater, and managing land uses in floodplains involve many complexities, albeit different ones than other aspects of water resources management. Due to their area-wide nature, flood services do not adhere well to single jurisdictions; that is why regional organizations such as the Corps of Engineers for large basins and regional districts such as the Urban Drainage and Flood Control District of Denver are especially valuable. With the massive Mississippi River floods of 1993, the Rhine River floods of 1994 and 1995 in Holland, and the persistent large-scale flooding in Bangladesh, our attention is riveted to the difficulty of permanent solutions to the flooding problem. The best thing we can do is to follow valid principles whenever possible.

In the realm of water quality, more attention is needed to the integrating aspects of urban sanitation, best management practices, stormwater management, and other public works practices that affect environmental quality.

Questions

1. Imagine yourself in the role of an expert witness representing a federal agency that is being sued for alleged improper operation of a flood control dam. You are meeting with your defense attorneys and rehearsing how to answer the inevitable challenges that will come from the plaintiff's attorneys. Answer the following questions.
   a. Which court system of the United States would you expect to handle the trial?
   b. What laws, case results, and legal theories might apply to the case?
   c. As an expert witness, might you be expected to use computer models to justify your opinions? If so, which ones might be useful in such a case?
   d. If you were questioned about the level of confidence you had in your hydrologic, hydraulic, or modeling computations, how might you present your analysis of level of confidence?

2. In a federally subsidized water project, who should pay for flood control benefits? Can you justify your answer?

3. If a regional flood control program included facilities with joint use such as recreation in floodplains, detention ponds, trails, and the like, how would you suggest the program for maintenance be organized?

4. Levees are flood control facilities that serve many people in dispersed locations. Who should have responsibility for maintaining them?

5. After a flood wave enters a reservoir, the hydrograph leaving the reservoir differs from that entering the reservoir. Why? What methods are available to compute the outflow hydrograph?

6. In urban areas, who should pay for stormwater services and programs? Is a stormwater utility a good idea? How should it be organized?

7. What is your position on structural versus nonstructural flood control programs? What are the issues that bear on the choice of measures?

References

Charney, Joel R., Bangladesh Needs Reform, Not Just Aid, *The Wall Street Journal,* May 22, 1991.

Cobb, Charles E., Jr., Bangladesh: When the Water Comes, *National Geographic,* June 1993.

Grigg, Neil S., *Water Resources Planning,* McGraw-Hill, New York, 1985.

Interagency Floodplain Management Review Committee, *Sharing the Challenge: Floodplain Management into the 21st Century,* Washington, DC, June 1994.

James, L. Douglas, and Keith Pitman, The Flood Action Program: Combining Approaches, *Natural Hazards Observer,* University of Colorado, March 1992.

Khondker, Habibul Haque, Floods and Politics in Bangladesh, *Natural Hazards Observer,* University of Colorado, March 1992.

Matthai, H. F., Floods of June 1965 in South Platte River Basin, Colorado, USGS Water Supply Paper 1850-B, 1969.

McCain, Jerald F, Storm and Flood of July 31–August 1, 1976, in the Big Thompson River and Cache la Poudre River Basins, Larimer and Weld Counties, Colorado, USGS Professional Paper 1115, 1979.

National Weather Service, *Great Mississippi River Flood of 1993,* Washington, DC, 1994.

Natural Hazards Research and Applications Research Center, *Floodplain Management in the United States: An Assessment Report, Vol. 1, Summary,* University of Colorado, Boulder, 1992.

Schilling, Kyle E., Claudia Copeland, Joseph Dixon, James Smythe, Mary Vincent, and Jan Peterson, The Nation's Public Works: Report on Water Resources, National Council on Public Works Improvement, Washington, DC, May 1987.

Tucker, L. Scott, Tucker Talk, in *Flood Hazard News,* Denver, December 1994.

Water, U.S. Agency for International Development, Asia Bureau and Near East Bureau, Winter 1993, Issue 3, prepared by ISPAN Project.

Wescoat, James L., Jr., Five Comments on the Bangladesh Flood Action Plan, *Natural Hazards Observer,* University of Colorado, March 1992.

12

Planning and Managing
Water Infrastructure

Introduction

Water system components in the "built environment" are capital assets that require careful attention to the engineering tasks of planning, design, construction, operation, and maintenance. This case study chapter reviews these planning and management tasks, and presents three case studies to illustrate different aspects and phases of water infrastructure. The case studies illustrate the variable context of project development.

The first case study, the Ft. Collins Water Treatment Facilities Master Plan, illustrates a modern process for planning water treatment upgrades, but with financial controversies and need for public involvement. The case study presents unique methods to identify and present important decision-related information. A case study of the Colorado Big Thompson Project shows the historical development of a major Western water resources project that gained approval during the era of federal water development. The case illustrates the dominance of political considerations during that era, and shows the important role of citizen advocacy and organization. The third case is the California Water Plan. It is presented to illustrate how a state government can plan and develop an extensive water project.

Planning and Development Process

Chapter 4 explained the generic steps in the problem-solving and planning and development process. One version of the process has conceptual planning, preliminary planning, preliminary design, and final design as increasingly detailed (and costly) stages of planning. These would be followed, for an implemented project, with preparation of construction documents, contracting, construction, and inspection. At that point, the project would be turned over to the operator for shakedown and then operation and maintenance phases, all part of the continuum of design–construct–operate.

In water project development, terminology is similar to other infrastructure categories. Common terms are reconnaissance, feasibility, definite project, and final plan phases. The World Bank's terminology illustrates the phases. It refers to the "project cycle," which consists of six stages: identification, preparation, appraisal, negotiation and approval, implementation and supervision, and ex-post evaluation.

The reconnaissance phase, related to conceptualization and preliminary planning, identifies possible projects that meet development goals. It leads to recommendations for further studies rather than to definite plans. After it is decided that a definite project is to be pursued, the process involves the owner, the general contractor and subcontractors, the architect, the engineer, the surveyors, and any other team members such as the attorney.

The feasibility stage, in some ways similar to the preliminary design phase, goes into more detail and establishes financial, technological, environmental, and political feasibility. It may result in documents that are costly to prepare, depending on the complexity of the project. Facts must be determined, sites must be selected, additional feasibility studies must be carried out, and financing must be arranged.

The definite project phase results in plans, specifications, and operating agreements—all the guidance needed to construct and operate the project. In the design task, final details must be set for all aspects of construction. This phase is quite costly and may involve complex teams of consultants and advisory organizations.

The United Nations (1964) has prepared several guideline documents for planning of water projects. One of them, *Manual of Standards and Criteria for Planning Water Resource Projects,* sets forth the traditional approaches to planning. It describes the phases as reconnaissance, feasibility investigation, and definite plan investigation.

Traditional planning processes of this kind have been fine-tuned to incorporate sophisticated environmental, social, and economic investigations, and are now much more complex than in the past. The U.S. experience with the National Environmental Policy Act, environmental impact statements, and the Principles and Standards of the Water Resources Council illustrate these concepts.

The change in planning had the effect of expanding the feasibility criteria beyond the traditional technological, legal, and financial categories to much greater emphasis on environmental and social criteria. Whereas technological, legal, and financial categories can be studied rationally and quantitatively, the social and environmental categories involve more subjective and political judgments. These are the subject of Chap. 4.

Design and Construction

After a project enters the definite project phase, all the guidance needed to construct and operate the project (plans, specifications, and operating agreements) is prepared. In general, design and construction are managed by the engineering staff of the organization. This staff handles functions from initial surveys and plans to construction management and, finally, record keeping. Keeping all of these phases going in the proper direction and order is the task of the project management process.

The design process involves creative decision making about the configuration and details of projects. There are many decisions to make that require experience and consultation. The results of design will be the drawings, documents, and plans necessary to initiate the construction process. With so much emphasis on liability and safety today, the design process has taken on new constraints, and the use of computers is increasing at the same time, adding to the complexity.

The initiation of construction involves numerous procedural steps. The construction process begins with preparation of contract documents, and it involves bidding, review, award, organization, construction itself, inspection, and acceptance. A recognized process for these steps has developed over many years to control costs, quality of construction, and the quality of the final product.

In the construction phase, legal aspects are very important. The legal instruments of construction contracting generally consist of bid advertisement, information for bidders, general and special provisions, measurement and payment information, proposal form, notice of award, notice to proceed, change order information, form of contract, detailed specifications, contract plans and drawings, bonds, insurance certificates, and other certifications. The adversarial environment of construction today demands that all parties avoid litigation. The problems with obtaining liability insurance greatly complicate the design-construction process. There is much emphasis on alternative dispute resolution (ADR) and techniques such as partnering to avoid litigation.

Operations and Maintenance Phase

After construction is completed, the project must be operated and managed. Operations management is, in effect, the final phase of problem solving and planning. Maintenance receives equal billing with operations in the term, "operations and maintenance," or "O&M." It must be supplied at different levels, and providing the proper levels has implications for both the operating and capital budgets. In effect, there are four separate maintenance-related functions: condition assessment, inventory, preventive maintenance, and corrective maintenance. The latter is sometimes called major and corrective maintenance, to stress the major nature of some of the corrective actions. The condition assessment activity is a link between the operation and the maintenance function, and illustrates why the two functions must be unified. If the condition starts to worsen, the operation will be affected, and it will be time to schedule repairs and maintenance.

The concept of the *maintenance management system* (MMS) has arisen in recent years as an attempt to bring together the disparate concepts of maintenance activities into a holistic approach to caring for a system. In effect, it is the systems approach applied to maintenance. A MMS is, in effect, a program for making sure that overall maintenance is managed adequately. It involves all the tasks of management—planning, organizing, and controlling—and it requires an effective decision support system. The MMS will include condition assessment, preventive maintenance, and corrective and major maintenance, and the decision support system will provide the information and data needed for these activities.

Large water systems such as dams and reservoirs are subject to special operating rules (see Chap. 13). The operation and maintenance considerations for urban water systems are discussed by Grigg (1986).

A special case of water systems operation and maintenance is the irrigation system, which involves complex social systems. Whereas irrigation systems in developed countries can be managed much the same as other infrastructure systems, in developing countries the social systems make the operation and maintenance problems more complex. Approaches to management include the development model proposed by Lowdermilk and others (1983), and the operations and maintenance learning process proposed by Skogerboe (1986).

Case Study: Ft. Collins Water Treatment Facilities Master Plan

The first case study is the Ft. Collins (Colorado) Water Treatment Facilities Master Plan (Water Treatment Master Plan). The plan includes both straightforward and political elements, the latter related to public acceptance of rate increases and dealing with growth versus antigrowth factions of the city.

As Ft. Collins grew, the city recognized the need for a water treatment facilities master plan, and the water utility managers retained a consultant. According to the consulting engineer, a comprehensive study of the city's treated water service system had never been attempted before, and the basic purpose of the plan was "to consolidate past efforts, assess existing conditions, project future conditions, and develop a comprehensive plan that can serve as a guide to the improvement and optimum operation of the City's WTF's" (Black & Veatch, 1994).

From a water manager's viewpoint, the challenge of the master plan is how to use it to gain approval for the optimum choice and sequence of plan elements. To pursue this question, utility managers worked hard to configure the plan for public review and approval. This meant that the plan elements had to be segregated into understandable categories, then evaluated in terms of costs and benefits—not to compute benefit–cost ratios, but to enable the public, and the city council, to evaluate the relative values and priorities of the recommended improvements.

The layout of Ft. Collins's water supply and treatment facilities is generally as shown in Fig. 12.1, which illustrates diversions of raw water from the Cache La Poudre River and from Horsetooth Reservoir, part of the Colorado Big Thompson Project (to be described next).

The result of the master plan process identified four categories of improvements: source of supply, treatment process improvements, support systems, and alternative elements. Costs for these elements, totaling around $70 million, were allocated to reliability, regulations, and capacity improvements, which enabled priorities to be set. Table 12.1 shows examples of this allocation, and illustrates how costs can be allocated to different purposes to aid in the decision-making process.

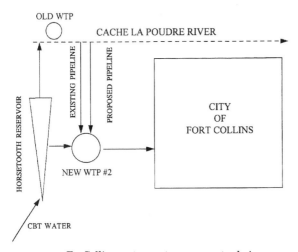

Figure 12.1   Ft. Collins water system: conceptual view.

TABLE 12.1   Ft. Collins Water Treatment Facilities Master Plan

Cost Allocation of Master Plan Elements

Improvement	Total cost ($ millions)	Reliability (%)	Regulations (%)	Capacity (%)
Source of supply:				
New Poudre River pipeline	28.4	100	0	0
Second connection to Horsetooth Reservoir	0.8	95	0	5
Treatment process improvements:				
Media and piping in new filters	0.5	66	17	17
Treatment train 5	8.5	34	51	15
Disinfection contact chamber	3.7	13	77	10
Support systems:				
Chemical feed and storage facility	4.0	60	40	0

Next, the improvements were scheduled according to priorities into three phases. Figure 12.2 shows the decision matrix that was worked out by the city's staff to illustrate the choices. Although the matrix looks quite simple, it is actually the product of several meetings with the city's Water Board and its Engineering Subcommittee, and of the joint City Council-Water Board Planning Committee. Notice how much information is in the matrix. Across the top are the problems that are addressed by the master plan elements. These include health, safety, reliability, economics, growth, customer service, efficiency, and environmental issues—quite a bundle of issues. These are generally aligned into priorities as health and regulatory, safety, and then items having more to do with customer satisfaction.

Next, notice that the master plan elements have been disaggregated into decision packages and scheduled into three phases. The items are generally set up so as to reflect highest-priority items first. In the columns at the bottom of the matrix are notes to show the alternatives that were considered. This helps the display because the city staff and their consultants have already picked what they consider to be the best plans, and the display shows the decision makers what other items were considered.

Finally, at the bottom of the chart are the total costs and schedules of the construction phases. Figure 12.2 contains information on recommended improvements, priorities, problems and benefits, secondary impacts, total costs, and alternatives considered. This is quite

Water Treatment Master Plan - Recommended Improvements

Problems Identified and Why Problems Need to Be Solved

Recommended Improvements	First Priority: Existing chlorine facility does not meet current standards for public or worker safety. • Health risks • Environmental • Building codes	Not enough contact time for disinfection of treated water. • Health risk • Future regs • Chemical use	Customers are exposed to health risk due to cryptosporidium and viruses. • Acute illness • Deaths • Future regs	Second Priority: Water supply is vulnerable to interruption. • Reliability • Restrictions • Fire protection • Economic loss	Electrical system's not reliable. • Safety issues • Interruptions • Reliability	Limited capacity of chemical feed and storage systems. • Safety issues • Chemical storage • Efficiency • Reliability	Third Priority: High levels of manganese in the distribution system. • Stained laundry • Revenue loss • Customer service	Customers dissatisfied with taste and odor. • Customer service • Health risk??	Need for environmentally responsible solid waste disposal method. • Environmental • Efficiency	Schedule and Cost
Phase 1										
Construct new chlorination facility $1,034,000										**Phase 1** Total Cost of $7,260,000 5% Rate Adjustment Design: 1995 Construction: 1996-97
Construct disinfection contact chamber $2,946,000										
Construct backwash water recovery facility $1,500,000										
Upgrade media in filters (1-8) $940,000										
Construct misc electrical improvements. $440,000										
Install media, piping & controls in new filters. $500,000										
Phase 2										
Construct new sedimentation basin (T5) $8,000,000 - $13,000,000										**Phase 2** Total Cost of $15-22 Million 10-12% Rate Adjustment Design: 1997 Construction: 1998-99
Upgrade settling process (T3) $3,000,000 - $5,000,000										
Construct chemical feed & storage facilities. $4,000,000										
Convert T1 & T2 to pre-treatment facilities. $2,440,000										
Phase 3										
Construct new Poudre raw water pipeline. $18,800,000										**Phase 3** Total Cost of $24-38 Million 12-20% Rate Adjustment Design: 1999 Construction: 2000-02
Construct additional Horsetooth connection. $800,000										
Construct waste solids handling facilities. $2,510,000										
Alternate Plan Element: Add ozone to water treatment process $13,870,000										
Other Options Considered:	• Modify the existing facility	• Baffles in reservoirs	• Ozone • Membranes • Chlorine dioxide	• Use existing canals, lakes, new pump station and pipeline • Second outlet from Horsetooth • Modify existing Poudre pipeline		• Modify existing facilities	• Chlorine dioxide • Ozone • Multiple level outlet in Horsetooth • Biological contactor • Reservoir treatment • Poudre pipeline	• Granular activated carbon • Ozone • Multiple level outlet in Horsetooth • Poudre pipeline	• Dewater solids and truck to ranch • Mechanical dewater and truck to ranch • Build facility to dewater at ranch	

Figure 12.2 Water Treatment Master Plan recommendations. (*Source: City of Ft. Collins, Michael B. Smith, 1995.*)

a bit of information to pack into one display, but it illustrates the complexities that must be considered in such a master plan.

Case Study: The Colorado Big Thompson Project

The next project selected for case study is the Colorado Big Thompson (CBT) Project, a set of reservoirs and facilities that diverts water from one side of the Rocky Mountains to the other. Numerous sources have been consulted to describe the project, but the primary one that outlines the sequence of project development is Dan Tyler's (1992) book, *The Last Water Hole in the West*. This book makes interesting reading, and is recommended for serious students of Western water history.

The CBT Project is also involved in the Platte River and Colorado River case studies of Chap. 19. It created a major interbasin transfer of water from the Colorado to the Platte River Basin, and thus affects water issues all the way from New Orleans to Los Angeles. Water management by the CBT's management authority, the Northern Colorado Water Conservancy District, is also described in Chap. 21.

Figure 12.3 shows the dimensions of the CBT project. Basically, it collects water on the western slope of the Rockies and diverts it to the eastern slope, with an original purpose of providing supplemental irrigation water.

As described by a publication of the Northern Colorado Water Conservancy District (1987), the CBT project has 1,011,490 acre-feet total water storage; diverts 231,301 acre-feet in annual deliveries to the East Slope; has 2428 entities or individuals who own water shares, and provides water to 23 cities or towns. It involves 12 reservoirs, 34.4 miles of tunnel, 95.5 miles of canals, conduits, or siphons, has 6 hydro plants, 3 pumping stations, and 183,950 kW in installed capacity of hydropower. The water flows at a maximum of 550 cfs through the 9.9-ft-diameter Alva Adams Tunnel, a length of 13.1 miles. The drop is 2800 ft from the tunnel to Flatiron Reservoir, one of the hydro facilities. The project took 19 years to complete.

The CBT Project is the largest interbasin transfer from the Colorado River system. Of total diversions from Colorado's East Slope to the West Slope of some 635,000 acre-feet in 1978, the CBT Project accounts for 264,000 acre-feet of it. All of these diversions are from the Colorado River Basin headwaters, and represent about 635,000/15,000,000 or 4 percent of total Colorado River Basin yield (see Chap. 19).

The Last Water Hole in the West chronicles the monumental struggles and political issues involved in the CBT Project as it unfolded, with the major action being in the New Deal era (see Chap. 4 for more

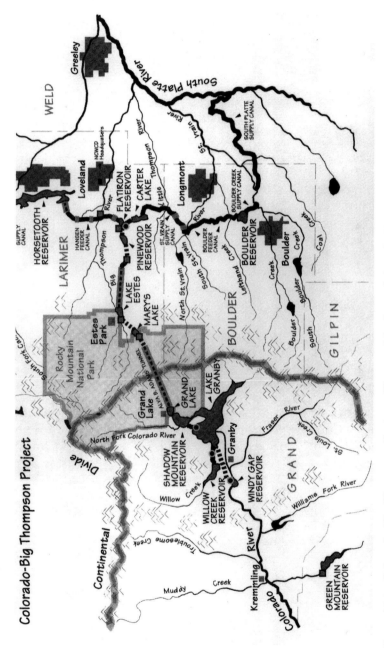

Figure 12.3 Colorado Big Thompson Project. (*Source: Northern Colorado Water Conservancy District.*)

detail about the planning environment of this era). Milestones, as summarized by the Northern Colorado Water Conservancy District, the NCWCD (1987), are:

1884	First preliminary survey
1902	Reclamation Act
1933	Northern Colorado organizes for Grand Lake Project
1933	Tipton report on feasibility
1935	Bureau of Reclamation surveys
1936	Name assigned: Colorado Big Thompson Project
1937	Colorado Conservancy District Act
1937	East Slope and West Slope agreement
1937	Senator Adams presents plans to Congress
1937	President Roosevelt signs first construction allocation
1937	First meeting of NCWCD Board of Directors
1938	Contract with United States for repayment costs
1938	Construction begins
1947	First water deliveries
1951	First deliveries from Horsetooth Reservoir
1956	Completion of original project
1957	Partial maintenance responsibility transferred to NCWCD
1962	Commencement of repayment
1986	Final maintenance responsibility transferred to NCWCD

The CBT case study illustrates a number of issues related to planning and management of projects.

The players involved civic leaders, the Bureau of Reclamation, state water agencies, governors, Congress, the President of the United States, water users, interest groups, and consultants. If the case took place today, it would involve environmental groups in a big way. In fact, the environmental groups are monitoring the project and its future actions.

The reconnaissance, feasibility, and definite project stages were followed, but over a long period of time and with a number of different actors.

The project has effective operation and maintenance by a local water district. Ownership and responsibility have been largely transferred from the federal government.

California Water Plan

The California Water Plan is one of the most significant nonfederal water schemes in the United States. It illustrates several issues relating to topics of this book: identification of water needs, schemes for large-scale water transfers, regionalization, construction of dams, reservoirs, canals, and related facilities, and continuing controversy over management of the facilities.

In California, most of the water is in the north and most of the people are in the south. The Metropolitan Water District of Southern California (MWD) was created in 1928 to provide wholesale water to the growing cities of the south. This provided a regional approach for meeting raw water needs, but did not provide the water itself.

Twenty-seven member agencies select directors for the MWD's board. These include cities such as Los Angeles, water districts, and water authorities such as the San Diego County Water Authority. Los Angeles had begun importing Owens Valley water early in the twentieth century, but the rest of the area was short of water to meet the growth of the 1920s and 1930s. In 1931, voters approved a bond issue and MWD constructed the California River Aqueduct, which opened in 1941 (Metropolitan Water District, 1988).

However, during the 1950s, southern California's growth accelerated, and new supplies were needed. In 1960 the state's voters approved the largest aqueduct program in history, known as the State Water Project and the California Aqueduct. This drew water from the Sacramento–San Joaquin Delta southward. (See Chap. 22 for a discussion of problems in the delta.) The MWD bears about two-thirds of the cost of the program and has contracted with the state for the ultimate delivery of about 2 million acre-feet per year.

Figure 12.4 shows the major features of California's State Water Project and other major water facilities of the state. Major water supply is developed from the north and center of the state from mountain runoff, and the south is mostly dry.

The California Water Plan, authorized in 1947 and first published in 1957, was to be the guide for the orderly development of the state's water resources, including the State Water Project (Imperato, 1991). A series of publications by the Department of Water Resources (DWR) outlined the plan: *Water Resources of California* (1951); *Water Utilization and Requirements for California* (1955); *The California Water Plan* (1957); *Implementation of the California Water Plan* (1966); *Water for California: The California Water Plan Outlook for 1970* (1970); *The California Water Plan Outlook for 1974* (1974); *The California Water Plan: Projected Use and Available Water Supplies to 2010* (1983); and *California Water: Looking to the Future* (1987). The latter publication, Bulletin 160-87 of the DWR, signaled a "shift in the Department of Water Resource's approach to long-range planning from one of system expansion to one of more efficient utilization of existing water supplies" (Imperato, 1991).

By 1993, California was facing growing conflicts in its water plan (Rosenbaum, 1993). The year marked the end of a six-year drought,

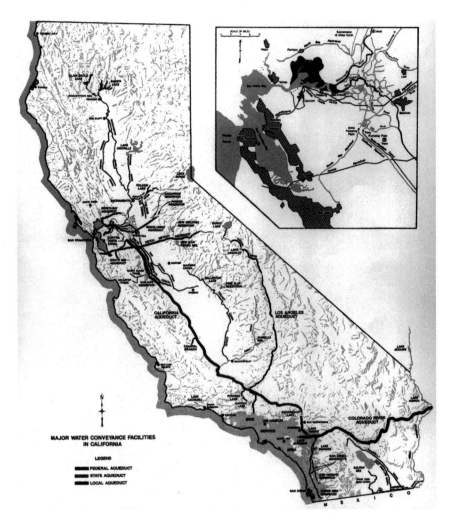

Figure 12.4 Major water conveyance facilities in California. (*Source: Metropolitan Water District of Southern California.*)

but in the new 30-year water plan (to the year 2020), the DWR warned that the state faces the loss of 3.5 million acre-feet by the year 2000. The reasons are allotment cutbacks on the Colorado River, environmental measures to protect rivers and wetlands, and the protection of the Bay-Delta system (see Chap. 22). This loss of water comes at a time that the state is facing a growing population and expects to have 49 million by the year 2020.

California's agriculture uses 26.6 million acre-feet per year, and cities about a fourth as much, but cities contain 91 percent of the state's population. This sets up a conflict between cities and agriculture (see Chap. 24 for more details on this conflict).

Figure 12.5 shows the extent of major water transfers in California. Clearly, the major movements of water are north to south, to serve both cities and agriculture. California's population of about 30 million

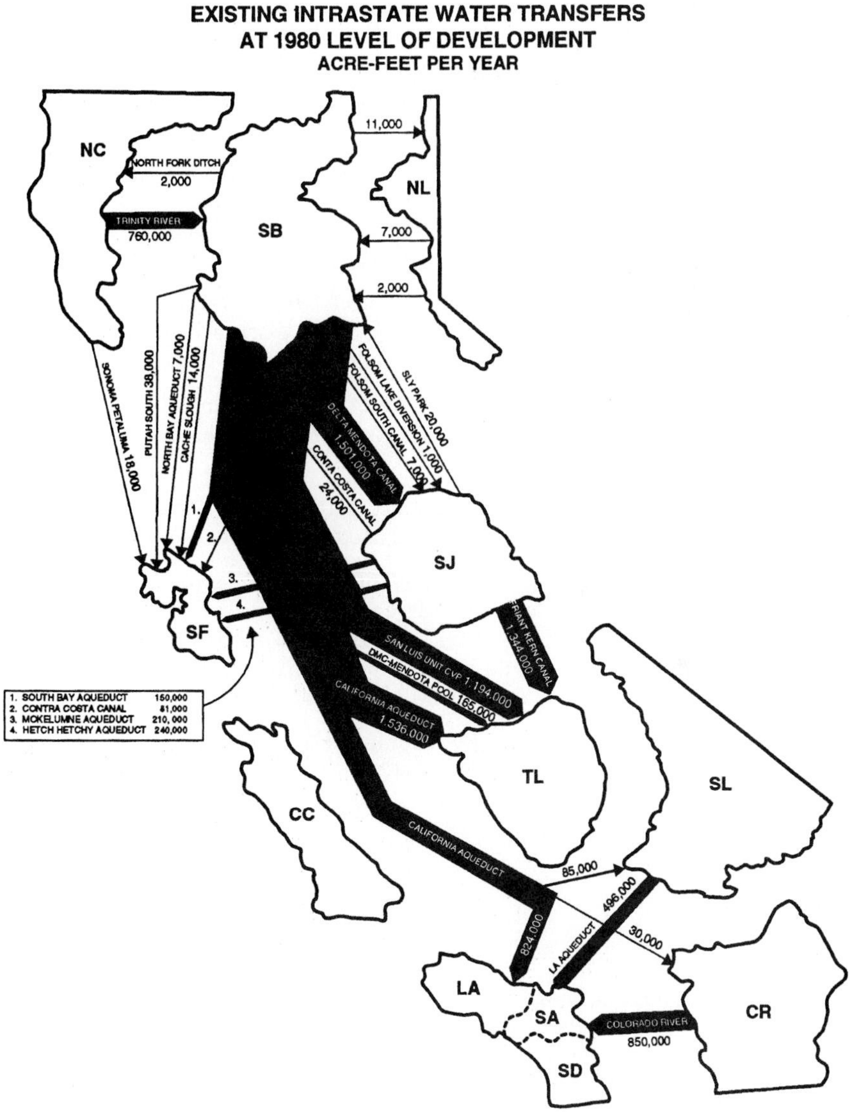

**EXISTING INTRASTATE WATER TRANSFERS
AT 1980 LEVEL OF DEVELOPMENT
ACRE-FEET PER YEAR**

Figure 12.5 Water transfers in California. (*Source: California Department of Water Resources, 1983.*)

in 1995 is growing, and providing water for future growth will be a challenge. The last major dam built in California was finished in 1979. Even with conservation programs, water managers figure that the state needs more storage. The stage is set for much more conflict in the future. Chapter 24 discusses the major aspects of the coming conflicts in the context of Western water management.

Questions

1. Earlier projects such as the Colorado Big Thompson Project were built with federal subsidies. Assuming that these projects could be built today, how should they be financed?

2. If more management measures like demand management and incentives from pricing were implemented, could fewer water projects meet water demands from society? What would be the implications of policies that implemented more aggressive use of nonstructural measures?

3. What is the current status of China's Three Gorges Dam? In your opinion, is a "megaproject" like this a good thing? What alternatives are available? Can you think of other "megaprojects" and describe their impacts?

4. Name the four accounts in the "Planning and Guidelines." How are these evaluation criteria used in project evaluation in the public sector? In the private sector?

5. Name the components of a maintenance management system and describe how they would apply to capital facilities used in a water resources system.

6. Table 12.1 shows several aspects of planning for expansion of the Ft. Collins Water Treatment Plant. The table is meant to illustrate decision criteria to decision makers and the public. A table showing the accounts from the Principles and Guidelines would have a similar purpose. Why does Table 12.1 not show the Principles and Guidelines "accounts"?

7. Do you think the Colorado Big Thompson Project could be built in the United States today? Why or why not? How about the California Water Project? If these projects had not been built, what would be the differences in the areas they serve?

References

Black & Veatch, Inc., City of Ft. Collins Water Treatment Facilities Master Plan, Draft, August 1994.

Grigg, Neil S., *Urban Water Infrastructure: Planning, Management, and Operations,* John Wiley, New York, 1986.

Imperato, Pamela R. Lee, In Dry Dock: Refitting the California Water Plan, *Jesse Marvin Unruh Assembly Fellowship Journal,* Vol. II, Sacramento, CA, 1991.

Lowdermilk, M. K., W. Clyma, L. Dunn, M. Haider, W. Laitos, L. Nelson, D. Sunada, C. Podmore, and T. Podmore, *Diagnostic Analysis of Irrigation Systems: Volume I, Concepts and Methodology,* Water Management Synthesis Project, Colorado State University, Ft. Collins, December 1983.

Metropolitan Water District, *Water for Southern California,* Los Angeles, June 1988.

Northern Colorado Water Conservancy District, *Waternews,* Loveland, CO, 1987.

Rosenbaum, David B., California Faces Growing Conflicts, *ENR,* August 2, 1993.

Skogerboe, Gaylord V., *Operations and Maintenance Learning Process,* International Irrigation Center, Utah State University, Logan, UT, 1986.

Tyler, Daniel, *The Last Water Hole in the West,* University of Colorado Press, Boulder, 1992.

United Nations Economic Commission for Asia and the Far East, *Manual of Standards and Criteria for Planning Water Resource Projects,* Water Resources Series No. 26, New York, 1964.

13

Reservoir Operations and Management

Reservoirs are the most important man-made storage elements in water systems because their capacity and operational schedules determine the rates and volumes of flows in streams. They have enabled humans to make the desert bloom, and to provide water supplies for large and concentrated populations.

A reservoir is usually created by the construction of a dam across a flowing stream. Storing water is necessary because when water occurs naturally in streams and is not stored, it is sometimes not available when needed. Reservoirs solve this problem by capturing water when it occurs and making it available at later times. However, with today's environmental awareness, reservoirs are sometimes negative symbols of "man taking dominion over nature," and work against sustainable development as they interfere with natural ecology.

While the concept of a reservoir may bring to mind a large body of water, many small reservoirs are also in service. These include urban water tanks, farm ponds, regulating lakes, and small industrial or recreational facilities. These small reservoirs can have important cumulative effects in rural regions.

In this chapter I describe the nature and characteristics (engineering and hydrologic aspects) of reservoirs, and how they are operated to respond to multiple demands. Key aspects of reservoir dynamics, such as lake water quality problems and annual turnover of reservoir water, are described briefly. Several case studies are presented to illustrate reservoir sizing, operation during drought or flood, operation in a region subject to the appropriation doctrine, operation of a complex system of reservoirs, and situations with environmental conflicts.

Purposes of Reservoirs

As a storage facility, a reservoir serves either to smooth out a variable inflow, as in flood control, or to provide reserves to meet variable downstream demands. Within these broad functions, reservoirs are built to meet eight economic and environmental purposes: flood control, navigation, hydroelectric power, irrigation, municipal and industrial water supply, water quality, fish and wildlife, and recreation (Johnson, 1990).

Flood control is provided by a reservoir or series of reservoirs by reserving space in the available reservoir pool to store flood waters so that they do not pass immediately downstream and flood vulnerable areas.

Navigation is actually a supply issue, to regulate downstream flows.

Hydroelectric power.   Two types of reservoirs provide for hydropower: the larger reservoirs that carry water over from season to season, and smaller reservoirs that provide run-of-river power generation.

Irrigation.   Providing irrigation water requires carrying over a supply from a wet season to a dry season. Examples are the Rocky Mountains, where snowmelt peaks in late May or early June and the main demand for irrigation water is in September, and Southeast Asia, where, depending on the location, a monsoon season might occur for four months, followed by eight months of drier weather when supplemental irrigation water is needed.

Municipal and industrial supply.   Municipal and industrial (M&I) surface water supplies are stored by reservoirs until needed and demanded by the users.

Water quality enhancement.   Downstream of reservoirs water quality changes may occur due to choices about the quantity or the quality of water releases.

Fish and wildlife enhancement is another environmental purpose of water releases from reservoirs.

Recreation is an important aspect of reservoir uses and may include a range of activities such as boating, swimming, fishing, rafting, hiking, viewing, photography, and general enjoyment of nature.

Reservoir Characteristics and Configuration

Generally, a reservoir is divided into zones of water that are reserved for different uses. As shown by Fig. 13.1, these zones are basically working storage, sometimes called useful storage; multiple-purpose

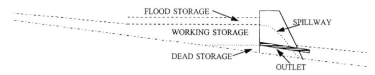

Figure 13.1 Components of a reservoir system.

capacity or operating storage; flood control storage, which might involve part of the operating capacity as well as surcharge storage; and dead storage, normally reserved to be filled with sediment.

Johnson (1990), in a report for the U.S. Army Corps of Engineers, divides storage into three zones: exclusive, multiple-purpose, and inactive. These correspond to flood, work, and dead storage in Fig. 13.1. Johnson subdivides multiple-purpose storage into seasonal flood control storage and conservation purposes that include navigation, hydropower, recreation, and supply of water for M&I, irrigation, fish and wildlife, and water quality purposes.

Planning of Reservoirs

Because both future inflow and needs can vary statistically, the planning of a reservoir's size is a statistical and operations analysis exercise. That is why *operational hydrology* is an important subset of the field of hydrology.

The reservoir sizing process carries the risks that the capacity will be too small to meet the purposes or too large for the reservoir to fill. To face these risks, the planner should compile as much historical data as possible and make studies of how the planned reservoir would have performed if it had been in place during the historical period. These are sometimes referred to as "what if" studies.

After the location and size of a reservoir are determined, many aspects of the construction process must be settled through the *design process,* an effort that involves engineers, geologists, hydrologists, financiers, attorneys, and other professionals. Among other considerations, it is necessary to make the dam as safe as possible to avoid placing people who live downstream at risk due to dam failure. While there have been few dam failures, when they occur they may cause unacceptable levels of damage.

Planning for reservoirs involves different scenarios. Some that are used as exercises in my class on water resources management include: an urban water supply storage tank, small, on the order of a million gallons, and sized to provide water to meet the high-demand periods from a rather constant inflow rate determined by a pump or treatment plant capacity; a multiple-purpose reservoir with the goal to practice computations and use basic storage and flow units for a system with a reservoir planned on a stream channel, a diversion for a city water supply, a diversion for irrigation uses, return flows, and requirements for in-stream flows both for water quality and for fish and wildlife minimum flows; and a raw water reservoir sized to deal with the risk of drought based on historical or simulated statistics of inflow.

Operation of Reservoirs

After a dam is built, it must be operated correctly. The key person is the *operator,* who makes decisions about when to release or store water. The operator may be a part-time worker who lives near a lake and who occasionally operates a gate, or a highly trained engineer working at a remote location who makes system decisions based on computer forecasts.

In the past, reservoir operating decisions were made by *rule curves,* which provided the operator with simple guidelines about how much water to release and what lake levels to maintain. As the science of forecasting and the use of computers has become more complex, however, reservoir operation has become more sophisticated. It is not uncommon to have a reservoir control center where operators use computers to monitor weather forecasts furnished from satellite data and simulate future demands for water to make decisions about water releases. They may also be bound by legal requirements to release water for downstream users, including fish and wildlife.

Basically, what is required in the operations problem is an analytical tool to illustrate the time rate of inflows, outflows, and change in storage. In other words, it is necessary to consider the storage equation both numerically and graphically. Recall that the storage equation relates inflow, outflow, and change in storage as functions of time, where for any time period such as an hour, day, month, or year,

$$Q_i - Q_o = DS$$

where Q_i is rate of inflow, Q_o is rate of outflow, and DS is change in storage.

A simple way to view the operations plan is to consider a rule curve such as the one shown in Fig. 13.2, which is taken from the Corps of Engineers operating plan for Lake Lanier in Georgia (see Chap. 19). A rule curve illustrates a typical operations time period, usually a year, and shows the boundaries within which operations should occur.

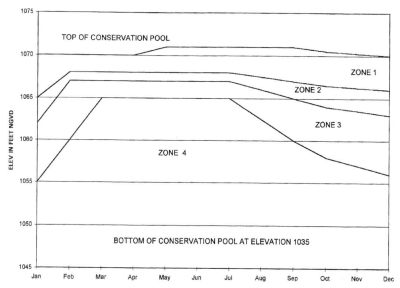

Figure 13.2 Rule curve for Lake Lanier, Georgia. (*Source: Adapted from U.S. Army Corps of Engineers, 1989.*)

Maintenance and Renewal of Reservoirs

Reservoirs require continuing attention from their owners and managers. Problems that may need attention include excessive deposition of sediments on the lake bottom that reduce the capacity to store water, pollution of lake waters, eutrophication (aging of waters with excessive growth of algae), shoreline protection, and associated issues such as dam leakage or settlement. Sedimentation is a particularly difficult problem because it occurs naturally and is impossible to control fully, but it may ruin a reservoir's capacity to store water. Some water management agencies are employing devices to flush sediment past dams.

Controversies over Reservoirs

The construction of new or altered reservoirs is often resisted by members of the public because of side effects that are considered to be negative. One of these side effects is reservoir evaporation, which can consume waters that could otherwise be used in the river environment for fish and wildlife or to flush salts through the stream system. As a result of evaporation, the microclimate around the reservoir might even be altered. Seepage caused by the reservoir impoundment can change the local patterns of underground water. On the valley floor, settlement can occur due to the weight of the water in storage. Ecosystems can be changed because the schedule of water release and the quality of the water can be altered. Also, when reservoirs are built, large numbers of residents may have to be resettled. Several of the case studies in this book, for example, Two Forks (Chap. 10), involve reservoir controversies.

Water Quality in Reservoirs

Water quality in reservoirs is an important consideration, both in operating decisions and in managing water quality in stream systems (see Chap. 14). A reservoir provides a different water environment than its associated stream, and water quality changes are influenced by currents, temperature, light, wind action, and other climatic conditions. For example, the annual water turnover in lakes is an important factor in oxygen content and associated parameters. In Horsetooth Reservoir, for example, one of the reservoirs in the Colorado Big Thompson system (Chap. 12), bacteria that are associated with manganese thrive at different water levels, depending on the temperature. The city water utility must consider these temperature variations in selecting water for withdrawals in order to control manganese problems.

Another water quality problem in lakes is eutrophication, or the aging of waters. This can result in algae blooms and dramatic changes in fish habitat. This is also a problem in estuaries, where resident times for water are long (see Chap. 22).

Lake sediments also play important parts in the overall quality picture. Chemical and biological contaminants can become trapped in sediments and remain for many years. For example, on the James River above Norfolk, Virginia, a chemical company was discharging the organic chemical Kepone and it became trapped in the sediments and affected fish populations. Estimates were that the fishery would remain closed for about 50 years.

Fish and Wildlife Issues

Although they are opposed by environmentalists, reservoirs play important roles in sustaining fish and wildlife. The ecology of a reservoir system is different from that of its associated stream system, but both an aquatic and a terrestrial ecology develop around the reservoir. Reservoirs are also popular spots for recreation, including fishing. A perspective on issues related to stream versus lake ecology can be gained from the Platte River case study (Chap. 19). The critical habitat was furnished by the stream system, which has been altered by the reservoirs and diversions, but one reservoir, Lake McConaughy, is a major recreational facility for the entire region.

Case Studies

Five brief case studies or situations related to reservoirs will be presented: reservoir sizing, reservoir operation during drought, reservoir operation during flood, operation of a reservoir in a water-short region subject to appropriation doctrine of water law, operation of a complex system of reservoirs, and planning or operations situations where there is conflict over water management objectives. Each will be presented in enough detail to illustrate the water management issues involved, but not with complete hydrologic data.

Reservoir sizing

As the United States has passed from an era of dam building to one of opposition to dams, most reservoirs have already been planned and built. In many cases, planning decisions on reservoir size were based on conservative, "don't run out of water" criteria rather than on scientific, multipurpose sizing procedures. The U.S. Bureau of Reclamation confronted this situation in 1990 as it sought new missions and developed a series of reports on sizing criteria for water projects (U.S. Bureau of Reclamation, 1990). These reports included four case studies to illustrate the methods. An overview of the method will be presented with a summary of the bureau's case studies.

The historic guidelines were: 50 percent shortage of irrigation water in the most critical year; cumulative shortage of 100 percent of annual irrigation demand in a 10-year period; and no shortage of M&I water. No criteria were developed for in-stream flow quantities.

Case studies of four bureau projects were studied. They included Lake Cachuma, on the Santa Ynez River north of Santa Barbara, California; the Tualatin Project, adjacent to the city of Portland, Oregon; the Cheney Division of Wichita Project, Kansas, focused on the Cheney Dam and Reservoir on the North Fork of the Ninnescah River, Kansas; and the Dolores Project, located in the Dolores and San Juan Basins, Colorado.

After applying their proposed methodology to these four case studies, the bureau concluded that their procedure could generate marginal benefit curves for alternative reservoir sizes, thus enabling a decision to be based on economics and to consider multiple criteria. They also concluded that the usefulness of older guidelines was limited to preliminary studies. Using the more sophisticated model approach, as opposed to guidelines, they found that when variability was a factor, as in Lake Cachuma, there can be severe impacts on high-value crops due to shortages. In view of differences in key variables, the study team recommended that the Bureau of Reclamation adopt the model approach for detailed studies of reservoir sizing. Key variables would include relationship of benefits to reservoir capacity, shortage criteria, water uses served, and selection of hydrologic period for study.

Reservoir operation during drought

The case study to discuss reservoir operations during drought is Lake Sidney Lanier, located on the Chattahoochee River above Atlanta, Georgia. Overall management of the basin is discussed in Chap. 20.

Lake Lanier is located in the headwaters of the Apalachicola-Chattahoochee-Flint (ACF) River Basin of Georgia, Alabama, and Florida. It is one of the major recreational reservoirs in the eastern United States, and has a surface area of 38,024 acres or almost 60 mi^2 at the top of the power pool, where the total capacity is 1,917,000 acre-feet. The 1040 mi^2 of drainage area includes regions of heavy annual precipitation. Lanier is the largest reservoir on the Chattahoochee River as well as the ACF system and drives a series of other mainstem reservoirs that culminate at Apalachicola Bay. By the time the river reaches Jim Woodruff Lock and Dam just below its confluence with the Flint, it drains 17,230 mi^2.

Operation of Lake Lanier water levels is a challenge during drought. Lanier contains 65 percent of the basin's conservation storage and is the major resource for release of stored water during drought (U.S. Army Corps of Engineers, 1989).

Figure 13.2 shows the Water Control Action Zones for Lake Lanier, and illustrates the thinking about operation of the reservoir to meet the authorized objectives of flood control, navigation, hydropower, water supply, and recreation. The problem faced by Lanier is how to satisfy all of these at times of low flow, especially considering water quality as well.

During droughts of the 1980s there were conflicts on the river at different points. Recreational users on the lake did not want their lake levels dropped too much, navigation interests found that they could not navigate, water supply interests became concerned about both quantity and quality of supplies, and environmental interests expressed concern about fish and wildlife. The Corps' report describes actions taken to deal with the drought conflicts. Suffice it to say that drought problems on this major facility have not yet been resolved. See Chap. 20 for more details.

Reservoir operation during flood

This case study deals with the operation of run-of-river reservoirs during flooding. It describes a case in the U.S. District Court that involved litigation of 1983 flood damages on the Black Warrior River near Tuscaloosa, Alabama.

Several lawsuits had been consolidated, and the cases were tried in September 1986 with a decision rendered in May 1987 (U.S. District Court, 1987). Judge Daniel H. Thomas, an expert in maritime law, stated that the case was one of the most interesting he had heard in a long career on the federal bench.

The general situation was that on the evening of December 2–3, 1983, towboats with coal barges were plying the Black Warrior River navigation system and hauling coal from strip mines near Birmingham to the Port of Mobile. The river system of interest involved two reaches: from Bankhead Dam to Holt Dam, and from Holt Dam to Oliver Dam, as shown in Fig. 13.3.

As a barge departed Holt Lock about 1:45 a.m. on December 3, 1983, the dam tender noticed that the pool was rising rapidly. Sometime in that period, he received complaints from boat owners moored at a marina in Holt Lake. He opened gates rapidly, even though the barge was just downstream. The dramatic rise in the outflow from Holt Reservoir beginning at 0140 hours effectively doubled the discharge in a short period of time.

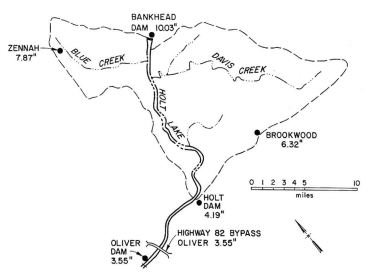

Figure 13.3 Holt Lake watershed: local area. (*Source: Adapted from U. S. Army Corps of Engineers, 1967.*)

At about 2:30 a.m., the towboat, with six loaded coal barges, hit a bridge pier at the Highway 82 Bypass located at mile point 341.5. The flood wave caused by increasing discharge apparently caught the barge, causing the operator to lose control of his tow and hit the bridge pier.

After hitting the dam, the tow broke up and careened downstream toward Oliver Lock and Dam. Towboats located downstream heard the distress call, tied off their barges, and went to provide assistance. Some of their barges broke loose due to the rapidly rising water, washed over Oliver Dam, and sank. The barges that washed over Oliver Dam struck other barges moored below the dam and broke them loose. One of the loose barges washed against an oil dock, and one or more barges apparently damaged submerged natural gas lines. Eventually 13 loaded barges sank, the oil dock was destroyed, and the gas lines were ruptured. Later in the month an explosion occurred when a barge struck a remaining gas line that had apparently floated up from the bottom of the stream.

The authorized purposes of Holt and Bankhead are navigation and power. The project manuals for Holt and Bankhead clearly state that there is no provision for flood control (U.S. Army Corps of Engineers, May 25, 1959).

To summarize the arguments, the government generally argued that the lack of flood control provisions in the project excused it from the obligations it would have for a flood control project. The plaintiffs generally argued that the Corps still had responsibility to anticipate the flood and prepare the flood pool, to warn the river traffic, and to avoid releasing flood waves onto river traffic.

The trial court expected run-of-river projects to be operated with consideration of the consequences to downstream river traffic of the flooding, even though there is no provision for flood control in the projects. However, this was reversed by the Court of Appeals, which ruled that it was an Act of God. The events and the trials illustrate important lessons for flood control operation of reservoirs, but it appears that the jury is still out on some of the rules and principles.

Operation of reservoirs under appropriation doctrine

When constraints on reservoir operation become severe, as in the appropriation doctrine of water law, decisions become more difficult and are placed under intense scrutiny (Eckhardt, 1991). This case study deals with a hypothetical but realistic situation in Colorado. Eckhardt classified facilities into subsystems to enable the identification of reservoirs and facilities belonging to different entities; in that way exchanges and transfers can be arranged from one subsystem to another, just as they must be agreed on, and adjudicated in water court, from one owner to another. The water right "owner" is the basic decision-making unit in this system.

Under Colorado water rights administration, there must be an operator–water commissioner exchange of information. The manual reservoir operations procedure in effect is quite complex and requires much coordination. Considerable information is shared every 24 hours, resulting in a final check to see if subsystem demands (legal entitlements) are being met so that adjustments can be made. In reality, a system like this has to have flexibility and some cooperation, because it is never possible to know with certainty all aspects of the system state.

Eckhardt proposed a system to work with modern, computer-based models and tools. His proposed system would introduce a continuous, real-time computation of natural flows and water available to release from reservoirs, so that water right owners could decide on the basis of reliable information about their current water accounts. It is in reality a complex, real-time water accounting procedure. The framework for the system, based on a large spreadsheet, includes system simulation, information management, and an operator interface.

A complex system: The Colorado River

While operating any reservoir is a complex exercise, it is mind-boggling to consider the complexities of operating a large, integrated system such as the Colorado River. The Colorado River system is discussed in Chap. 19, and here I will present just a few aspects of river operational decisions.

Given that a number of mainstem reservoirs are to be operated, and that numerous compact and treaty requirements are involved, it is no wonder that operational decisions cannot be reduced to a simple table or chart. In fact, the seven basin states and the Bureau of Reclamation follow the Criteria for the Coordinated Long-Range Operation of Colorado River Reservoirs, a set of rules first promulgated in June 1970 and which have been reviewed four times, each time with greater involvement and more attention to environmental issues (Gold, 1991). The Law of the River that generally governs the operation is described in Chap. 19.

Each year, an Annual Operating Plan (AOP) is developed for the individual reservoirs. These AOPs are developed in consultation with each state, federal agencies, and representatives of environmental, recreational, water user, and Native American Tribal groups. In other words, the final development of an AOP follows the model described throughout this book: final decisions made by an authority (the Bureau of Reclamation) after coordinating with the stakeholders.

Because of environmental concerns, operation of the system started to receive a lot of scrutiny in the early 1980s. Starting in 1982, the Glen Canyon Environmental Studies sought to determine the effects of dam operation on downstream environmental resources. The coordination process involved a number of entities that were following the process intensely. A few mentioned by Gold were the National Park Service, the Fish and Wildlife Service, the Bureau of Indian Affairs, the Western Area Power Administration, the Arizona Game and Fish Department, the Navajo Nation, the Hopi Tribe, the Hualapai Tribe, and the Havasupai Tribe. This partial list illustrates the difficulty of coordinating a complex issue such as this.

In the final analysis, what is sought in the changes to reservoir operation is to move back toward the natural pattern of river hydrographs. That tends to negate some of the reasons for water storage, to hold peaks back for later release to meet water supply needs.

Reservoirs with environmental conflict situations

Two cases, both on the Platte River system, illustrate the high degree of conflict that can arise in reservoir planning or operations over environmental issues. The Two Forks case, described in Chap. 10, illustrates the conflict over reservoir planning and siting. Lake McConaughy, located on the North Platte in Nebraska, illustrates the conflict that can arise over reservoir operations (Chap. 19).

Questions

1. For each basic purpose of reservoirs, discuss how the purpose could be met without a reservoir. The purposes are flood control, navigation, hydroelectric power, irrigation, municipal and industrial water supply, water quality, fish and wildlife, and recreation.

2. What methods might be used to reduce the buildup of sediment and extend the life of a reservoir?

3. What is a rule curve for a reservoir, and how can it be used to negotiate water management schedules with different interest groups?

4. List the alleged side effects of reservoirs and describe how serious they might be in different situations.

5. In general, what will be the effect of a reservoir on habitat in a stream, and how would you expect the reservoir to impact the fish species?

6. Given that many major reservoirs in the United States were built in the twentieth century, what major issues would you predict for them during the twenty-first century?

7. You will design a reservoir for urban water supply, irrigation, flood control, hydropower, navigation, and fish and wildlife enhancement. To plan the reservoir you must know the *demand* for each of these water management purposes. Describe briefly how the demand function for each of these purposes is found and give examples of quantities that are commonly experienced in the United States. As an example, for industrial water the demand function must be found for each industrial process that would use the water, and a typical quantity might be x acre-feet per ton of the industrial process.

References

Eckhardt, John R., Real-Time Reservoir Operation Decision Support under the Appropriation Doctrine, Ph.D. dissertation, Colorado State University, Ft. Collins, Spring 1991.

Gold, Rick L., Environmental Protection and the Operation of the Colorado River System, Bureau of Reclamation Working Paper, July 23, 1991.

Johnson, William, *A Preliminary Assessment of Corps of Engineers' Reservoirs, Their Purposes and Susceptibility to Drought,* Hydrologic Engineering Center, U.S. Army Corps of Engineers, Research Document No. 33, Davis, CA, December 1990.

U.S. Army Corps of Engineers, *Reservoir Regulation,* EM 1110-2-3600, May 25, 1959.

U.S. Army Corps of Engineers, Mobile District, *Post Authorization Change Notification for the Reallocation of Storage from Hydropower to Water Supply at Lake Lanier, Georgia,* October 1989.

U.S. Bureau of Reclamation, Summary Report on Sizing Criteria for Water Projects, SAC24229.AO, Denver, May 1990.

U.S. District Court for the Southern District of Alabama, Southern Division, Warrior & Gulf Navigation Co., et al., vs United States of America, et al., Civil Action Nos. 84-0632 T, 84-0672-T, 84-1341-T, 85-0574-T, and 85-0983-T, 1987.

14

Water Quality Management and Nonpoint Source Control

This case study chapter discusses the evolution and issues of the U.S. approach to water quality management, and two specific cases: one that illustrates in a practical manner how the Clean Water Act is administered, and the other to illustrate the findings of a comprehensive study on water quality policy.

The United States has spent more than $500 billion on water quality management since the 1970s (Intergovernmental Task Force on Monitoring Water Quality et al., 1992). Like other complex areas, water quality programs were developed incrementally, but a coordinated design for the overall program has evolved. It consists of research and studies, standards, permits, monitoring programs, reporting, and enforcement. The program is based on regulations rather than incentives and market forces, but in the future, limitations of the approach will force other mechanisms into play, including quasi-market methods to negotiate and coordinate in other than command-and-control formats.

Point and Nonpoint Source Control

Clearly, the multiplicity of point and nonpoint sources makes integrated management a real challenge. Point sources are difficult enough, but nonpoint source pollution is recognized as the single largest contributor to water quality problems.

The Clean Water Act sought to deal with the nonpoint problem, and Section 208 of the act provides for studies of area-wide water quality management. The philosophy is to consider nonpoint sources and point sources in a comprehensive framework. Unfortunately, both the technical and institutional aspects of nonpoint source control are more complex than those for point sources. Therefore, not much progress has been made on nonpoint sources. However, the emerging "watershed approach" may offer fresh hope (see Chap. 16).

Point sources are individual points where wastewater is discharged from pipes into receiving waters. Except for underwater discharges, these are visible and can be identified, inventoried, and controlled. Nonpoint sources result from land uses, and are prime targets for the watershed approach. Agricultural sources are the largest category of pollutant by volume. These include sediment, animal waste, fertilizers, pesticides, and herbicides. Logging is a closely related activity. Industrial and transportation activities add residuals, air pollutants, and solid waste sites. Construction adds to the sediment load, and can include oil and grease, gasoline, chemicals, and various industrial products resulting from the building process. Mining can produce sediments and chemical pollution in the form of acid mine drainage. Roadways produce sediment and chemicals. Landfills and solid waste sites of all types contribute to nonpoint source problems through leakage, runoff, and leaching. Urban runoff can be a source of substantial pollutant loads, ranging from sediment to bacteriological contaminants to chemicals and heavy metals. Also, hydrologic modifications drain aquifers, add sediment, and alter local ecology.

Management measures fall into three categories (Brooks et al., 1994): regulatory (zoning, regulations, land and water rights, controls, permits, prohibitions, and license); fiscal (prices, taxes, subsidies, fines, and grants); and direct public investment and management (technical assistance, research, education, land management, installation of structures, and infrastructure). The management measures that have evolved to protect watersheds from nonpoint sources are generally called *best management practices,* or BMPs. Novotny and Chesters (1981) classify BMPs into three categories: source control of hazardous lands and land uses, collection control and reduction of pollutant delivery to receiving waters, and treatment of runoff.

Regulatory-Based and Market-Based Approaches

During the 1960s, as the U.S. system was evolving, there were advocates of both the regulatory-based and market-based (or economics-based) approaches. In the regulatory approach, the regulations are set, and dischargers must meet them (command and control). The economics approach creates a system of charges and incentives.

The regulatory approach is like a speed limit that applies to each vehicle: The limit is set, and if you violate it, you pay a penalty. One vehicle cannot agree to go slow and trade its extra speed to another. The parallel situation for water quality is that each discharger has to meet the standards set, but the standards are supposed to reflect the stream carrying capacity, as a speed limit is supposed to reflect highway capacity. Under the regulatory approach, government authorities set rules, and dischargers must comply. The authorities (federal and state governments) command and control the dischargers (local government, individuals, and industry). While this seems to go against the grain of free enterprise, the approach was adopted only after a period of careful study, and evolved over a number of years.

The reasons for the United States' selecting the regulatory-based approach are reviewed by Grigg and Fleming (1980). Basically, three goals were pursued: the water quality objectives, the goal of state primacy and decentralized administration, and a goal to equalize water quality standards across the country. In this latter goal, one federal role was to prevent one state from setting standards low just to attract industry and jobs from another state.

The economics approach tries to emulate the marketplace. It offers the promise of economic efficiency, it might stimulate innovation, and it seems administratively simple compared to the regulatory approach. These arguments have been made repeatedly by economists, and they have apparent merit. However, the economics approach has received little support from lawmakers, regulators, or industry. The regulatory system seems more practical, and it involves no new taxes on industry.

Limitations of the command-and-control approach have become evident. Evidence is that "pollution trading" and quasi-market approaches may gradually emerge. The new watershed approach taken by the Environmental Protection Agency (EPA) signals such a possibility (see Chap. 16). According to Glaze (1994), in the years ahead we will see "a combination of command and control, self-regulation and a very large number of economic instruments such as pollution taxes, tradeable permits, the removal of distorting subsidies...."

Evolution of Water Quality Management in the United States

The U.S. system evolved over about the last 90 years, mostly during the twentieth century. Prior to about 1900, waterborne diseases such as typhoid and cholera were prevalent in the United States, and there was no organized approach to water quality management. Public health management as a profession was just beginning, and the links between waterborne diseases and death rates were just becoming apparent.

About 1900, things began to happen. First, in 1899, the Rivers and Harbors Act prohibited discharge of refuse into waterways that would interfere with navigation, unless a permit was obtained from the U.S. Army Corps of Engineers. Although this regulatory authority did not seem to extend to release of wastewaters, it set the stage for later programs. Then, in 1914, the U.S. Public Health Service (PHS) was established. Part of it evolved into the Federal Water Pollution Control Administration (FWPCA) and then into the EPA. The PHS's water activity concentrated initially on safety of drinking water.

Several federal water pollution control acts and amendments were passed in the period 1945–1965. The 1948 act provided limited federal financial assistance for the construction of municipal wastewater treatment facilities, and the 1956 and 1961 acts increased this assistance. In 1965 the Water Quality Act required states to develop water quality standards for interstate waters, created the FWPCA, and increased financial assistance. The EPA was created in 1969, and the FWPCA was merged into it. The Federal Water Pollution Control Act of 1972 (the Clean Water Act) created the system of water quality management we have today. It became the centerpiece of U.S. programs and strategy.

Quite a few other environmental laws affect water quality. Especially noteworthy is the Safe Drinking Water Act (1974) and amendments (see Chap. 6 for a discussion of environmental statutes).

Clean Water Act

The Clean Water Act contains many features and has been amended several times. It initiated a national permit system for all point source dischargers, it created a uniform system of technology-based effluent standards for industrial dischargers, and it greatly increased federal financial assistance. It also designated the Corps of Engineers as the authority to administer the Section 404 permitting program over dredge and fill activities.

Other aspects of the 1972 act included the statement of water quality goals, the Section 201 regional planning program, Section 208 water quality planning, and enforcement programs. The authorization for the Wastewater Treatment Construction Grants Program was perhaps the most dramatic part of the program.

The 1977 Clean Water Act Amendments added control of priority toxic pollutants to the federal program and encouraged states to take control of the permit system.

The 1981 Municipal Treatment Construction Grant Amendments reduced federal financial support, and the 1987 Water Quality Act phased out the Construction Grants Program, but provided initial capitalization grants to states for revolving funds. See Chap. 7 for a discussion of the difference between grants and loans programs.

The 1981 act also required the EPA to develop regulations for stormwater runoff control, and it required states to prepare nonpoint source management programs. At this time, phasing in a stormwater permit program is one of the major policy issues on the water quality docket.

At this writing, reauthorization of the act is, as usual, a contentious political issue. The conclusions to this chapter outline some of the issues to be faced.

How the System Works

The overall water quality management system includes control of drinking water quality, control of in-stream and groundwater quality, and control of effluents from wastewater plants and nonpoint source discharge points. While the system has been put together a piece at a time, it is meant to be a "cradle-to-grave" approach to water quality.

The laws that regulate different aspects of the hydrologic cycle are sometimes administered by different agencies and seem quite separate, but they utilize common water quality measurement parameters, and the output of one process may be the input to another; thus, the overall approach is integrated, as shown on Fig. 14.1.

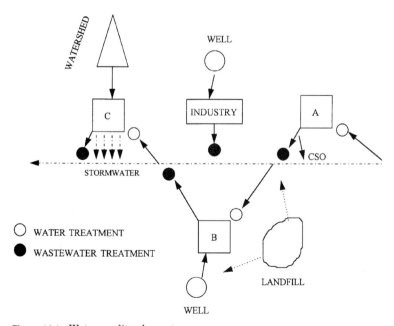

Figure 14.1 Water quality elements.

Note the successive use of water as it moves through its cycle. The withdrawal of surface water by all three cities is shown, with City B having a supplemental well supply that is threatened by seepage from a landfill, which also threatens the main stream. City C has a supplemental source from a protected upland watershed. The industry uses self-supplied water and furnishes its own treatment. City A has a problem with combined sewer overflow (CSO), and City C has significant stormwater discharges, which are not regulated for water quality.

The withdrawals for drinking water are regulated by the Safe Drinking Water Act. The effluents from the wastewater treatment plants are regulated by the Clean Water Act, which also specifies the guidelines for setting stream water quality standards. The nonpoint sources are not yet effectively regulated. Nor is regulation of groundwater quality very effective yet.

Monitoring of ambient and effluent waters is necessary to assess water quality, and to determine if the limits of National Pollutant Discharge Elimination System (NPDES) permits are being met. Wasteload allocation decisions are made using water quality models. To test the extremes of the system, dry weather levels are used. In addition to riverine waters, the quality of reservoirs, lakes, estuaries, and coastal waters is assessed. It is difficult to assess the impact of nonpoint sources, however.

Describing and measuring water quality is a challenge because it has physical, chemical, and biological dimensions. Chapter 2 gives descriptions of water quality parameters. Not only must these be measured against standards, they are expensive and difficult to measure, and data protocols are not standard.

Criteria and Standards

The water quality management system depends on criteria and standards. It is a top-down approach.

Based on research, the EPA sets overall criteria for specified pollutants. The EPA's Office of Science and Technology, a part of the Office of Water, is responsible for developing the criteria (U.S. General Accounting Office, 1994). Water quality criteria were called for in the original Clean Water Act to protect human health and aquatic life. A 1976 consent decree required the EPA to publish criteria for a specified set of pollutants by 1979. These were later designated by Congress as toxic pollutants under Section 307(a) of the act, and the EPA selected 126 key chemicals or classes of chemicals for priority status. By 1994, 99 of the 126 criteria had been published, but activity had slowed. The EPA feels that it has released criteria for the most serious pollutants, but some groups would like it to publish more criteria and revise others. Many feel that the approach of setting criteria is flawed, and that other problems such as destruction of habitat and loss of biodiversity are more important than individual pollutants (U.S. General Accounting Office, 1994).

Based on the water quality criteria, states are required to set stream standards which recognize unique characteristics of the waters and their uses and which protect human and aquatic life. A typical stream classification system begins with waters suitable for a drinking water supply, and works down to waters of lower quality. There will be unique regulations for each category, and they will be revised from time to time. Stream standards have significant economic and social implications, and revising them is a difficult process. Effluent standards set under the authority of the Clean Water Act prescribe the levels of contaminants that can be discharged. They are set after an analysis of the ability of the stream to assimilate the oxygen-demanding wastes and with a view to prohibiting other harmful discharges. These become part of the permit conditions that place requirements on individual dischargers.

Water Quality Monitoring and Assessment

Monitoring is necessary to assess water quality and to make sure that management programs are working, from the earliest stages of assessment to the final phases of compliance with a discharge permit. How to design a monitoring system has been the subject of courses at Colorado State University (Sanders and Ward, 1993).

The basic water quality parameters are oxygen levels, bacterial levels, and concentrations of chemicals and nutrients. Many parameters are involved, and some are not easy to measure. For example, in estuary monitoring, a list of characterization parameters includes land use data, freshwater distribution, inflow and drawdown, shoreline development, erosion rates and frequency, severity of storm events, nutrient enrichment, dissolved oxygen, phosphate, total nitrogen, inorganic nitrogen, nitrate, ammonium, organic nitrogen, toxic metals, pesticides, organics, fish landings, catch per unit, nursery areas juvenile index, spawning areas, plant data, and pollutant data (Davies, 1985).

The terms "monitoring" and "assessment" sometimes mean different processes. The National Academy of Sciences' Water Science and Technology Board (1986) defined them this way: Monitoring is the "repetitive collection of water quality data for some specific purpose, e.g., compliance and enforcement, or establishment of a management strategy. Assessment, on the other hand, uses monitoring data and other information to make an evaluation or interpretation of the data in terms of ambient conditions, identification of water quality problems, sources of pollutants and their impact, trends and effectiveness of control programs."

The U.S. EPA (1985) publishes guidance for monitoring programs. Three basic purposes of monitoring are listed: to conduct water quality assessments, to develop water quality-based controls, and to assess compliance with and effectiveness of controls. Data uses are given as national water quality assessment, state-wide water quality assessment, regional oversight, program management and wasteload allocations, construction grants design, and checking of water quality standards.

Both ambient water quality and releases of contaminants to the waters (effluents) need surveillance. In the United States, ambient monitoring is carried out through a combination of fixed stations operated by various government entities, intensive surveys, and biological monitoring. Most effluent monitoring is done by dischargers in conjunction with requirements of their discharge permits, supervised by state regulatory agencies with oversight by the federal government. Monitoring of nonpoint sources is, for the most part, done only on a special study basis.

It would be convenient to specify water quality with a single water quality index, but attempts to develop an index fail without agreement about objectives or the parameters to include.

The Intergovernmental Task Force on Monitoring Water Quality (1992) concluded that in spite of large investments, the nation is "unable to document adequately the effectiveness of these investments in achieving the objectives of the Clean Water Act and other Federal and State legislation related to water quality." Recommendations included an integrated national strategy; more focus on multimedia, geographically based activities, biological and ecological information, and nonpoint sources, wetland, and sediment concerns; voluntary and cooperative efforts at integration using technologically advanced tools; establishment of a coordination partnership for monitoring; establishment of a permanent information standards and compatibility council; further evaluation of technologies; and a training program to implement the plan.

Although water quality assessment is difficult, it does provide information that can be compiled to gain ideas about trends and progress (see chapter conclusions). Section 305(b) of the Clean Water Act requires states to submit a biennial assessment. For example, in 1992, the Wisconsin Department of Natural Resources (1992) released the Wisconsin Water Quality Assessment Report to Congress.

Because the technology for water quality assessment has not produced reliable overviews of the nation's water quality, the U.S. Geological Survey launched a National Water Quality Assessment

(NAWQA) Program in 1986 (U.S. General Accounting Office, 1993). Obstacles include that data is often unable to meet the NAWQA needs as a secondary user, common data standards and definitions are missing, the quality of data is uncertain, and there is a wide variety of sampling and analysis procedures. NAWQA efforts are leading, however, to greater coordination of data management technologies and management.

Water resources assessment is necessary for any water planning and management process and goes beyond water quality alone. It was featured by Agenda 21 as one of seven major program themes for the freshwater sector (Grigg, 1993). The unique feature of assessment is its focus on activities that relate to data collection, management, and evaluation, and to other information-related activities such as research and development, and planning studies.

Chapter 18 of Agenda 21 stresses the data aspects of assessment. The writers probably considered that all other information activities would be handled under integrated water planning and management.

Using Agenda 21's terms, there is a close relationship between "assessment" and "integrated management." Integrated management includes activities to formulate national action plans and investment programs; integrate measures including the inventory of water resources; develop databases, forecasting, and other models; optimize allocation; implement demand management, pricing, and regulatory measures; implement risk management for flood and drought; promote rational water use through education and pricing; mobilize water resources in arid areas; promote international scientific cooperation; develop innovative sources; promote water conservation; support local water user groups; develop public participation; strengthen cooperation at all levels; and increase education and dissemination of information.

Sophisticated concepts of water resources assessment have been around for over 50 years. An experiment that aided in this was the work of the National Resources Planning Board (NRPB), a New Deal program (Clawson, 1981). Later, the Water Resources Planning Act of 1965 included a requirement for water resources assessment as one of the functions of the U.S. Water Resources Council (1968):

> To maintain a continuing study and prepare periodically an assessment of the adequacy of supplies of water necessary to meet the water requirements in each water resources region in the United States and of the national interest therein.

James, Larson, and Hoggan (1983) concluded that while the national assessments were useful, they never really became an integral part of the national water planning process.

It will be important that monitoring be designed to complement the use of mathematical models in the future. Monitoring needs can be so expensive and extensive that they may in the future need to go beyond governmental programs and include citizen volunteers. Meanwhile, some monitoring techniques are becoming more standardized (Hall and Glysson, 1991).

Treatment Plants

There are several different kinds of treatment plants, including water treatment, wastewater treatment, pretreatment, industrial wastewater treatment, wet weather treatment, and so on. Chapter 3 outlines treatment plants as part of overall water resources systems, and Chap. 2 discusses physical, chemical, and biological aspects of water quality. Figures 14.2 and 14.3 illustrate two examples of the basic processes of water and wastewater treatment. Figure 14.2 is an example of the processes used to treat drinking water. It illustrates different physical and chemical processes that convert raw water into pure water that can leave the clearwell to the distribution system. Figure 14.3 illustrates typical processes of a wastewater treatment plant, showing progressive conversion of raw wastewater to final effluent and sludge.

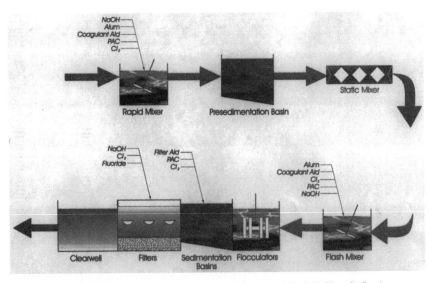

Figure 14.2   Typical water treatment processes. (*Courtesy Black & Veatch, Inc.*)

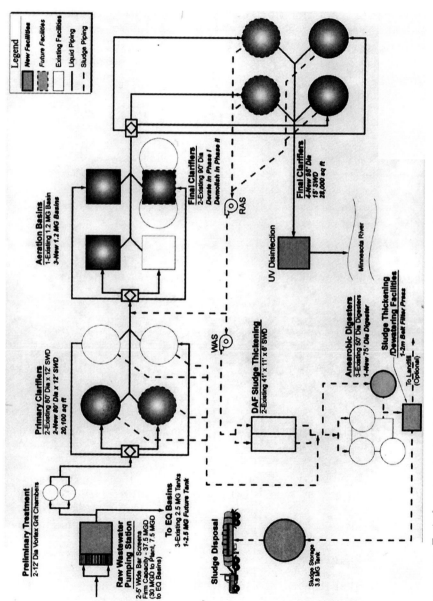

Figure 14.3 Typical wastewater treatment processes. (*Courtesy Black & Veatch, Inc.*)

Water Quality Databases

Water quality data is particularly important to monitoring and assessment. Monitoring, in turn, is a key to management actions such as treatment plant planning and siting, permitting, and enforcement. A great deal of improvement is needed in the complex arena of water quality data management.

Saito (1992) and Saito, Grigg, and Ward (1994) review the status of water quality data management and conclude that more integration is needed at the state government level. Specifically, with regard to Colorado, they recommend that the state develop a coordinated water quality and quantity database, with an interagency and intergovernmental task force organized to help design the database.

As should be clear from this chapter (and from Chap. 19 on river basins), design of databases is quite complex. It may be too early to fully integrate water quality and quantity databases, but it is not too early to launch integrated databases for each category and provide for relationships between them. That is what relational databases are about anyway.

Water Quality Modeling

In addition to data on water quality, modeling is used to set permit conditions and assess water quality.

Why model streams? There are two basic reasons. First, to improve our understanding of the system. Models can help our understanding of complex relationships that we cannot measure. The more urgent reason, however, is to provide information for decision making about management actions.

Models are complex because there are many scenarios and constituents. An EPA report of the 1970s placed them in six categories (Grimsrud et al., 1976). These include steady-state stream models (simplified stream models); steady-state estuary models (simplified estuary models); quasi-dynamic stream models (such as QUAL II, which seems to be the way stream models are going); dynamic estuary and stream models (which are more difficult to run); dynamic lake models; and near field models (which give detail on issues such as outfall plumes).

Of the current models, QUAL-IIE may be the best known. It is widely used and accepted, relatively inexpensive and easy to apply, and can model diurnal temperature, dissolved oxygen, and algae.

As models simulate actual systems, they must begin by simulating the hydraulic interrelationships. This requires the ability to simulate unsteady flows in nonuniform channels. Then the model must be able to handle mass transport phenomena, an order of magnitude more difficult than hydraulics. Next there are complex biological and chemical changes to be handled, such as algal sequences and oxygen transfer. Then there is the relationship between the benthic layer and water column to be evaluated. Finally, living things in the food chain must be considered.

How useful have models been? The results are mixed. In the Chesapeake Bay Program, a research effort costing millions of dollars developed models, but due to the complexity of the situation, none were in use by 1985 for bay-wide management.

The EPA's experience with managing and maintaining water quality models shows that their use is increasing (Bouchard et al., 1993). The Center for Exposure Assessment Modeling (CEAM) at the EPA's Athens, Georgia, laboratory was established in 1987, and it deals with scientific and technical exposure assessment to support environmental risk-based decisions.

In the United States, experience with using models for management has shown that some principles should be observed. These include standard models, open structure, documentation, support, and training. Standard models simplify review and evaluation of results by management. Open structure models allow the inspection of the inner workings of the models and enable documentation to be understood.

Case Study: Water Quality Management by State Government in North Carolina

The first case study is water quality management in North Carolina. My own experience there adds some perspective, and the background of the agencies involved is described in some detail in Chap. 23.

Briefly, water quality management began to receive attention about 1877, when the North Carolina Board of Health was organized (Howells, 1990). In 1928, the State Stream Sanitation and Conservation Committee (SSSCC) had been formed, and it was formalized through legislation in 1945, but no appropriation was made until 1951–1952. In 1959 the Department of Water Resources was established, and the SSSCC and Division of Water Pollution Control (transferred from the Department of Health) were made part of it. Later it became the Department of Water and Air Resources, and still later it merged into the Department of Natural and Economic Resources (1971), which became the Department of Natural Resources and Community Development (1977) and later the Department of Environment, Health, and Natural Resources (1989).

My experience with the agency was from 1979 to 1982, when the Clean Water Act was administered by the Division of Environmental Management and the Safe Drinking Water Act was administered by a separate agency in the Department of Health, which also handled solid wastes. Our main activities were permitting, monitoring, and enforcement as they related to cities and industries. The agency had a lot of legal activity, and had an Office of Legal Affairs with several attorneys who specialized in environmental law. Also, the Attorney General's Office had an attorney assigned to help with cases involving the AG's role.

Permitting

Permit authority is under Section 401 of the Clean Water Act, delegated to the state by the EPA. Most states have this authority, because they do not want the federal government controlling their industries. Generally, a new or existing industry will submit a permit application to the state environmental agency. If the applicant is not satisfied, the permit conditions can be appealed. If the EPA does not feel that the state is following the law, it can intervene.

North Carolina, like many states, has numerous small industries, or "mom-and-pop" plants. For both small and large plants, the permit conditions are an important part of the cost of doing business. In one case, a textile plant in the coastal region was applying for renewal of a discharge permit. One of the constituents discharged was phosphorus, a nutrient linked to estuary algal blooms (see Chap. 16 on estuary management, Albemarle-Pamlico case). Ideally, the plant could cut out all phosphorus and help the estuary. However, the plant was small, and to modernize it would essentially put the company out of business. On the other hand, at the time little was known about the link between phosphorus and eutrophication. What should be permit conditions be?

Our action was to meet with the staff members and scientists involved in permitting to review the knowledge base. We requested the opinions of university scientists who had studied the estuary. Based on all of the advice we received, we set a standard at the background level of phosphorus in the estuary. While we believed it would be tough on the plant owners, it was not unreasonable, and if it would really hurt them financially, they could appeal to the Environmental Management Commission. To the best of my recollection, the plant was able to comply with the permit conditions, and we were able to support estuary cleanup goals at the same time that jobs and a family business were protected.

Monitoring and enforcement

Monitoring and enforcement are the next steps in environmental administration. Monitoring is done both by the permittee (effluent monitoring) and by the agency (ambient monitoring). Enforcement is necessary when a permit condition is violated.

A case dealt with a situation located in the coastal water region of North Carolina, where an industry was suspected of violating permit conditions and of contributing additional nutrients through airborne and nonpoint source routes. This was potentially a serious problem and could even have been linked to the deteriorated health of the estuary.

The industry was filing the normal effluent monitoring reports, which showed that wastewater discharges contained nitrogen, but were within limits. However, local residents reported that there might be illegal discharges taking place through storm drains. Also, we suspected that airborne nutrients were going up the stack, being deposited into the watershed, and carried to the estuary during rainfall events. Finally, it was suspected that waste fertilizer had been buried on the site and was leaching into the estuary by groundwater flow.

Monitoring was a challenge due to the different routes that could be taken by the nutrients. Also, as the estuary flow ranged back and forth due to tidal action, it was difficult to sample the surface waters to gain a picture of actual releases. We tried all routes: measuring nonpoint sources, measuring the stream, taking samples on the site, and sampling the water discharge.

Finally, a picture emerged that this plant was responsible for significant nitrogen discharges to the estuary. As nitrogen is a key nutrient linked to nuisance algal blooms, we decided that enforcement action was necessary. However, the enforcement action could not be straightforward for a permit violation, because such a violation had not been detected. The enforcement action was ultimately unsuccessful, as the industry appealed and was upheld by a special panel.

Enforcement

The final situation is a pure enforcement scenario. The enforcement staff reported that a city had violated its wastewater permit conditions. The city was repairing its sludge dewatering centrifuges and had taken several out of service. However, it did not make provisions for handling the sludge during the repair period, and excessive amounts were being discharged into the river. As this was sludge and not wastewater, the violation did not show up in the regular reports of ambient monitoring. However, recreational boaters reported to the agency that the streambed downstream of the plant was becoming caked with sludge. Upon investigation, we found that this was indeed the case. The enforcement staff prepared their case and made a recommendation for a penalty under North Carolina statutes.

When this occurred in 1979, environmental enforcement was not as well developed as it is today. After receiving the recommendation, I called several officials to learn about current practices and experiences in determining the amount of civil penalties. Finally, I determined that few precedents were available, and that we had to set a precedent with a civil penalty. The penalty was appealed to the Environmental Management Commission by the city, but it was upheld, and the city paid. The precedent of a state agency fining a city did not sit well with all involved. However, I believe that firm enforcement of the rules is the only way to run an environmental program; otherwise there will be double standards, one for industry and another for government, and it will lead to sloppy management.

Lessons from the case study

The administration of a complex management system like the Clean Water Act involves a lot of behind-the-scenes work. The case situations illustrate a little of it, but do not reveal the full range of conflict inherent in different staff opinions, interest group pressures, scientific disputes, and the politics of influence.

There are examples of the basic elements of the U.S. system: the standards, permitting, modeling, monitoring, and enforcement. With these elements, the range of activities involved from research to law enforcement becomes apparent.

Any part of the scene involves the players. In the case studies we see the roles of cities and industries that apply for permits, the state agency that administers the program, and the relationships between the state agency and the oversight agency, the EPA. Also, the players include interest-group members such as local boosters, university professors, and environmentalists. Politics plays an important part in program administration, because ultimately environmental administration, like other public-interest programs, involves a lot of balancing of interests, such as finding the balance point between the most ambitious environmental goals and economic development.

From the case situations, one can see issues that still plague water quality management today. The command-and-control aspects of environmental administration place a lot of power in the hands of agency officials. This power must be balanced by avenues of appeal and negotiation. One can see the unresolved and important role of nonpoint sources, as compared to the simpler point sources. Finally, one can see that unless an effective program for law enforcement is in place, none of the programs make any difference.

The North Carolina case illustrates typical experience with the Clean Water Act in a state government. Now, let us look at another case that deals with the future—Water Quality 2000.

Water Quality 2000

The Water Quality 2000 case illustrates a number of the themes of this book: complex issues, multiple players, coordinated studies, and policy recommendations. In 1988, a consortium of more than 80 public, private, and nonprofit organizations began a four-phase effort to identify a national policy for water quality management (Water Quality 2000, 1992).

Water Quality 2000's vision statement is "society living in harmony with healthy natural systems," and the corresponding goal is "to develop and implement an integrated policy for the nation to protect and enhance water quality that supports society living in harmony with healthy natural systems."

Process of Water Quality 2000

Water Quality 2000 represents a broad, jointly funded effort to reach consensus on what is needed to develop and implement an appropriate water quality policy. As such, it is an example of the collaborative approach to policy setting that is needed to deal with policy issues.

The process was developed through the initiative of individuals who were able to coalesce a broad effort among public and private organizations, with the organizational focus located at the Water Pollution Control Federation (later the Water Environment Federation).

As a process, the steering committee determined that the policy should seek broad representation; a long-range, visionary, and holistic perspective; maximum consensus on national principles; a focus on water quality, not quantity, but with a balanced view of surface, ground, and atmospheric waters; and should lead to a specific agenda for action. The process would assure healthy aquatic, estuarine, and marine ecosystems, healthy drinking water supplies and adequate water quality for other uses, and protection of human health from water quality hazards associated with recreation, fish, and shellfish consumption and other uses. Thus, the goals were consistent with coordinated water management actions aimed at an integrated approach to water management.

Condition of the nation's waters and aquatic habitat

The condition of the nation's waters and aquatic habitat is generally determined and reported through semiannual reports of the states. These reports are required by Section 305(b) of the Clean Water Act and are intended to provide assessments of progress toward the two interim goals of the act. According to Water Quality 2000, the information provided by the reports is useful, but not fully adequate to assess the condition of the water bodies.

Sources of pollution

Water Quality 2000 lists nine categories of sources of water contamination. The list is in alphabetical order, not in order of impact, because different interest groups will insist that they not be apportioned the major blame for water pollution unless the scientific evidence is definitive. The source categories are agriculture, community wastewater systems, deposition from atmospheric sources, industrial dischargers, land alteration, stocking and harvest of aquatic species, transportation systems, urban runoff, and water projects.

Causes of water pollution

Water Quality 2000 explained that the sources of water pollution are driven by decisions about how society lives; farms; produces and consumes; transports people and goods; plans for the future; and how it acted in the past.

U.S. commitment to clean water

Since the Clean Water Act was passed in 1972, the United States has made quite a financial commitment to clean water. According to Water Quality 2000, the United States spends more on pollution control per capita and more per unit of output than most other industrialized nations, including Great Britain, Japan, and Germany. Since 1970, investments by all levels of government and by industry totaled a minimum of $239 billion in capital facilities and $234 billion for operation of facilities and programs in water pollution control (1986 dollars). Many of the water pollution control programs are not included in these totals.

Impediments to clean water

Water Quality 2000 listed the following categories of impediments to clean water: narrowly focused water policy; institutional conflicts; legislative and regulatory overlaps, conflicts and gaps; insufficient funding and incentives; inadequate attention to the need for trained personnel; limitations on research and development; and inadequate public commitment to water quality.

Emerging issues in water quality

Water Quality 2000 identified 12 emerging issues as critical to the years immediately ahead in water quality management. These include preventing pollution, controlling runoff from urban and rural lands, focusing on toxic constituents, protecting aquatic ecosystems, coping with multimedia pollution, protecting groundwater, increasing scientific understanding of water quality issues, promoting wise use of resources, setting priorities, providing safe drinking water, managing growth and development, and financing water resource improvements.

Conclusions about Water
Quality Management

The Water Quality 2000 study group concluded that the condition of the nation's waters has improved since 1972, but many problems remain. One issue in this conclusion is whether the data enable us to reach such a conclusion. The Association of State and Interstate Water Pollution Control Administrators (ASIWPCA) reached a similar conclusion (Savage, 1994). However, data and assessment problems inhibit analysis and conclusions, and it is apparent that more attention is needed to data management.

The nation has made a tremendous investment in water quality, some $500 billion since 1972, with about $85 billion coming from the federal government.

According to Water Quality 2000, the fundamental causes of water pollution lie in the way we live, farm, produce, consume, transport people and goods, and plan for the future. This fact has far-reaching implications for our ability to achieve sustainable development.

Rather than narrowly focused water policies, we need more integrated solutions such as watershed-based approaches and win–win plans. Other problems include conflicts among water quality institutions; legislative and regulatory overlaps, conflicts, and gaps; funding and incentives for clean water programs; inadequate attention to the need for trained personnel; research and development programs that are insufficient to meet the challenge; and inadequate communication.

With any approach, there are formidable challenges to meeting the goals, including prevention, control, focus on toxics, protecting ecosystems, coping with multimedia pollution, protecting groundwater, increasing understanding, promoting wise use, setting priorities, providing safe drinking water, managing growth, and financing improvements.

Common themes among study groups include the need for new national water resources policy; preventing pollution in the first place; developing both individual and collective responsibility for water resources; and focusing on the watershed for planning and management.

The tools of change should include securing public commitment through education and training; preventing pollution; promoting wise use of resources; managing growth and development; increasing scientific understanding and improving technologies; and eliminating, resolving, and filling regulatory and legislative overlaps, conflicts, and gaps.

Almost all parties believe that new frameworks for problem solving are needed. These should add funding, flexibility, better science, and broad-based participation. They should strengthen existing state and federal programs, provide incentives for collaborative efforts, and feature improvements in public education to change the way we live.

In the 1990s, several attempts have been made to reauthorize and revise the Clean Water Act. These show the specific legislative approaches to reforming water quality management. In 1995, for example, the House of Representatives passed H.R. 961, the Clean Water Act Amendments of 1995, but the Senate did not act on it. The act would increase flexibility for pollutant trading and end-of-pipe versus pretreatment; encourage watershed approaches; revise nonpoint and stormwater management approaches; redefine wetlands; require greater use of cost–benefit analysis and risk assessment; and increase the funding available to the state revolving loan fund. While these seem to be mostly "tune-ups" to the existing program, they are controversial. The Majority Staff of the House Transportation and Infrastructure Committee felt it necessary to issue a document entitled "Setting the Record Straight: A Response to the Myths Regarding the Clean Water Act Amendments of 1995." The act illustrates how each provision of the law becomes a site for battles between interest groups.

As a final word, the goals of Water Quality 2000 and sustainable development are really the same—"society living in harmony with healthy natural systems," and "to develop and implement an integrated policy for the nation to protect and enhance water quality that supports society living in harmony with healthy natural systems."

Questions

1. Discuss the advantages and disadvantages of a regulatory versus market-based approach to water quality management.

2. What is your philosophy on the need to subsidize wastewater treatment? Do you believe the funds spent by the United States on the Construction Grants Program were well spent?

3. Describe the working features of the Clean Water Act. If you were to redesign it, what would your design be?

4. What are the largest sources of pollution and the major pollutants in the United States?

5. Explain what the NPDES program is and how it relates to monitoring of water resources.

6. Explain the Section 404 program in water resources management and how it relates to the siting of new reservoirs.

7. How does the Water Quality 2000 vision statement, "society living in harmony with healthy natural systems," relate to the concept of "sustainable development"?

8. Water Quality 2000 concluded that fundamental changes in institutions, business, government, and individual lifestyles will be required to enhance water quality. Do you agree, and what would be some of the changes needed?

References

Bouchard, Dermont C., Robert B. Ambrose, Thomas O. Barnwell, and David W. Disney, *Environmental Software at the U.S. Environmental Protection Agency Center for Exposure Assessment Modeling,* U.S. Environmental Protection Agency, Center for Exposure Assessment Modeling, Athens, GA, 1993.

Brooks, Kenneth N., Peter F. Folliott, Hans M. Gregersen, and K. William Easter, *Policies for Sustainable Development: The Role of Watershed Management,* EPAT/MUCIA Policy Brief No. 6, Arlington, VA, August 1994.

Clawson, Marion, *New Deal Planning, The National Resources Planning Board,* Johns Hopkins Press, Baltimore, 1981.

Davies, Tudor, Management Principles for Estuaries, Environmental Protection Agency, unpublished, 1985.

Glaze, William H., Training the Next Generation of Environmental Professionals: Problems at the Academy, *Update, Universities Council on Water Resources,* Winter 1994.

Grigg, Neil S., Water Resources Assessment, USCID/USCOLD Earth Summit Workshop, Washington, DC, June 28, 1993.

Grigg, Neil S., and George H. Fleming, *Water Quality Management in River Basins: U.S. National Experience,* Progress in Water Technology, Vol. 13, International Association of Water Pollution Research, Pergamon Press, London, 1980.

Grimsrud, G. Paul, E. J. Finnemore, and H. J. Owen, *Evaluation of Water Quality Models: A Management Guide for Planners,* EPA 600/5-76-004, July 1976.

Hall, V. W., and Glysson, G.,D., eds., *Monitoring Water in the 1990's: Meeting New Challenges,* American Society for Testing and Materials, Philadelphia, 1991.

Howells, David H., *Quest for Clean Streams in North Carolina: An Historical Account of Stream Pollution Control in North Carolina,* Water Resources Research Institute, Raleigh, NC, November 1990.

Intergovernmental Task Force on Monitoring Water Quality, Interagency Advisory Committee on Water Data, Water Information Coordination Program, *Ambient Water Quality Monitoring in the United States: First Year Review, Evaluation, and Recommendations,* Washington, DC, December 1992.

James, L. Douglas, Dean T. Larson, and Daniel H. Hoggan, National Water Assessment: Needed or Not, *Water Resources Bulletin,* Vol. 19, No. 4, August 1983.

National Academy of Sciences, Water Science and Technology Board, National Water Quality Monitoring and Assessment, Report of a Colloquium, Washington, DC, May 21–22, 1986.

Novotny, Vladmir, and Gordon Chesters, *Handbook of Nonpoint Pollution,* Van Nostrand Reinhold, New York, 1981.

Saito, Laurel, Water Quality Data Management, Technical Report No. 59, Colorado Water Resources Research Institute, Ft. Collins, July 1992.

Saito, Laurel, Neil S. Grigg, and Robert C. Ward, Water Quality Data Management: A Survey of Current Trends, *Journal of Water Resources Planning and Management, ASCE,* Vol. 120, No. 2, 1994.

Sanders, Thomas G., and Robert C. Ward, How to Design a Water Quality Monitoring System, Colorado State University, unpublished, 1993.

Savage, Roberta H., Clean Water Act Reauthorization: The States' Perspective, in *Update, Universities Council on Water Resources,* Winter 1994.

U.S. Environmental Protection Agency, *Guidance for State Water Monitoring and Wasteload Allocation Programs,* EPA 440/4-85-031, Washington, DC, October 1985.

U.S. General Accounting Office, *National Water Quality Assessment: Geological Survey Faces Formidable Data Management Challenges,* GAO/IMTEC-93-30, Washington, DC, June 1993.

U.S. General Accounting Office, *Water Pollution: EPA Needs to Set Priorities for Water Quality Criteria Issues,* GAO/RCED-94-117, Washington, DC, June 1994.

Water Quality 2000, *A National Water Agenda for the 21st Century,* Alexandria, VA, November 1992.

Wisconsin Department of Natural Resources, Wisconsin Water Quality Assessment Report to Congress, Madison, April 1992.

15

Water Administration: Allocation, Control, Transfers, and Compacts

Introduction

With conflict over water use growing, practical systems of regulating water delivery are needed. These systems must be based on law, and for practical purposes we shall call them "water administration." They are based mostly on state government programs to regulate diversions, permit systems, water accounting, and interstate compact administration. In some cases, federal courts, including the U.S. Supreme Court, administer water systems. Water administration systems have been in existence for a long time in Western states, but some states in humid regions are just beginning to implement them.

Interbasin transfers are a special problem because of the contentious issues involved. They are a way to facilitate water transfers, but some consider it an ecological and social insult to transfer water from basins of origin.

This case study chapter explains water administration as a body of tools and techniques that can be used to regulate water quantity uses. It illustrates the tools and issues with specific references to Eastern and Western situations, and it discusses promising management techniques such as ag-to-city transfers.

With population growth, competition for water increases. This is obvious in the West, where the climate is semiarid, but similar competition is arising in humid areas because industrialization and environmental demands increase competition. As a result, effective and equitable systems for control of water quantity are necessary. There are many issues, and management systems for water allocation, control, and transfers are being worked out incrementally. In most cases, it is not possible simply to let the "laissez faire" doctrine apply—that is, to let nature take its course and people take what they need. Disputes require court cases, negotiation, complex studies, and large expenses.

For the most part, control of water quantity is a matter of law, but it involves a great deal of quasi-legal administrative work which involves managers and engineers. Regulatory systems that arise from federal and state statutes and agency rules form a large part of the control system. While law is central for allocation and control, not all disputes can be handled by courts, and administrative systems and dispute resolution mechanisms are also important.

The beginning point in a system for water administration is a procedure to make and record water allocations. Then, a system to monitor available water is required, followed by daily decision making about who can use water and who cannot. Water administrators are needed to turn on and off diversion structures and to "police" the system. An after-the-fact reporting system records the historical experiences so that future policies can be set. Finally, a dispute resolution mechanism is required. These steps in the process are required for surface and groundwater allocations, for proposals to transfer water from place to place, including from one basin to another, and for interstate issues, which are more complex.

Management systems seek efficiency and efficiency in use of water, and to reduce the conflict of water management. Equity in water ownership and use should be provided by the law. Issues such as providing water to those who need it most, such as during drought, will arise. Also, conflicts arise due to differing perceptions of water needs and requirements, mainly in environmental disputes.

Water Accounting

In Chap. 2, we outlined how hydrology is, to a large extent, water accounting. In tracking issues related to water transfers, controls, and administrative systems, the accounting for water is a crucial topic. Some of the details of systems are outlined in the text by Rice and White, *Engineering Aspects of Water Law* (1987).

Surface Water Administration Systems

It would be convenient if a system of water control could be implemented whereby each person on a stream would obey and take only his or her share. However, people do not act like that, and it is necessary to administer control systems with permits, adjudications, monitoring, and enforcement actions. Even if all water users had perfect intentions, they still would not know how much water was available and whether all legal water users were being satisfied; thus they could not make the right decisions all the time. The need to "police" water users, and the need to evaluate information to determine how much water is available, give rise to the need for a system of water administration.

When water is scarce, a system is essential. When water is more abundant, it may be necessary to regulate only in times of shortage. Varied systems operate around the nation.

In the West, states developed systems to be administered by "state engineers." These state engineers have considerable authority to decide how to allocate water, given the system of water rights and the flows in the stream.

In the East, comparable systems did not develop, but recently, a number of states have implemented permit systems that have the same goals. However, they lack large irrigation users for the most part, and water administration is simpler, but becoming more complex.

Groundwater Administration

Until recent years, there was little administration of groundwater. Groundwater was, for the most part, administered separately from surface water. If a person wanted to drill a well, it was freely allowed, subject to health department rules.

In recent years, states have realized that groundwater systems are not separate from surface water, and they have passed laws and regulations to govern pumping. The general approach to regulating groundwater is similar to that of surface water; that is, law or permits are used to specify where, when, and how much water can be pumped.

Both tributary and nontributary aquifers are covered. A tributary aquifer is clearly connected to a stream; a nontributary aquifer has no apparent connection to the surface water. However, most aquifers have a hydraulic connection somewhere, although we may not know where it is due to lack of hydrogeologic data. In the case of tributary wells, the pumping limits will be specified according to the effects on streams. Nontributary wells can be pumped up to their rate of recharge, or *safe yield,* or if the water involved is fossil water, with low or zero rates of recharge, then how long the aquifer will last must be specified, and that will determine the allowable rate of pumping.

Systems for River Basin Control

Permit systems or administrative systems deal with water users one at a time. For an entire river basin, different approaches are needed. During the twentieth century, systems for basin-wide control have evolved. States which lack management systems do not really have the capability for basin-wide management.

Systems for river basin control vary in design, but water masters, water commissioners, ditch riders, river basin engineers, and other arrangements are common.

Water Transfers and Marketing

General issues

Today, water transfers and water marketing are becoming "hot" topics. Lund and Israel (1995) have published an excellent summary of principles and issues of water transfers.

In the West, where supplies are short and unpredictable, transfers make a lot of sense. The National Research Council (NRC) (1992) has published a study on water transfers in the West, and defined them as "a change in the point of diversion or a change in the type or location of use, ranging from simple internal adjustments to actual sales." The NRC concluded that voluntary water transfers "are the most significant mechanisms available today for responding to the West's changing water needs but that broad 'third party' participation is essential if transfers are to be both efficient and equitable."

Water marketing means simply to sell the right to use water, either through permanent sales or by leasing the right to use water on a temporary basis. Water marketing is common in Colorado, for example, as will be discussed later in the chapter.

It is interesting to compare water and electricity transfers. Electric power can be "wheeled" from place to place over connected transmission lines. Water for the most cannot be moved easily, although in some cases, as in the California water project, it is moved over large distances. Water can be moved on paper, however, through systems of voluntary exchanges.

The U.S. Geological Survey (USGS) devoted an article to voluntary transfers in its 1985 National Water Summary (Wahl and Osterhoudt, 1985). The USGS classified types of transfers as isolated, negotiated transactions; short-term water exchanges to alleviate drought; organized water banks and exchanges; and established water markets. As case studies, they described a 40-year lease of water by a Utah power utility from the Bureau of Reclamation, a purchase by another Utah power utility of pooled irrigation water from several districts, a transaction between the city of Casper, Wyoming, and an irrigation district, negotiations between California's Imperial Irrigation District and the Metropolitan Water District of Southern California (MWD), federal water banking in California during the 1976–1977 drought, and several others.

Water transfers and reallocations have become more important in recent years because of competition for limited supplies, and tight regulatory standards on developing new supplies. As a result, interest in transferring water that has already been developed has increased.

While the legal and political complexities of water transfers and re-allocations have been recognized for a long time in states that follow the appropriation doctrine of water law, states in more humid regions are catching up with these complexities due to competition for water, environmentalism, and related political issues.

The two terms, "transfer" and "reallocation," have similar meanings. Transfers can be from one person to another (transfer of ownership); from one point of use to another; or from one type of use to another; or from one schedule of use to another. Each of these can involve complex engineering and legal determinations.

If a transfer of place of use is from one basin to another, it is called an *interbasin transfer,* and involves more complexities than a transfer within the same basin.

The term "reallocation" differs from the term "transfer" mostly from the standpoint that someone must be doing the allocating and reallocating; in other words, it implies an authority that can allocate, as opposed to a legal system that confers ownership. Following that line of reasoning, it follows that the right to use water or storage space that has been allocated from an authority might be reallocated to a different party. This sounds much simpler than it really is, because there are multiple legal and political issues that must be faced in almost any reallocation issue, such as reallocation of storage in federal reservoirs, which will be discussed later in the chapter.

Transfer of ownership

As explained in Chap. 6, ownership of a right to use water is a creature of a water rights system. Change in ownership can best be seen in the appropriation rights system, but the concept is certain to increase in reallocation issues that will occur as development occurs and changes in the nonappropriation states.

To illustrate the concept, consider the simple case shown in Fig. 15.1, where a farmer with a diversion at stream point C chooses to sell water rights to a city with diversion at point A. The city discharges waste-water downstream of point B, where another farmer takes water. Unless some arrangement is made, the farm takeoff at B will have less water than before because of the change in point of diversion. This

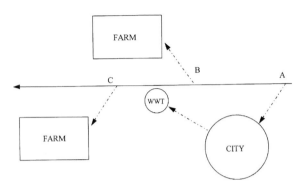

Figure 15.1   Water rights layout.

must be worked out before the transfer takes place. Of course, most transfers of ownership involve far more than this simple situation; the place, time, and type of use are usually all different.

As an example of change in ownership, consider the controversial water rights purchase by the city of Thornton, Colorado, from farmers in northern Colorado. The city of Thornton, a northern suburb of Denver, found in the 1980s that it needed added water supplies for projected growth. After studying its options, it decided to purchase irrigation water from farmers in another basin, and proceeded with a complex deal to buy water and farms in northern Colorado. Working through an agent who would not specify who the client was, Thornton purchased 300 farms and about 60,000 acre-feet of water for eventual transfer to its use. At the time of this writing (1994), legal challenges have been mostly settled, and the project appears to be on its way to implementation. This example should be considered in the light of the Two Forks case (Chap. 10), where the regional aspects of the Denver supply were reviewed.

In the future, change in ownership will occur in permit systems as well. Suppose a user has a 25-year permit to use water in a humid state, but no additional water is available for a new industry that desires to locate nearby. If the original user could be convinced—say, by a reasonable payment—to discontinue a portion of the use so that the new industry can locate, what is wrong with that? These kinds of situations are already arising, but since most states are still developing and refining their permit systems, legal challenges and principles for such transfers remain for the future.

Transfer of point of diversion, schedule, type of use

In the water-short West, water accounting is practiced, and it is common to give attention to the schedule and type of use and points of diversion and return. It is essential that these be known so that equity for all water right owners can be maintained.

If the point of diversion is changed, water that might have been available downstream may not be available, and even the "carriage duty" of the water, or the duty to share stream losses, may be affected.

Type of use affects consumptive use. An irrigation use will consume more than a wintertime urban use, for example. Computing the consumptive use is a key requirement of water administration.

Schedule of use is important because users depend on their water at specific times. If a user who normally has been using more water in the wintertime (say, a snowmaking operation at a ski area) changes its demand schedule to use more in the summer (say, for a new fishing pond with evaporation potential), then downstream users will be affected.

City–ag transfers

One of the methods to resolve water problems in the West is the temporary transfer of water from irrigation uses to urban uses during dry periods. This concept makes a lot of sense; in effect, the farm uses are considered a "reserve" for the cities. There are a few examples, but the rigidity of the appropriation doctrine makes it difficult to implement the systems in a general sense.

Clark (1992) studied this subject for a thesis at Colorado State University and concluded that temporary water transfers make an important water supply alternative in Colorado and California. In California, for example, Clark found that the 1980s drought had stimulated transfer activity. The Yuba County water agency had six one-time transfers totaling 572,000 acre-feet of water that were quite complex and involved numerous parties. Another California example cited by Clark is the Arvin-Edison/MWD exchange. In this plan, up to 200,000 acre-feet of the MWD's water goes to Arvin Edison Water District (A-EWD) for underground storage in normal years, up to a limit of 800,000 acre-feet cumulative. In dry years, MWD takes about 93,000 acre-feet for a cost of $20 million, or about $215 per acre-foot.

In Colorado, the situation is somewhat more complex due to the numerous small water users who transfer water. Temporary transfers are facilitated when they can occur within a large area such as a water district, but between water users, the water court system of Colorado imposes significant transfer costs. A notable exception is the Northern Colorado Water Conservancy District, which administers water from the Colorado Big Thompson Project (CBT). This water, originating on the West Slope, is transbasin diversion water and is generally available to move around more freely than East Slope surface water.

Clark developed a concept to implement temporary transfers in Colorado through what he called the "Water Rights Option Agreement" that is basically a drought water supply plan. The concept still has to be proved out.

Interbasin Transfers

Interbasin transfers of water are popular and logical proposals for those needing water, but fiercely resisted by those losing the water. In interbasin transfers the "emotion" of water can be seen clearly.

Transfers can be large scale, as a proposal to divert Mississippi River water to West Texas, or small scale, as a proposal to divert water from a stream for a small town, and return the wastewater to a different creek, which may ultimately reach the same main stream.

I have encountered interbasin transfer issues in several states, and they are always controversial. In the West, they are generally permitted under water law, as a method to transfer water from point to point. In the East, they are handled on an ad-hoc basis, usually under the rules of the permit system or authority of the state government or courts. Local or federal rules might also become a factor.

There are several examples of interbasin transfer among this book's case studies. In the Colorado River Basin, the CBT project (Chap. 12) features the largest interbasin transfer from the Colorado River to the Mississippi River Basin. This transfer has impacts all the way from the Rocky Mountains to the Gulf of California. The Apalachicola-Chattahoochee-Flint (ACF) system (Chap. 19) has interbasin transfers resulting from Atlanta's water use. This is a common mechanism of interbasin transfer—a city withdraws water supply from one basin and discharges wastewater to another. In North Carolina's Yadkin River Basin (Chap. 19), residents have opposed interbasin transfer and introduced bills into the legislature to prohibit it. This is discussed in more detail later in this chapter. In the Virginia Beach water supply case (this chapter), the city seeks to pipe water from North Carolina's Roanoke River Basin to a coastal city. The diversion would impact Albemarle Sound (see Chap. 22).

In Colorado, the term "basin of origin" has been adopted to refer to interbasin transfer issues. A basin of origin is the home basin for waters, and the legal and political question is if its residents and landowners have rights to continued benefits of the water. In Colorado's San Luis Valley, for example, this was a major issue in a private company's proposal for massive diversions of groundwater from the basin. Basin of origin issues are reviewed in more detail later in the chapter.

Interstate and International Streams: Issues, Coordination, Compacts

When water flows across a state line, it creates a "transboundary issue." Normally, such issues must be handled within the context of river basin management, and when more than one state is involved, the complexity grows. There are several examples of interbasin river basin problems in Chap. 19: the Platte River involving Colorado, Wyoming, and Nebraska; the Delaware River involving New York, Pennsylvania, Delaware, and New Jersey; the Colorado River, which involves seven states and Mexico; and the ACF, which involves Georgia, Alabama, and Florida.

Problems of interstate or international streams involve both water quantity and quality. The most visible issue is water quantity because the most urgent problems are supply related, for both cities and agriculture. For example, an important issue in the Middle East peace talks of 1994 was division of the waters of the Jordan River. Except in some cases, water quality issues are more subtle. In the United States, interstate water quality problems are supposed to be handled by uniform federal stream standards. However, stream standards are set by states, so the potential for problems exists. In international rivers, a few treaties, such as, for example, on the Rhine River, have been signed, but the countries in Europe (East and West) are just beginning to develop uniform water quality standards.

Chapter 19 reviews institutional coordination and enforcement mechanisms for interstate rivers. This chapter's case study for interstate compacts is the Pecos River Basin, where this author has served as River Master.

Pecos River Compact

As shown in Fig. 15.2, the Pecos River heads in the mountains of north-central New Mexico and flows about 900 miles to join the Rio Grande near Langtry, Texas. The history of the Pecos River Compact illustrates a range of issues related to interstate compact negotiation, challenge, and administration.

The Pecos drains about 25,000 mi^2 in New Mexico and about 19,000 mi^2 in Texas. The river can be viewed as having three major reaches: above Alamogordo Dam, between the dam and the New Mexico and Texas state lines, and the reach in Texas. Precipitation in the basin is in the general range of 11 to 14 in annually, and extensive irrigation is practiced, both from surface and groundwater diversions. There were about 210,000 irrigated acres in the two states in 1948, about 75 percent in New Mexico.

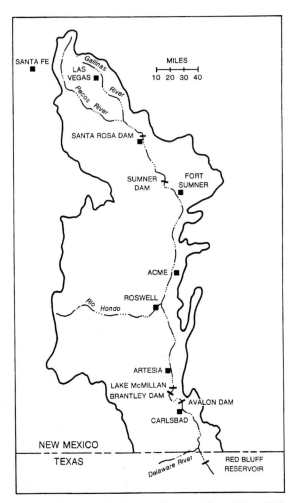

Figure 15.2   Pecos River Basin. (*Source: Adapted from U. S. Geological Survey, 1988.*)

Water development in the basin dates to before the Spanish conquest, as Coronado found irrigation practiced among Indians in the upper reach in 1540. The Ft. Sumner project was initiated in 1863 and rehabilitated by the Bureau of Reclamation in 1906. Irrigation in the Roswell area began from surface waters about 1889, from artesian waters about 1891, and from shallow wells after 1927. Irrigation in the Carlsbad area started about the same time, with the use of McMillan and Avalon Reservoirs (constructed in the 1890s) for storage. Alamogordo Reservoir was constructed in 1937 to replace the deteriorated capacity of these reservoirs.

The controversy about interstate allocation of water dates back to about 1914, when a report by the Bureau of Reclamation reported on the need for a state-line reservoir to regulate Texas's share of the water. Red Bluff reservoir was completed in 1936 as a Public Works Administration project. Details of the controversy can be traced in Supreme Court documents (Breitenstein, 1979).

A Compact Commission was organized in 1923. The compact it drafted failed due to a veto by the New Mexico governor. In 1931 the Texas legislature authorized a suit, but it was not filed. The National Resources Planning Board completed a study called the Pecos River Joint Investigation in October 1942, and this report figured prominently in future compact negotiations (National Resources Planning Board, 1942).

The Pecos River Compact Commission was formed in 1942 and held its first meeting on February 9, 1943. The Compact was ratified by New Mexico on February 9, 1949, by Texas on March 4, 1949, and by the U.S. Congress on June 9, 1949. Two key features of the Compact were the organization of the Pecos River Commission and a statement of New Mexico's obligation under Article III(a) of the Compact.

New Mexico's Article III(a) obligation has been the subject of considerable litigation before two special masters and the Supreme Court. It is stated this way in the Compact: "Except as stated in paragraph (f) of this Article, New Mexico shall not deplete by man's activities the flow of the Pecos River at the New Mexico–Texas state line below an amount that will give Texas a quantity of water equivalent to that available to Texas under the 1947 condition."

The Supreme Court approved the Special Master's elaboration of the 1947 condition, which is stated this way: "The 1947 condition is that situation in the Pecos River Basin which produced in New Mexico the man-made depletions resulting from the stage of development existing at the beginning of the year 1947 and from the augmented Fort Sumner and Carlsbad acreage" (Breitenstein, 1984).

The Engineer Advisory Committee to the Compact Commission recommended and the negotiators accepted an "inflow–outflow" method of apportioning waters. This is quoted by Special Master Breitenstein this way: "The inflow-outflow method involves the determination of the correlation between an index of the inflow to a basin as measured at certain gaging stations and the outflow from the basin" (Breitenstein, 1979).

The Pecos River Commission had difficulty resolving disputes due to the lack of a tie-breaking vote, and this problem led to a Supreme Court case (*Texas v. New Mexico*, Supreme Court No. 65 Original). Over the years there have been numerous hearings and activities to resolve the case. The case was divided into two parts by the Court after a July

1986 report by Special Master Charles J. Meyers (Meyers, 1987). One part concerns remedies for past shortfalls set at 340,100 acre-feet for the period 1950–1983. This was resolved by a financial judgment of about $14 million against New Mexico.

For the future enforcement of the Compact, an Amended Decree was issued by the Court on March 28, 1988. It set forth the schedule for the administration of the Compact and the duties of the River Master. The following are some of the principal features of the Amended Decree:

- The concept of the water year and accounting year
- Adoption of the Pecos River Master's Manual
- Annual calculation by the River Master of a shortfall or overage by New Mexico using the inflow-outflow method described in the Manual
- Procedures for modifying the Manual
- Requirement for New Mexico to submit a plan for making up shortfalls

The first calculation was made under the Amended Decree in accounting year 1988, which covered water year 1987. As of this writing, after seven years of administering the Amended Decree, New Mexico has adopted an apparent policy of maintaining a net overage to avoid having to make up deliveries at the state line. This is, in some ways, a type of water banking operation. The other aspect of administering the Amended Decree has been to determine modifications of the River Master's Manual. Basically, these involve changes to the formulas for water accounting in the reaches. Proposals for changes can generate contention, or in some cases they can be resolved with simple negotiations.

Mediation to Solve Conflicts

Unfortunately, too many disputes over water use end up in court. However, an important goal of water planning and management is to reduce these conflicts and work out solutions. Solving water use conflicts is one of the main needs in collaborative water planning, and one of the principal subjects of this book.

The use of negotiation to arrive at coordinated solutions to problems is a valuable took when properly applied. There are a number of avenues, ranging from one-on-one bargaining to the use of formal mediation processes. In legal parlance, these are referred to as *alternative dispute resolution,* or ADR. Chapter 4 reviews some of the techniques of ADR.

Now, we will turn to several case studies about water allocation and control.

Colorado's System of Water Administration

Colorado's system of water administration stems from its water laws. An interesting aspect of the Colorado system is that it is court-based, whereas most Western states have administrative systems. The difference is that in the court-based system, judges hear the evidence about water transfers, and issue decrees. In an administrative system, the state engineer normally decides on these matters. The decisions are pretty much the same, but who makes them is different.

As I understand it, the decisions by the Colorado Water Courts are actually decisions of the state's court system, although many rulings are based mostly on the work of a "referee" who works for the water court judge. In effect, the rulings are binding as court decisions, having the force of law. In an administrative system, the state engineer's decisions would also have legal implications and go beyond mere "administrative" decisions, and they would be subject to appeal.

Colorado has more activity in water transfers than most states, and some people joke that the state has too many water lawyers and engineers who are employed in studying all of the water transfers.

After the decrees are issued, it is up to the state engineer to administer them on a daily basis. To do this, the Office of the State Engineer, the Division of Water Resources, part of the Department of Natural Resources, operates seven regional offices in water divisions that correspond to the river basins of the state. Each of these is headed by a division engineer who oversees a group of water commissioners who work with local water users to administer the water diversions and records. The water commissioners actually make the on-the-ground decisions about who gets water and who does not.

Surface water rights are prioritized according to the dates of the decree. A water user might have an "1890 direct flow right to 5 cfs," for example. This would mean that the user could divert up to 5 ft^3/sec, in priority, during times when he or she had historically used the water. However, the user could not change the time, place, rate, or schedule of use without going to water court. It is up to the state engineer's forces to determine when the water user can divert.

Water users may be senior or junior users, depending on the level of priority. A junior upstream user may have to forgo diversion so that the water will travel downstream to meet a senior's needs, for example.

Colorado operates on a system of "calls." A call occurs when the division engineer determines that there is enough water in the river to meet rights to a certain date—say, 1910 rights—then all juniors must stop diverting. Determining when a call is necessary is the principal technical challenge facing the division engineer. If the determination is not right, water may be wasted or a downstream user may not get water to which he or she is entitled. (Of course, water users in downstream states are glad for water to be "wasted" in Colorado, because then they get more.)

Groundwater was integrated into the surface water system only in 1969. Now, to pump wells, the user has to be in priority. To provide for users who drilled wells prior to 1969, but at junior dates in the 1930s to 1950s when most wells were drilled, a provision for plans of augmentation was included in the law. A plan of augmentation enables the user to provide replacement water at times when the call is on the river so that the well can continue in operation. Plans of augmentation are complex arrangements. A special organization, Groundwater Appropriators of the South Platte (GASP), was organized specifically to cooperate in plans of augmentation.

Theoretically, Colorado's system looks relatively neat. However, working out problems on the ground, in the face of differing needs and imperfect information, is a real challenge. One experienced water attorney said that there are three ways that one can view the system of administration: the way it is supposed to work, the way people think it works, and the way it really works on the ground. What this means is that there must inevitably be deals, trades, special arrangements, and other compromises that take place in real time among local water users, and these are too complex to follow the theory. The role of the water commissioner is thus to serve as coordinator and facilitator as well as administrator.

In Colorado's system, decisions are made according to users in an entire river basin. Thus, the system recognizes basin-wide priorities. However, the lack of a method to allocate pain and share shortages is an obvious deficiency of this method.

There are many issues that arise as a result of Colorado's system and water laws. One important issue is interbasin transfer.

Basin of origin issue

In Colorado, interbasin transfers are called "basin of origin" or "area of origin" issues. The name seems to have come about from legislation introduced to provide a way to manage proposals for interbasin transfer, which is legal under Colorado's version of the appropriation doctrine.

The unresolved issue is that while interbasin transfer may be legal under the appropriation doctrine, the law itself was developed in a time when economic and social issues were simpler. When there was no economy or society in the "basin of origin," people did not depend on the continued availability of the water to sustain their societies, but now, with whole regions depending on the water, although they do not own it, when water is transferred, it might wipe out lives and family businesses.

On the basis of pure economics, this may seem acceptable, because the "water may flow uphill to money." After all, the small family businesses in the rural area of origin may not be able to compete with financial power of the big city buying the water. Also, if natural systems are depending on the water to sustain the ecology, they have no money to buy the water. In earlier days the appropriation doctrine, in its pure form, said "tough."

Today, attempts to include the "public interest" in decisions is the avenue to provide better decision making on issues such as these. However, defining the public interest is a challenge filled with conflict. See Chap. 24, which discusses the public-interest issue and related issues as they relate to Western water management.

Several examples of basin of origin issues were discussed earlier in this chapter. One, the city of Thornton's purchase of northern Colorado water, involves a transfer from the Poudre River Basin to Denver. Another, the San Luis Valley groundwater case, would result in a transfer of closed-basin groundwater supplies to Denver. Still another, the CBT project, involved transfer of water from the West Slope to the East Slope of Colorado.

North Carolina's System

In contrast to the Western model, states in the East use administrative systems based on permits, information systems, or capacity-use systems. These management systems are created by state law and differ from state to state.

A permit system requires each user of water—say, over 100,000 gal/day, to get a permit. A use of 100,000 gal/day, or 0.1 mgd, would correspond to a small community of 667 people (at 150 gpcd) or a small industry. When water was short, as in a drought, there would need to be a system of curtailing uses so that the pain could be shared. If this system provides for uniform sharing of shortages, then it differs from the appropriation doctrine, which does not allocate shortages but takes users out completely, according to priority.

Information systems on water use are preludes to permit systems, enabling the state to gather data on the users so that when systems become overstressed, arrangements can be made to deal with it.

North Carolina and a few other states have chosen "capacity-use" systems. These provide the state with authority to curtail or regulate water use when a "capacity-use" designation is declared by a state authority such as North Carolina's Environmental Management Commission. The designation can apply to either surface water or groundwater uses.

Experience shows that users will go a long way to avoid a capacity-use designation. They do not like being regulated. Such management systems are controversial, and not easy to implement. Nevertheless, with droughts and water shortages, states in the East are beginning to realize the necessity for such management systems.

Eastern states face the same issues as those in the West, but with less intensity. For example, in North Carolina I recall a proposal by a local government to divert practically all of the water from a small trout stream for a water supply. It was in a rural area, and there was no apparent authority to protect in-stream flows. The only permit authority involved was a Corps of Engineers Section 404 permit. The state agency placed permit conditions for low flows as part of their comments on the Section 404 application, but these were only suggestions and lacked authority to protect the stream. The state agency was ignored, the permit was granted, and it provided the community with a low-cost addition to its water supply, but it seriously impacted a trout stream.

Another issue involved proposals to divert Yadkin River water to cities along the urbanizing crescent of North Carolina. An opposition group has been working in political channels for years to prevent this. This issue illustrates the intensity of opposition to interbasin transfers.

Virginia Beach Water Supply Case Study

The Virginia Beach, Virginia, water supply case study, presented in an earlier book on water planning (Grigg, 1985), is still unfolding. This interesting and important case study illustrates several issues that affect water supply planning, including interbasin and interstate transfer. I began to follow it in 1977 and at this writing have 18 years of information about it. Most of the following background on its early development is taken from Walker and Bridgemen (1985).

The case study reports an interbasin, interstate water transfer issue that has been going on for over 17 years at the time of this writing. It also describes issues relating to an environmental impact statement (EIS), political issues, metropolitan cooperation, and water conservation.

Virginia Beach was a sleepy beach town of 22,584 in 1940, but after World War II it grew rapidly. It continued to grow faster than projections, and by 1985 it had 324,000 residents. By 1994, Virginia Beach's population was up to 410,000.

Virginia Beach's water supply situation has been precarious for years; the town relied on Norfolk for raw water and on desalting and wells for supply; thus, it lacked a large-scale, dependable supply at the beginning of the study period. The contract with Norfolk was quite unfavorable to Virginia Beach. Going back to 1923, Virginia Beach contracted with Norfolk to build transmission lines in Virginia Beach and to supply surplus water. Virginia Beach later purchased the distribution system, to take title in 1993, but it was entitled only to "surplus water" from Norfolk, and was not entitled to introduce additional water into the city. In 1985 Virginia Beach had about 30 mgd available from Norfolk, but was using only about 25 mgd. Droughts in 1977 and 1980–1981 made Virginia Beach aware of its water supply vulnerability. As Norfolk's supply is mostly unregulated surface water, the security of its supplies is less than desirable, and Norfolk's efforts to drill wells was resisted by other local governments.

The need for wells was resolved by agreements with local governments, but the imposed conditions did not provide much security or water supply independence for Virginia Beach. These conditions compelled Virginia Beach to continue to seek its own independent water supply.

Virginia Beach issued a position paper in 1981 to spell out its alternatives (Virginia Beach, 1981). It listed 24 projects ranging from an interbasin transfer of water to desalting plants. At that time, it identified a diversion from the Appomattox River as the preferred alternative. Virginia Beach was looking at "go it alone" alternatives, and did not get into institutional alternatives for cooperation. Cox and Shabman (1983) identified some of them, such as demand reduction, regional interconnections, groundwater, and combined strategies.

After a period of analysis, the alternatives changed, and were reduced to four: Lake Genito, Assamoosick Reservoir, Lake Gaston, and an in-town alternative. Virginia Beach developed a plan for a Lake Gaston interbasin diversion from Lake Gaston in the Roanoke River Basin (Fig. 15.3), and decided to implement it in 1983. This set into play a set of opposing forces and regulatory mechanisms.

The Lake Gaston alternative would take the water from an existing reservoir owned by Virginia Power and transport it via a 60-in pipeline for about 85 miles to Norfolk's raw water system, located in the city of Suffolk. The use of the water would increase from a planned quantity of 10 mgd in the early 1990s to 57 mgd by 2030. The line would also supply 10 mgd for Chesapeake and one mgd each for Frankin and Isle of Wight County by 2030.

In the beginning, the projected cost was $176 million for 60 mgd of raw water. That translates into about 67,300 acre-feet per year, or a capital cost of about $2615 per acre-foot to develop the raw water.

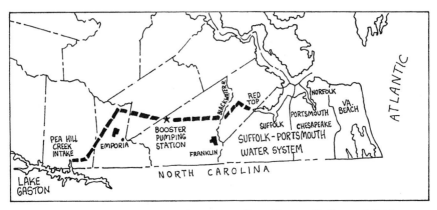

Figure 15.3 Virginia Beach's proposed water pipeline. (*Source: Walker and Bridgemen, 1985.*)

Virginia Beach needed the cooperation of North Carolina to gain permission for the pipeline. The two states had organized a bi-state committee to discuss alternative solutions to border problems involving river water quality, groundwater withdrawals, and interbasin transfer of water supplies. The committee originally met in 1974–1976, and was reorganized in 1978. A number of meetings were held, and progress was apparent, but in 1983, the Helms–Hunt Senate race interfered, and on August 25, 1983, the Secretary of Natural Resources for North Carolina announced that North Carolina would oppose the project. That ended the cooperation between the states.

Virginia Beach applied for a Section 404 permit on July 15, 1983. At the same time, the city decided to contract with the Corps of Engineers for storage rights to 10,200 acre-feet in the Buggs Island Lake pursuant to the Water Supply Act of 1958, thus providing the capability to keep the water level in Lake Gaston constant. This would be a reallocation of storage from hydropower to water supply in a Corps reservoir.

On December 7, 1983, on the basis of their environmental assessment (EA) under NEPA, the Corps issued a Finding of No Significant Impact (FONSI) under both Section 404 and Section 10 of the Rivers and Harbors Act, and they issued the permit on January 9, 1984. This set into motion a series of legal and political challenges. North Carolina began a process to file suit. Virginia Beach started a countersuit process. Also, North Carolina's congressional delegation introduced to bill to halt the project, and southside Virginia counties also entered the fray through a group called the Roanoke River Basin Association (RRBA).

Virginia Beach's legal strategy was to stop all future suits and to gain a declaratory judgment that the riparians on the river had no rights to future use of the waters it would divert. They were apparently trying to halt future suits that would tie them up endlessly. North Carolina's basic suit was to have a full EIS performed. They were joined by the RRBA in the suit.

A federal court ruling of July 7, 1987, remanded the case to the Corps and ordered the Corps to do a reevaluation of the project's effect on striped bass in the Roanoke River. Also, the Corps was to reevaluate the need for the full 60 mgd. In 1987 Virginia Beach was saying that it was still committed to Lake Gaston, but it was also looking toward near-term alternatives such as deep wells and desalting. Also, regional cities and counties were starting to make requests to be included in the project, seeing it as a regional alternative. This introduced the power factor into the equation: Those who control the water control the land use and growth cards.

The Corps replied on December 21, 1988, and reaffirmed the earlier FONSI in both the environmental and the water needs categories. I learned, however, that North Carolina has raised issues that related to the FERC relicensing procedures and of consistency with the Coastal Zone Management Act.

Water consumption had been reduced by a mandatory conservation program to a per-capita consumption of just 78 gpcd, with overall consumption at 30 mgd. The city is still seeking approval of the Lake Gaston pipeline project, but North Carolina continues to present legal challenges. North Carolina is using the Coastal Zone Management Act to seek a review of the project from two federal agencies, the Federal Regulatory Energy Commission and the National Oceanic and Atmospheric Administration. North Carolina's stated concerns include that the project would hurt the state's farmers, affect the striped bass population, diminish hydroelectric energy production, and harm wetlands (Virginia Beach, 1994).

In 1994, the Federal Energy Regulatory Commission announced that it would require an in-depth environmental impact statement before it could rule on the project. Virginia Beach thus faces more legal and technical hurdles before it is able to proceed.

Virginia Beach's quest for Lake Gaston water bears some similarities to the Two Forks case, but involves a different set of actors and issues. The two cartoons in Figs. 15.4 and 15.5 illustrate, perhaps more clearly than words, some of the feelings and attitudes about the issue.

There must be a more rational way of managing the public water resources of the state so that the concerns of both the exporting and importing areas are balanced for the benefit of all

Figure 15.4 Players and conflicts in Virginia Beach water case. (*Source: Walker and Bridgemen, 1985.*)

Figure 15.5 Editorial cartoon, Virginia Beach water dispute. (*Source: Reprinted by permission of the News and Observer of Raleigh, North Carolina.*)

Questions

1. The need for state engineers and water administration programs emerged in the West before the eastern part of the United States. What factors argue that similar programs are now needed in the East?

2. Should groundwater withdrawals be regulated in the same way as surface water withdrawals, or how should they be handled?

3. How does water marketing compare to marketing of electric power? Should water marketing be allowed?

4. How would reallocation of water or transfer of ownership be handled in the states that follow the appropriation or riparian doctrine?

5. Explain how a city–agriculture transfer of water would work and give your opinion whether it is a good concept to meet water needs in urban areas.

6. What is meant by the controversy over "basin of origin" protection in Colorado? Should Colorado have basin of origin protection in its water policy?

7. How should third-party effects be handled in water marketing?

References

Breitenstein, Jean S., Report of Special Master on Obligation of New Mexico to Texas under the Pecos River Compact, Supreme Court of the United States, No. 65 Original, filed October 15, 1979.

Breitenstein, Jean S., Report and Recommendation of Special Master, Supreme Court of the United States, No. 65 Original, filed February 27, 1984.

Clark, John R., Temporary Water Transfers for Urban Water Supply during Drought, Ph.D. dissertation, Colorado State University, April 1992.

Cox, William E., and Leonard A. Shabman, Institutional Issues Affecting Water Supply Development: Illustrations from Southeast Virginia, *Water Resources Research Center Bulletin 138,* Blacksburg, VA, March 1983.

Grigg, Neil S., *Water Resources Planning,* McGraw-Hill, New York, 1985.

Lund, Jay R., and Morris Israel, Water Transfers in Water Resource Systems, *Journal of Water Resources Planning and Management,* Vol. 121, No. 2, March/April 1995.

Meyers, Charles J., Report of Special Master, Supreme Court of the United States, No. 65 Original, filed November 1987.

National Research Council, Committee on Western Water Management, Water Science and Technology Board, *Water Transfers in the West: Efficiency, Equity, and the Environment,* National Academy Press, Washington, DC, 1992.

National Resources Planning Board, *Regional Planning: Part X, the Pecos River Joint Investigation in the Pecos River Basin in New Mexico and Texas,* U.S. Government Printing Office, Washington, DC, 1942.

Rice, Leonard, and Michael D. White, *Engineering Aspects of Water Law,* John Wiley, New York, 1987.

Virginia Beach, *Water Position Paper: A Commitment to the Future,* Virginia Beach, VA, September 1981.

Virginia Beach Looks to Cut Deeper into Water Consumption, *Waterweek* (American Water Works Association), January 31, 1994.

Wahl, Richard W., and Frank H. Osterhoudt, *Voluntary Transfers of Water in the West,* National Water Summary, U.S. Geological Survey, Reston, VA, 1985.

Walker, William R., and Phyllis Bridgemen, Anatomy of a Water Problem: Virginia Beach's Experience Suggests Time for a Change, Virginia Water Resources Center, Special Report No. 18, Virginia Polytechnic Institute and State University, Blacksburg, VA, August 1985.

16

Watersheds and Riverine Systems

Introduction

Watersheds and riverine systems, including wetlands, streams, riparian zones, and aquifers, provide critical support for ecosystems as well as producing and delivering water supplies for human use. Preserving these natural water systems is the key to sustainable development. This chapter presents concepts for linking the natural elements in a water resources system and illustrates how they should be managed. It includes specific cases on watershed management, wetlands, and in-stream flow issues.

Terminology is important to provide a unifying concept for an integrated natural water system—the conduits for the hydrologic system and rules for how the water behaves in them. Doppelt, Scurlock, Frissell, and Karr (1993) used "riverine systems," which they define as "the entire river network, including tributaries, side channels, sloughs, intermittent streams, etc.," and "riparian area," which they define as the stream-side vegetation buffer zone, or the "transition zone between the flowing water and terrestrial ecosystems and . . . a very important part of the riverine-riparian ecosystem." The ecosystems can be captured in another term, "riverine-riparian ecosystems," which was first used, according to Doppelt et al., by the National Research Council in their 1992 report, *Restoration of Aquatic Ecosystems.*

Terms can be confusing, and I believe that the overall concept of the *riverine system* best fits this chapter's perspective. Thus, the chapter's title, "Watersheds and Riverine Systems," is meant to frame the natural water system and underscore the need to manage the natural facilities involved in water management decisions.

Every management action should strive for ecological integrity in watershed and riverine systems management. That is the only way we can hope to achieve sustainable water systems. However, "ecological integrity" may be difficult to define. The term appears in the Clean Water Act reauthorization as an attempt to integrate biological criteria into the water quality program. At Colorado State University, a task force defined ecological integrity as existing when "an ecosystem where interconnected elements of physical habitat, and the surficial processes that create and maintain them, are capable of supporting and maintaining the full range of biota adapted for the region" (Colorado Water Resources Research Institute, 1995).

Riverine systems need a comprehensive approach—sort of a "total river management" program. The same type of goal is stated by Doppelt et al. (1993, p. xxi) as "a comprehensive new approach to the crisis facing America's riverine systems and biodiversity: a strategic national community- and ecosystem-based watershed restoration initiative . . . a coordinated federal initiative."

Figure 16.1 illustrates a conceptual view of a riverine system that appeared in a publication of the U.S. Forest Service (1993).

Figure 16.1 Watershed and riparian area in Alaska. (*Source: U.S. Forest Service, 1993.*)

From the Past to the Future

Programs are evolving with goals of restoring ecosystems and developing sustainability in water management. The increased emphasis on managing water in watersheds is a result of this, and the U.S. Environmental Protection Agency (EPA) established an office of wetlands, oceans, and watersheds in 1991. The first watershed conference in 1992 drew a large crowd. It is a good concept, but the challenge is how to implement it. It would be ideal if the "ecosystems" approach could be taken to water resources management, but that has not been possible due to political problems. In the real world, the tools we use to coordinate and manage natural systems are laws, politics, and administrative mechanisms.

Watershed Management

The term "watershed management" is meant to capture the sum of the actions taken to preserve and maintain watersheds. The watershed, the land area draining to a point on a stream or river (or drainage basin, catchment, or river basin on a larger scale) is nature's production unit for water supplies. According to Brooks et al. (1994), watershed management is the "process of guiding and coordinating use of land and water resources in a watershed."

When watersheds remain in pristine condition, water quality and quantity are normally protected. The only exceptions would be conditions of natural disaster, such as mudslides, fire, avalanche, volcanic eruption, or drought. The watershed is an important source of drinking water. Ideally, a protected watershed can provide water that is pure enough to drink without any additional treatment. We know, however, that this ideal is not often obtained. However, a combination of a protected watershed and aquifer system can provide safe drinking water. Land use threats to watersheds can come directly, from human activities in the watersheds, or through indirect transport by air pollutants. Direct land use activities that cause most of the problems include cropping, pasture and rangeland use, logging, construction, mining, urbanization, roadways and transportation, feedlots and animal husbandry, hydrologic modifications in streams, and landfills.

When watersheds are not protected against poor land use practices, negative impacts can include excessive erosion, higher flood peaks, and water quality deterioration. Some of the land use practices that threaten watersheds include logging, urbanization, recreation, vehicles, farming, mining, cattle grazing, and even wildlife habitat. If the watershed is allowed to deteriorate, then the water must be cleaned up before use. It is important to protect watersheds from poor and harmful land uses before contamination occurs.

Watershed management includes all measures that can be taken to protect, manage, and conserve water and related land resources. Measures that can be used to protect watersheds include land use controls, zoning, monitoring, restoration, and land use treatment.

In a policy brief, Brooks et al. (1994) make several summary points about watershed management: Watershed management provides a means to achieve sustainable land and water management; poor watershed management is a major cause of land and water degradation and rural poverty in the world; the main cause of poor watershed management is lack of good policy to encourage known beneficial practices; watershed boundaries conflict with political boundaries; the main policy challenge is to integrate watershed and political boundaries; and this requires policies that make people responsible for effects outside their sphere (internalize the externalities).

Effects of Land Use Practices

The effects of improper land use practices appear in nonpoint source pollution and in increased or altered flood peaks. Since the inception of the Clean Water Act, studies have sought to quantify the impacts of nonpoint source pollution. Many of these studies resulted from Section 208 of the act.

Agricultural sources are probably the largest category of pollutant by volume. These include sediment, fertilizers, pesticides, and herbicides. Fertilizers add excess nutrients in the forms of nitrogen and phosphorous to streams and lakes. Pesticides and herbicides can contain toxic chemicals and may be lethal to wildlife. Runoff from animal husbandry, especially cattle feeding, can add greatly to nonpoint runoff. Logging is a form of agricultural activity. It also can contribute sediment, nutrients, and chemicals to runoff.

Industrial activities add contaminants in the form of transportation and manufacturing residuals, air pollutants, and solid wastes. Construction adds primarily to the sediment load, but if careless management practices are followed, contamination can also include oil and grease, gasoline, other chemicals, and various industrial products resulting from the building process.

Mining can produce sediments and chemical pollution in the form of acid mine drainage. Acid mine drainage is primarily in the form of sulfuric acid that results from water leaching through sulfur-bearing materials such as pyrite (FeS_2).

Roadways and transportation can produce sediment and chemicals that leak from vehicles or result from the combustion process, such as lead, oil and grease, gasoline, antifreeze, paint residue, particulates, and carbon compounds.

Landfills and solid waste sites of all types contribute to nonpoint source problems through leakage, runoff, and leaching. These can be quite serious and supply a toxic blend of pollutants to streams and aquifers.

Urban runoff can be a source of substantial pollutant loads ranging from sediment to bacteriological contaminants to chemicals and heavy metals. Urban areas are combinations of industrial, farming, and transportation sites. A great deal of study was given in the 1980s to urban runoff.

In general, any hydrologic modification can drain local aquifers, add sediment, and alter the ecology of a watershed.

Evolving Strategies for Watershed Management

Significant changes are taking place in the use of the watershed as a planning unit. However, we do not know if the revolution will succeed. If it does, one might say that it has been a "watershed" event in water resources management (pun fully intended).

The challenge will be to reconcile the watershed and political viewpoints because although the emphasis is to be on the "watershed approach" to water resources management, as shown in other chapters, political problems make the watershed difficult to use as a focal point for decision making. The changes are due to the focus on the watershed as a management unit by water quality advocates, including the possible passage of legislation that will require states to take a watershed view of water quality problems.

At the same time, water policy advocates have called for the introduction of the watershed viewpoint into management decision making. The Long's Peak Group report is an example of that thinking (see later in this chapter and Chap. 9).

Case Studies in Watershed Management

The case studies in watershed management consist of two examples of ongoing studies and activities. The first is how the United States is now considering the watershed approach as an organizing principle for protection of water quality. The second is a study of how states and local governments have studied the protection of water supply catchments with land use restrictions.

Watersheds as management units
in the United States

The first case in watershed management is the move toward water quality management by watersheds in the United States. Proponents of the approach are seeing the watershed as the forcing mechanism to create integration and the systems approach, but it will not be easy. The case enables us to focus on this effort and, over the years ahead, to trace whether progress is made with the approach.

New initiatives would enable states or groups of states to voluntarily designate "watershed management units" and identify impaired waters within them. Then the EPA would approve the designations and require the states to manage the units. The states would submit watershed management plans, and could be eligible for incentives such as use of state revolving funds and use of longer-term discharge permits in the watersheds. There would also be more attention to nonpoint source reductions in the designated watersheds (*ENR,* 1993).

According to Wayland (1993), director of the EPA Office of Wetlands, Oceans and Watersheds, comprehensive watershed management entails recognizing that all resources within watersheds are part of interconnected systems and dependent on ecosystem health; identifying priorities and tailoring solutions to system needs; building partnerships and integrating federal, state, tribal, regional, territorial, local, and private programs within watersheds; and obtaining local commitment to implementing solutions.

North Carolina's water quality management agency has already taken a watershed approach to water quality management (Clark, 1993). The state's Division of Environmental Management (DEM) decided to organize the state into 17 river basins for the purpose of analysis of permitting, monitoring, waste load modeling, nonpoint source assessments, and planning. The intent is to use the concept of the total maximum daily load (TMDL) to evaluate the total waste that a stream can assimilate while still maintaining its standard of quality. Once the TMDL is determined, then point and nonpoint sources can be allocated. This should make possible innovations such as pollutant trading, agency banking, industrial recruitment mapping, and consolidation of wastewater discharges.

To achieve sustainability in water management, the Long's Peak Working Group recommends the use of the watershed as the basic unit of analysis and activity (Long's Peak Working Group, 1992). The group sees ecological integrity and restoration as a needed priority national policy objective. Their recommended principle is "Watersheds should form the basic unit of analysis and activity in order to protect and sustain aquatic biological diversity, including instream, wetland, riparian, and related upland resources. Watershed restoration priorities should, however, reflect the role and importance of these resources as components of larger regional, interstate, or even international ecosystems."

Protection of water supply catchments by state and local governments

This case illustrates how local and state governments can work together to provide watershed management approaches to protect urban water supplies.

Burby and others (1983) propose a comprehensive framework for protecting drinking water supplies that would incorporate an intervention program as the heart of the protective measures. The intervention program would be carefully planned on the basis of the need to protect water quality. It would include regulatory measures such as zoning, subdivision regulations, erosion, sedimentation and stormwater controls, regulation of septic tanks and on-site waste disposal systems, regulation of solid waste management, regulation of wastewater discharges, regulation of impoundment surfaces and shorelines, acquisition of property rights, and various techniques of incentives and education about the importance of watershed protection (Burby et al., 1983).

The American Water Works Research Foundation (AWWARF) sees watershed protection as an important element in protection of drinking water quality. In 1993 it approved a new research project to study watershed planning practices and impacts because it saw land use controls and land management practices as two of the most effective means for controlling water quality in watersheds. Some of the control and management programs it reviewed included the evolving national water policy, the surface water treatment rule of the Safe Drinking Water Act, the Clean Water Act and amendments, national wetlands policy, Forest Service guidelines, the Soil Conservation Service watershed protection program, and Bureau of Land Management grazing policies. A previous AWWARF project reviewed a number of utilities and published a report about effective watershed management for surface water supplies. This report includes brief case studies of the utilities' experiences.

While these measures are not official state policy in each state, they are used by some local and state governments to protect water supply catchments. In particular, the state of North Carolina, and the local governments in the Research Triangle area around Raleigh-Durham and Chapel Hill, have given attention to watershed protection. As with any land use measures, there has been a lot of controversy and battles are fought one by one, but the general ethic of watershed protection is present.

Some of the elements of watershed management that are used to protect water supply sources in North Carolina and other states include state rules that septic tanks located in water supply watersheds must be placed on lots larger than a minimum size, say 40,000 ft^2; subdivision regulations; zoning regulations; sediment control programs; and acquisition of watershed lands. Each of these techniques and others are in use in North Carolina and in other states. Burby et al. provide a detailed listing of measures and frequency of use by state and local governments in a number of sampled areas.

Riverine Systems

Riverine systems are literally rivers of life. In this section, we shall delineate their parts and describe the management issues briefly.

The riverine system, or the entire river network, includes tributaries, side channels and flood plains, associated wetlands and riparian zones, and tributary aquifers. Floodplains and associated wetlands literally teem with life and ecological activity.

In most systems one can see upland tributary watersheds that drain both surface and groundwater into the system. Along the stream corridor there will be riparian zones and tributary aquifers. There may be wetlands and lakes, sometimes created geologically.

The study of natural water systems falls into the fields of geology, geomorphology, aquatic ecology, limnology, and, of course, civil engineering. At Colorado State University, several courses attempt to pull together the various fields: courses in river mechanics, fluvial geomorphology, and aquatic ecology. No single discipline has a complete view of the overall picture.

Maintaining flows in streams within watersheds that have adequate quantity and quality is a challenge. Stream water quality is a matter of standards, and is discussed in Chap. 14. In-stream water quantity has been rather neglected until recent years, but quantity is related to both water quality and to the health of ecosystems.

The riparian zone along streams is the focal point of the water system–ecosystem interface. The riparian corridor is the strip of surface water, groundwater, and floodplain wetlands that sustains the aquatic ecosystem. Maintaining it in healthy condition is obviously critical to the functioning of natural systems. The overall management of the riparian zone is a rather new area of concern that has been approached in a fragmented way. In the 1960s the "blue-green" concept for planning stormwater systems was developed. Flood control planners have recognized for a long time the natural and recreational values associated with floodplains. For example, in Raleigh, North Carolina, the Raleigh Greenway Commission relied heavily on stream corridors for nature trails, which were very attractive. As discovered by Denver's Urban Drainage and Flood Control District, all channel bottoms qualify as wetlands (DeGroot and Tucker, 1985). This concept is giving way today to new thinking about how to design multiple-purpose drainage corridors to reconcile maintenance practices with protection of natural areas. The term adopted to identify the issue is "wetland bottom channels."

Management issues of riverine systems include a wide range of water quantity and quality issues, but many of them can be included in the "in-stream flow" concept.

Case on in-stream flow scheduling

In-stream flow is an emerging issue of importance in both arid and humid regions, but with greater urgency in dry regions where many streams may dry up completely. The basic issue is how to maintain an adequate flow in the stream at all times for all intended uses and to water the natural systems. An important technical issue is how to determine the *needs* for water in the stream. One way to begin is to picture a scene of a healthy aquatic habitat, and imagine as well the fish life cycle in the stream, somewhat as shown in Fig. 16.2.

The in-stream needs for water are the sum of the individual needs to meet the water purposes, with full consideration of the carriage needs. The concept of carriage needs is meant to take into account that water flowing in a stream may meet several purposes. For example, if x cfs are needed for fish and wildlife, that same x cfs might be on its way to a diversion point downstream to be used for another purpose, provided that fish and wildlife needs are still met further on. Another example would have an increment of flow provided to meet stream losses to deliver water to meet a certain withdrawal demand.

The problem of determining needed in-stream flows has come to focus on what is needed to sustain populations of fish and wildlife.

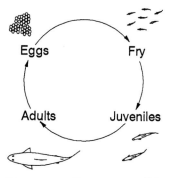

Figure 16.2 Brown trout life cycle. (*Source: Waddle, 1992.*)

For example, the flows needed to sustain a trout fishery can be estimated for a given stream reach. However, the exact flows needed at different times cannot be known with accuracy: Only estimates can be made. The issue boils down to the opinions of experts.

The U.S. Bureau of Reclamation has described the many ways that water projects and water management can affect in-stream flows (U.S. Bureau of Reclamation, 1992):

> Flow releases are usually designed to support a single species or life stage. Instream flows may be implemented to flush the system each spring or fall, to aid up- and downstream migration and access to spawning sites, to provide attraction flows for fish using fish ladders and assist with fish passage, to provide good spawning, incubation and rearing habitat, to provide feeding and resting habitat, to improve water quality parameters, to supply fish hatcheries constructed as mitigation, and other fish related applications.

Efforts to establish the quantity of flow needed have been made by the U.S. Fish and Wildlife's National Ecology Center, located in Ft. Collins, Colorado. Waddle (1992) has analyzed the problem in a Ph.D. dissertation aimed at showing how reservoir water management can optimize water available for species.

Waddle shows how several Western states have enacted laws to protect in-stream flows, and that: "The general purpose of all in-stream flow programs is to set aside water in selected streams, unavailable for consumptive appropriation below a specified level, for the protection of instream values." While there is no federal in-stream flow law, other federal laws can be used to protect in-stream flows, including permit constraints, the Endangered Species Act, and other environmental laws (see Chap. 6 for more discussion of these laws).

In most Western states, the in-stream flows that occur in stream reaches are there to satisfy downstream water rights, and benefit species by accident, not by design. This is changing now with the enactment of in-stream flow laws and programs meant to protect specific species.

Methods for determining in-stream flows are still under development. Waddle illustrates the range of methods and how they could be used to find the best water management scenarios.

The Platte River case study (see Chap. 19) can be used to illustrate in-stream flows and riparian areas. The following discussion is taken from the Bureau of Reclamation's in-stream flow manual, in which the Platte is used to illustrate the stream corridor problem.

> The Platte River flows through central and eastern Nebraska and has long been a major stopover for migratory birds traveling the Central Flyway. Six endangered species may be found on the Platte at different

times of the year: whooping crane, least tern, piping plover, bald eagle, Eskimo curlew, and peregrine falcon.

Historically, the Platte River is a wide, shallow, braided stream. Since 1865 the Platte width has decreased as much as 79 percent. In some extreme reaches, the width has decreased from 4,000 feet to 500 feet. These dramatic changes are due to a combination of reduced discharges, decreased sediment loads, and the establishment of woody riparian vegetation. In some places the river morphology is changing from the original braided condition to a meandering channel, which does not provide the open water and wet meadow habitat preferred by many of the endangered species.

For more than 20 years the Platte River has been the subject of a wide variety of scientific studies addressing the hydrology of the river system and how it affects the channel morphology. Presently, studies are being coordinated by the Platte River Management Joint Study which is directed by representatives of Federal, State, local, and environmental resource agencies. The goal of these studies is to monitor changes in the Platte River morphology and to develop operational schemes which will maximize the habitat benefits within the present developmental and institutional constraints. Reclamation is an active participant in these studies and is developing hydrologic and hydraulic models that will help assess the potential of alternative flow scenarios.

In this case maintaining desirable river channel characteristics will almost certainly be a combination of flow management and mechanical vegetation management. The Platte River Whooping Crane Trust is an active leader in the vegetation management research ongoing at the trust site along the Big Bend Reach of the river. Others are attempting to develop ways to predict the relationship of sediment transport and vegetation encroachment to channel morphology (U.S. Bureau of Reclamation, 1992).

A number of issues are being pursued to improve riparian conditions along the Platte River. The riparian conditions include practically all aspects of hydrology and ecology. Steady flow in the stream is necessary to sustain forage fishery that is used as food for birds and related wildlife. Flushing flows are seen as necessary to clear out seedlings to maintain adequate channel width to provide security for species such as whooping cranes. Timing of flows is an issue for bird nesting. In other words, hydrologic conditions are the integrating factor in the ecology of the riparian zone. In the stream habitat itself, there is concern about species preservation. As I was preparing this in 1995, the *Denver Post* (Obmascik, 1995) reported new concern about maintenance of rare fish species such as the suckermouth minnow, the brassy minnow, the stoneroller, common shiner, northern redbelly dace, lack chub, stonecat, orange-spotted sunfish, plains minnow, and plains top-minnow. The reason for the problems is water quality—city sewage, fertilizers, irrigation runoff, and other sources of nutrients.

Wetlands

Wetlands, or the parts of natural water systems that mostly remain wet, have many valuable functions, but managing them is a fractious problem in the United States because it pits developers and farmers against environmentalists and regulators.

Definition and scope of wetlands

That wetlands are part of the riverine environment is shown by the definition: "Those areas that are inundated or saturated by surface or ground water at a frequency and duration sufficient to support, and that under normal circumstances do support, a prevalence of vegetation typically adapted for life in saturated soil conditions. Wetlands generally include swamps, marshes, bogs and similar areas" (U.S. Army Corps of Engineers, undated). Also, wetlands include "sloughs, potholes, wet meadows, river overflows, mud flats and natural ponds (U.S. Environmental Protection Agency, 1978).

Figure 16.3 is a depiction of wetlands by the EPA.

Freshwater wetlands account for over 90 percent of U.S. wetlands. They may be sustained by rainfall, springs, or floodplain flow. Freshwater marshes are characterized by diverse kinds of grasses, whereas swamps are often dry in summer and may be characterized by woody plants including trees. Saltwater wetlands, although comprising less than 10 percent of U.S. wetlands, have received relatively more attention than their freshwater cousins because their alteration has been more visibly damaging.

In the presettlement United States, wetlands were abundant, but now the total area of wetlands has been greatly reduced. Draining

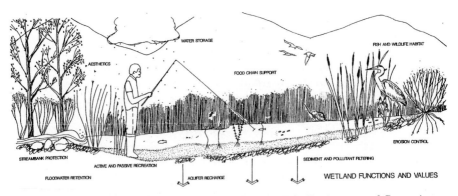

Figure 16.3 Wetland functions and values. (*Source: U.S. Environmental Protection Agency.*)

and filling of wetlands became serious with the 1850s Swamp Lands Act. Dahl (1990), of the U.S. Fish and Wildlife Service, estimated in 1990 that the United States had 221 million acres of wetlands in the lower 48 states during colonial times. By the 1980s we had lost 53 percent of that area, or over 60 acres of wetlands for every hour since colonial times. California lost the greatest percentage of its original wetlands (91 percent), while Florida lost the most acreage (9.3 million acres). Ten states lost 70 percent or more of their original wetland acreage: Arkansas, California, Connecticut, Illinois, Indiana, Iowa, Kentucky, Maryland, Missouri, and Ohio.

Functions and value of wetlands

An overview of wetland functions is shown in Fig. 16.3. Wetlands are feeding, spawning, and nursery grounds for more than half the salt-water finfish and shellfish harvested annually in the United States, and most of the freshwater gamefish as well. They constitute habitat for a third of our resident bird species, more than half the migratory birds, and for many of the plants and animals listed on the federal registry of endangered and threatened species. Wetlands also lock up peat and prevent it from being discharged into the atmosphere.

Wetlands have impressive economic and environmental values, including providing habitat for diverse species of fish, birds, and other wildlife; protecting groundwater supplies; purifying surface water by filtration and natural processes; controlling erosion; providing storage and buffering for flood control; and providing sites for recreation, education, scientific studies, and scenic viewing.

Saltwater marshes and swamps serve as habitat areas for a wide variety of saltwater fishes and coastal wildlife. Mangrove swamps, which cover much of southern Florida, are of great ecological importance because they can tolerate the undiluted salinity of the open sea and their root systems are areas of attachment for large communities of estuarine and marine organisms.

Wetlands face a variety of natural catastrophes, including flood, drought, ice damage, high wind, waves, and fire. They are able to buffer ecological systems from damage due to these catastrophes, although not without stress. For example, when Hurricane Andrew hit South Florida in 1992, the wildlife community of the Everglades suffered significantly.

Regulation of wetlands

Wetlands have been under attack for many years, and it is only in the past few years that protection has been provided through regulatory programs.

Wetland protection programs, evolving since the early 1970s, have been controversial, and the intensity of conflict seems to have grown with the years. As a land use issue, wetlands regulation involves a number of interests, and a number of federal agencies are involved in regulating wetlands, including the Corps of Engineers, the Fish and Wildlife Service, the Soil Conservation Service, and the Environmental Protection Agency. Also, more than 20 states have wetlands protection programs. The EPA is basically a regulatory agency, but wetlands sometimes casts the other agencies into unfamiliar roles—especially the Corps of Engineers, because it is responsible for administering Section 404 of the Clean Water Act, which deals with wetlands, whereas the Corps is primarily a water development agency.

The basic rule for regulating wetlands is the *Federal Manual for Identifying and Delineating Jurisdictional Wetlands* (1989), which defines the hydrologic, vegetation, and soil characteristics that qualify a site as a wetland subject to federal regulatory protection.

Conflict over wetlands centers around definitions. A 1991 Administration-proposed change would have removed from protection around half of the remaining 103 million acres of wetlands in the United States. As one example of complaints, environmentalists in Colorado objected that willow wetland areas near Rocky Mountain trout streams would no longer be protected. These willow areas trap sediment and provide food for insects, which in turn are important food sources for the fish.

The wetlands regulation game might go something like this. A farmer decides to drain some land for cropping, or a developer tries to get more usable area out of a parcel of land by draining part of it. A complaint is filed by someone, and the Corps comes out to inspect. The inspector finds that the farmer or developer failed to apply for a "404 permit." This seems incongruous to the farmer or developer, because his actions seem to have nothing to do with clean water. Conflicts continue, and eventually the matter is resolved, but with great difficulty. This is the front line of the battle for sustainable development.

One way to let the developer have what he or she wants is to require a mitigation plan. This is a plan whereby the developer agrees to create a certain amount of wetlands in exchange for some that is destroyed. Many plans include such mitigation plans, and the extent of the required mitigation is a negotiable item.

Case study on wetlands

On a continental scale, the wetlands issue is illustrated by the fates of Florida's Everglades and South America's Pantanal, two subjects of case study at the 1993 Interamerican Dialogue on Water Management in Miami (Wade et al., 1993). The two giant wetlands and their associated ecosystems were selected for study because they both illustrate current dynamics of fragile and important ecosystems that needed careful management in a time of rapid development.

Both areas involve large, internationally significant freshwater wetlands. The Pantanal is several times larger than the Everglades, has more elevation difference, and receives inflow from several tributaries. The Everglades receives inflow from a single river as well as sheetflow. Its ecosystem is known as the Kissimmee/Lake Okeechobee/Everglades (KLOE) ecosystem. The Pantanal supports a greater diversity and abundance of wildlife. The Everglades were partially drained beginning in the 1880s with drastic effects on the ecosystem. The Pantanal has not yet been affected by large-scale development, but a planned waterway, the Paraguay-Paraná Waterway, could affect it heavily.

The familiar problems of coordination and political will affect both ecosystems. While the KLOE system now has partial protection through the Everglades National Park, other areas are not protected. However, in 1993, a $465 million government–industry plan was announced that would distribute most of the cost to sugar growers, state government, and the South Florida Water Management District via a property tax (Gutfeld, 1993). The Brazilian economic crisis, together with a general ambivalence toward environmental regulation, has hurt efforts to protect the Pantanal.

The scales of the Everglades and the Panatanal problems illustrate the strategic importance of programs to protect wetlands. Local wetlands skirmishes can be thought of as parts of larger systems like them, and also illustrate why comprehensive water programs must include wetlands protection.

Questions

1. Several terms are used in this chapter to describe natural water systems. These include riverine systems, riparian areas, and riverine-riparian ecosystems. Can you think of a better term to describe the chain of features in the natural water system?

2. Explain the link between land use regulation and maintaining natural water systems.

3. The program area of floodplain management has become increasingly linked to wetlands and ecosystems. Who should take overall responsibility for this program area? The federal government? State government? Local government? Explain.

4. Managing and preserving wetlands is obviously a complex enterprise. What management framework would you suggest to reconcile the conflicting goals and views of wetland management? Should it be purely regulatory? Voluntary? Who should enforce it?

5. How does "watershed management," as a field of work and an organizational framework, compare to "river basin management"? What are the compelling reasons to adopt it, and what are the obstacles?

6. What responsibility should be taken by water supply agencies for protecting the natural water features within their jurisdictions? Who should pay for the protection, and how should charges be levied?

References

Brooks, Kenneth N., Peter F. Folliott, Hans M. Gregersen, and K. William Easter, Policies for Sustainable Development: The Role of Watershed Management, *EPAT/MUCIA Policy Brief No. 6*, Arlington, VA, August 1994.

Burby, Raymond J., Edward J. Kaiser, Todd L. Miller, and David H. Moreau, *Drinking Water Supplies, Protection through Watershed Management*, Ann Arbor Science Publishers, Ann Arbor, MI, 1983.

Clark, Alan R., The Big Picture, *APWA Reporter*, February 1993.

Colorado Water Resources Research Institute, Ecological Integrity, *Colorado Water*, June 1995.

Dahl, T. E., *Wetlands Losses in the United States, 1780's to 1980's*, U.S. Department of the Interior, Fish and Wildlife Service, Washington, DC, 1990.

DeGroot, Bill, and Scott Tucker, Wetland Bottom Channels—An Emerging Issue, *Flood Hazard News*, December 1985.

Doppelt, Bob, Mary Scurlock, Chris Frissell, and James Karr, *Entering the Watershed: A New Approach to Save America's River Ecosystems*, Island Press, Washington, DC, 1993.

ENR, A "Whole" Lot of Planning Going On, September 20, 1993.

Gutfeld, Rose, Agreement is Reached on Framework of $465 Million Plan to Save Everglades, *The Wall Street Journal*, July 14, 1993.

Long's Peak Working Group, *America's Waters: A New Era of Sustainability, Report of the Long's Peak Working Group on National Water Policy*, Natural Resources Law Center, University of Colorado, Boulder, December 1992.

National Research Council, *Restoration of Aquatic Ecosystems*, National Academy Press, Washington, DC, 1992.

Obmascik, Mark, South Platte's Rare Fish in Trouble, *Denver Post*, March 16, 1995.

U.S. Army Corps of Engineers, *Federal Manual for Identifying and Delineating Jurisdictional Wetlands*, Washington, DC, 1989.

U.S. Army Corps of Engineers, *Recognizing Wetlands*, Washington, DC, undated.

U.S. Bureau of Reclamation, *An Implementation Plan for Instream Flows, Reclamation's Strategic Plan*, Denver, December 1992.

U.S. Environmental Protection Agency, *Our Nation's Wetlands: An Interagency Task Force Report, including Executive Order 11990, Protection of Wetlands*, 1978.

U.S. Forest Service, *Forest Service Fish Habitat and Aquatic Ecosystems Research*, September 1993.

Wade, Jeffry S., John C. Tucker, and Richard G. Hamann, Comparative Analysis of the Florida Everglades and the South American Pantanal, Interamerican Dialogue on Water Management, Miami, October 27–30, 1993.

Waddle, Terry Jay, A Method for Instream Flow Water Management, Ph.D. dissertation, Colorado State University, Summer 1992.

Wayland, Robert H., III, Comprehensive Watershed Management: A View from EPA, in *Water Resources*, The Universities Council on Water Resources, No. 93, Autumn 1993.

17

Water Use Conservation and Efficiency

Introduction

With today's difficulty in finding new water supplies, Ben Franklin's adage, "A penny saved is a penny earned," certainly applies to water conservation. In fact, there is a multiplier effect: Conserved and salvaged water, along with water reuse, can save on treatment costs, on pipe sizes, and in other ways. No wonder water conservation is receiving a lot of attention.

During the last two decades, water conservation has become a key option for augmenting supplies. By the same token, demand management is considered a parallel management measure to supply augmentation.

The goal of this chapter is to outline and illustrate principles as well as advantages and techniques for water conservation and improvements in efficiency. Several policy issues are explored, and case studies illustrate recent discoveries and experiences in attempts to apply urban and agricultural water conservation in practice.

Urban and industrial water conservation are relatively easy to understand, but agricultural water conservation and efficiency is a misunderstood topic. One of the principal goals of the chapter is to clarify the efficiency concepts related to agricultural water use.

Philosophies of Conservation

The concept of sustainable development captures the conservation ethic which has been on center stage at many recent global and national environmental and water conferences.

Most people agree with the basic concepts of the ethic: Don't waste resources. It is in the specifics that we get hung up—what may be waste to you could be a productive use to me. The specifics include the difference between a "requirement," a "need," and a "demand," which are different concepts. As I understand the terms, a *need* is an expressed statement of a perceived requirement, and is not always a minimum requirement. *Requirement* has been adopted by agricultural water management to express the actual water that plants must have for productive growth. A *demand* is a request—what people will use when they are free to choose.

Earlier in this century, "conservation" meant to store water and save it for later productive uses. One study referred to this as "supply side conservation" (Western Governor's Association, 1984). Today, to store water is the opposite of "conservation" to some, as they see stored water to be a vehicle to promote waste. They recognize only "demand side conservation," that is, reducing the use of water.

Basically, water conservation today means to save water—to use as little as possible for a particular productive use. Thus, urban water conservation means to use as little as possible for washing, for flushing toilets, and for other domestic uses. Industrial water conservation means to use as little as possible to produce a product. Agricultural water conservation basically means to apply as little as possible, although as this category is more complex than the others, it requires further elaboration.

Urban and Industrial Water Conservation

Water for urban use is generally provided in the United States by public and private utilities. These include local governments, water districts, and private water companies. They vary greatly in terms of size, ownership, and type of operation. Water consumption, including residential, commercial, industrial, and urban irrigation use, ranges from a low of 40 gallons per capita per day (gpcd) in some small communities to more than 200 gpcd in some Western communities. These figures include some of the industries served by urban providers, but much of industrial water supply in the United States is self-supplied.

Bruvold (1988) analyzed the determinants of urban water use conservation, with an emphasis on the residential component. He posited a model that considered both the individual's knowledge about water use and conservation and the utility's actions to encourage conservation through program variables such as water use restrictions, rate structures, marginal price, and conservation education. Figure 17.1 illustrates the determinants.

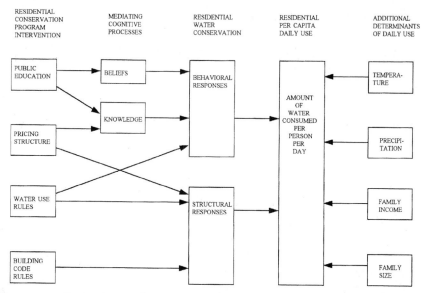

Figure 17.1 Determinants of residential water conservation. (*Source: Bruvold, 1988.*)

After assessing the experiences of a number of California utilities during a drought, Bruvold concluded that a combination of an inclining rate structure and education of consumers about their exact consumption, coupled with the cost, would be effective in curbing consumption.

Water use and price

Water use in the United States varies considerably. Data on actual use can be drawn from the periodic surveys of the American Water Works Association (Grigg, 1988).

For a surveyed population of 104 million in 1984, annual U.S. withdrawals (water utility production) were 176 gpcd. Of these withdrawals, 66 percent were from surface sources, 25 percent were from groundwater, and 9 percent were from purchased sources. There was substantial variation of water production among regions of the country, ranging from 138 gpcd in the northeast to 199 gpcd in the mountain states and heavy use in Florida, 286 gpcd.

U.S. utilities averaged raw water withdrawals of 44.8 million gallons per day (mgd). Average retail deliveries were 33.5 mgd, of which 56 percent were to residential customers ($\frac{5}{8}$–$\frac{3}{4}$ in connections) and 44 percent to other customers (connections 1 in and larger). Average peak deliveries were 61.8 mgd for the maximum day in 1984, and

69.8 mgd for the peak day of the past 10 years. The 1984 value represents a ratio of 1.38 to the average day's production. Average reported peak hour values were 69.8 mgd for 1984 and 85.0 mgd for the last 10 years. The average report of maximum production capacity was 81.0 mgd, a ratio of 1.3 to the average day.

The average values of unaccounted-for water (UAW) in the different regions were as shown in the table below. These data show the general tendency toward higher values of UAW in the older parts of the United States.

EPA region	1	2	3	4	5	6	7	8	9	10
Percent UAW	15.1	14.3	13.7	14.6	13.5	12.7	10.8	9.6	6.4	3.3

UAW also varied according to size of utility. The smallest and largest utilities had the highest losses, about 17 percent, and the medium-sized utilities ranged around 12–13 percent.

Thus the average utility in the survey withdrew about 45 mgd of raw water: 30 mgd from surface sources, 11 mgd from wells, and 4 mgd from purchased sources. It delivered about 33.5 mgd to retail customers and another 5.7 mgd to wholesale customers. Free metered distribution was 0.5 mgd. This leaves another 5.3 mgd or about 12 percent of raw water in the unaccounted-for category.

Responding utilities reported 20.66 million customers. Of the 18.45 million residential customers, 17.14 million (93 percent) are metered. Some 650,000 of the unmetered residential customers are in New York City. Of 2.21 million other customers, 2.16 million (98 percent) are metered and only about 50,000 are unmetered. An average of 95.6 percent of all distributed water is metered. Nationwide, 6.88 million meters were indoors, and 8.07 million outdoors, with 2.57 million having remote-reading capability. The climate accounts for most of the indoor–outdoor distribution variability: Hawaii has all outdoor meters; Minnesota has almost all indoor meters.

Utilities reported that the average residential bill was $143 for a water use of 113,000 gal, indicating an average cost 1000 gal of $1.27. Due to inflation and cost increases, this figure is higher at the time of this writing (1994).

Agricultural Water Conservation

One of the realities of water conservation debates is that agricultural water conservation is more complex and difficult to understand than urban water conservation, but the arguments against agricultural water "waste" are easy to sell.

The Council for Agricultural Science and Technology (CAST), a council of scientific organizations involved in agriculture, commissioned a report on effective use of water in irrigated agriculture, which was published in 1988. This report revealed the disparity in views among irrigation professionals about agricultural water use efficiency (Council for Agricultural Science and Technology, 1988).

The council's task force on irrigation, which prepared the report, noted that it tried to avoid the use of irrigation efficiency terms in the study because of the potential for misunderstanding and confusion. The task force did, however, include an appendix on "irrigation efficiency," and the following are the main points.

Generally, what is known as "irrigation efficiency" is the relationship between the water used in evapotranspiration (ET) by the crop and the total water applied. ET results from both crop water uptake by transpiration and evaporation from the soil surface. The ratio of the water entering the root zone of the soil and becoming available for ET and the total applied is called "application efficiency" or "water application efficiency."

The water that is not stored in the root zone after an irrigation is not necessarily wasted. It might serve other purposes, such as leaching of excess salts from the root zone.

Another set of terms has been developed to represent the volume of water stored in the root zone relative to the volume the root zone is capable of storing. These terms are "water storage efficiency" and "water requirement efficiency." Another set of terms describes distributional efficiency.

"Conveyance efficiency" is used to describe the fraction of water diverted or pumped that reaches the farm to be irrigated.

The task force pointed out that the effectiveness of irrigation cannot be described by a single efficiency term. What is needed are high values of application and storage efficiencies and uniform distribution.

The task force also recognized the need to measure irrigation return flows being used by downstream diverters in a river basin. Marion Jensen, chair of the task force, defined "effective irrigation efficiency" to represent this efficiency concept.

Thus, as the task force pointed out, "low values of irrigation efficiency do not necessarily result in losses of water to further economic use by the public, and that changing irrigation practice to increase efficiency will not necessarily result in substantial savings of water."

Keller and Keller (1995) present an overall summary of agricultural water use efficiency and explain how classical concepts ignore the return flow of water to the system. This makes computations of efficiency difficult to quantify and explain, and is probably the single factor inhibiting public understanding of agricultural water efficiency. To try to explain it better, the Kellers and their associate, David Seckler, introduced the term "water multiplier effect."

The CAST task force made three concluding observations. First, policy debates about agricultural water conservation are often hopelessly confused by lack of agreement about what constitutes conservation and who benefits from it. Second, water savings at the farm level usually do occur as a result of financial incentives, but basin or regional reuse possibilities are ignored. Finally, public policies that attempt to account for regional reuse may be costly to implement and futile.

Case Study: Water Meters in Ft. Collins

Issues surrounding water meters were the focus of a 1980s skirmish over whether to meter Ft. Collins, Colorado, water customers. Without meters it is impossible to use pricing to allocate and ration water.

For a number of months the city water board had debated meters as a method to extend water supply. This arose as a result of a water supply policy study which was presented to the city council on June 23, 1987. The study included meters as one of the policy options for extending the existing water supply (Ft. Collins City Council, 1987). The recommended option presented to the city council was to require meters on all new construction and to retrofit existing housing on a voluntary basis. The water board voted on August 21, 1993, by a 6-to-3 split vote, to recommend this policy to the council. After passing the measure on the first reading, at the second reading on December 1, 1987, the council deadlocked at 3–3 on the proposal, and by its rules the proposal was defeated. The local newspaper editorialized against the vote with a headline that read "Council Inaction on Water Meters Will Haunt City" (*Ft. Collins Coloradoan*, 1987). In 1991, the state of Colorado passed a mandatory water conservation bill that included metering, and Ft. Collins had to comply anyway. Why did the city not go ahead with the metering on its own?

Arguments in favor of metering were that meter-induced water conservation would reduce raw water requirements for the future; meters would reduce peak day demands and reduce need for treatment plant capacity; meters would allow the use of cost-of-service rates for customers; meters would enable the management of system demands, monitoring of system losses, and greater efficiency in the system; metering would place choice in the hands of the customer to decide how much to use and pay for. In addition, the conservation ethic argument arose frequently.

Arguments against meters were the initial cost of installing meters (cost argument); the cost to read and maintain meters (hassle argument); and that the quality of lawns would decline if owners saved money at the expense of lawn irrigation (green lawn argument). Another argument was that meters would discriminate against low-income customers. This argument split the liberal wing of the council, which normally voted as a block for environmental and low-income issues. Another argument against meters was that water waste actually constituted a "reserve" and meters could be installed at any time—in other words, develop all the water possible, then play the meter card last.

Research on the water meter and conservation issues in Ft. Collins shows articles on both sides going back many years. Generally, however, it was only in the 1980s that any serious thought was given to meters. Commercial customers in Ft. Collins had been metered since about the 1920s, but residential customers were assessed without meters on the basis of lot size. Ft. Collins was the only major city along Colorado's Front Range without a metering program. Several others, including Denver, had been in that category until the 1980s (Denver installed its last water meter in November 1993). The first meter-related action was in 1977, when the city council, on the recommendation of the water board, began to require meter yokes on new housing, just in case meters became required in the future.

The city staff had prepared a study in 1980 on the metering issue (Davis, 1988). It showed that the cost to meter 14,210 existing homes would be $3.9 million. The operation and maintenance cost was estimated at $7.40 per year per meter. After a study of nearby cities, the utility staff was estimating in 1987 that a savings of 20 percent of annual water use would result from a metering program. This translated into a savings in water rights needed of $10–12 million, expressed as a present value. This savings in capital cost was the principal argument in favor of metering advanced to the city council by the utility staff.

The anti-meter vote by the council did not mean that Ft. Collins was anticonservation. On the contrary, after embarking on the metering program in response to state legislation, and due to the general community support for water conservation, the Ft. Collins City Council and the water board initiated a "demand management" policy with 12 elements: a leak-detection program; an audit and efficiency effort at city-owned facilities; meters on all city water taps, along with 100 percent assessment of cost of water to city government customers; an aggressive public education program on water conservation; conversion of city landscape irrigation to raw water where feasible; training of city and commercial landscape irrigators on proper methods; institution of a voluntary certification program on efficient use for sprinkler contractors; amending the land Development Guidance System to add points for water-saving actions; setting standards for irrigation system efficiency for developments reviewed by city; implementing central irrigation control for city landscapes where feasible; development of conservation guidelines for city landscapes; and development of a zero-interest loan program for qualified water conservation projects (Clark and Bode, 1993).

Case Study: Agricultural Water Conservation and Efficiency

South Platte River water efficiency

One region where efficiency in irrigation use has been studied is Colorado's South Platte Basin (South Platte Team, 1990).

The first diversion in the basin occurred in 1858, during the Colorado gold rush. By 1859 the first 10 water rights had been established in the basin, and by 1875 ditches were going dry and a new stage of water development, including off-stream storage, had been initiated. By 1889 some 430,000 acres were under irrigation, and by 1909 the figure had reached 1,140,000 acres, nearly full development.

Today, the river is fully developed, with some 5800 water rights "decrees" (about 4500 for direct flow and 1300 for storage). The average annual supply of native streamflow is about 1.4 million acre-feet, joined by about 0.4 million acre-feet of imported water. Storage in the basin by some 370 reservoirs totals about 2.2 million acre-feet. During the 1970 water year, the total surface diversions were nearly 3 million acre-feet.

With a native plus imported supply that totals about 1.8 million acre-feet, and with total water withdrawals at about 4.5 million acre-feet, the river has a "reuse factor" of nearly 3. This factor is to some extent a measure of efficiency. The measure of efficiency, computed as output over input, might relate the water consumed to water supplied, but high efficiencies would rob the environment of in-stream water and downstream neighbors of their share (see Chap. 15 for a discussion of the South Platte Compact). Also, water quality changes with uses. In the final analysis, efficiency must consider all of the necessary uses of water—consumption, environmental, and water quality—and seems almost impossible to compute.

With the general public belief that agriculture "wastes" water, several Colorado politicians have sought to introduce water "salvage" bills. The idea is that if a water right owner salvages water by using less to irrigate, he or she should be able to sell the water, thereby gaining a financial advantage from the savings and making the water available for other uses. It sounds good, but the salvage issue presents several problems. First, what would keep a farmer from keeping the same acreage of crops in production but applying less water, increasing the application efficiency but not really reducing consumptive use? That would rob Peter to pay Paul. What would happen if the farmer increased conveyance efficiency by lining a ditch, but did not reduce consumptive use? Same outcome.

Attorneys pointed out that what was needed was not a salvage bill—the law already existed for salvage. The difficulty was in showing that downstream water right owners would not be injured. The complexity of the hydrology and water rights system inhibited a successful application of water salvage.

At this writing the only real case where water in Colorado is taken from agriculture and given to other uses is when a farm or crop is actually dried up. Even then the matter has to go through an expensive court process to transfer the point and time of use of water.

The Colorado legislature took up agricultural water efficiency in the 1989 session. The Colorado Water Conservation Board investigated what it would take to do basin studies of how efficiently water was used in each basin. One bill introduced was the water use efficiency act of 1989, but it was postponed indefinitely. The issue continues to emerge. In 1992 another bill sought the same goal, this time by proposing to make more specific the procedure that the state engineer had to approve the savings plan and it had to pass the court like any another change in use. This bill also was defeated.

The issue of agricultural water use efficiency is thus seen to be more complex than simply urban water conservation. I suspect that agriculture will continue to be criticized for excessive water use, but the issue is not so much irrigation practice as it is whether irrigation is an appropriate use of water in water-short regions where environmental goals are also sought. Chapter 24 includes a more detailed discussed of the conflicts among agricultural, urban, and environmental uses of water.

Water Pricing for Conservation:
A Task Committee Report

Water pricing can be a powerful tool to foster water use efficiency, in both urban and agricultural settings. To study the issue, the National Water Policy Committee of the American Society of Civil Engineers requested a task committee study on water pricing policy. The committee began its work in 1989 and submitted its report in 1991 (Task Committee, 1991).

Throughout history, water demands have been modest by modern standards, but increasing uses have created scarcities where none existed before. As scarcities increase, there is a need to allocate the resource to the most productive and valuable needs. The practice of "cost recovery" leads to artificial subsidies and misallocations.

Federal government water pricing policies have had a major impact on the development of the western United States and may be important in the East in the future. Some issues that need attention include: the length of long-term contracts; whether historical subsidies should be continued; collection and distribution of power and irrigation revenues; how to charge user fees for recreation; mitigation and public trust; selection of interest rates; how to do basic resource allocation; how to deal with Indian Trust responsibilities; and reallocation of storage. Dealing with these issues raises questions about the roles of federal agencies, ownership, and compensation.

Public and private utility water service pricing refers to water supply purveyors dealing with all uses other than agriculture; these include local governments, water districts, and private water companies. They vary greatly in terms of size, ownership, and type of operation.

Water rates vary greatly across the United States. The water industry is one of the United States' most capital-intensive industries in terms of asset requirements per dollar of revenue. The traditional method of determining price of water is cost recovery, including operation and maintenance, depreciation, taxes or payments in lieu of taxes, and cost of capital. Most rate structures use a declining rate basis as a function of usage, a concept that discourages conservation.

Issues for utilities include rate making, return on investment, replacement funding, paying for water quality improvements, pricing for conservation during drought and others. Policies considered include full-cost pricing, truth in pricing, lifeline rates, universal metering, time and space pricing and delivery, subsidizing conservation devices, and point-of-use storage and treatment.

Clearly, water is an underpriced resource in utilities. However, with the Safe Drinking Water Act and other cost pushes, the price is certain to rise.

Water policy is linked to agricultural policy, and numerous issues are involved, such as whether agriculture is special; what is the U.S. responsibility or interest in meeting the global demand for food and fiber; whether it is necessary to subsidize agriculture to compete globally; whether agriculture is linked to national security and health; whether we should have a cheap food policy; whether double subsidies should be allowed; how to control overproduction; how to deal with agriculture and wetlands issues; and what to do about development of marginal lands.

The committee took the position that prices for agricultural water should be set to recover the cost of providing water on a long-term basis, including capital repayment, operation and maintenance including costs of delivery, rehabilitation and improvement costs, and maintenance of adequate reserve funds. Its rationale was that pur-

veyors of agricultural water often set the price with little or no reserve funds for rehabilitation, modernization, emergencies, or
drought years. Revenues may include water and power sales; flood
control, drainage, and other services; recreation fees; fees for services
such as permits and plan checking; and taxes. Expenses may include
operating charges and costs; flood control, drainage, and other services; engineering and other services provided to farmers; and others.

The committee recognized that pricing should consider the general
well-being of the economy of the local area. Rates should be reviewed
on a recurring basis—say, every five years—and stabilized. Rates
might vary within an agricultural service area to reflect the cost of
service. They should promote water conservation, water quality protection, and wise use of the water resource. Promoting wise use might
involve tiered pricing, incentives or disincentives to pump groundwater, and other water management considerations.

Water transfers can help to improve water use efficiency by moving
water from place of surplus to place of need. The committee recognized that water allocation adjustments are needed between and
among agriculture and municipalities, and that policies should promote or prevent needed changes according to the prevailing public interest. Management, by government agencies, must balance gains
and losses in dealing with winners and losers. Many gains and losses
are economic and can be assessed in terms of willingness to pay.
Management reviews of potential transfers should address economic
impacts on people other than water users; social impacts on rural
areas that lose water; water use efficiency in urban areas that gain
water; issues of stimulating growth in arid regions; environmental
impacts of changing times and places of water use and of transporting
water through river and canal systems.

In dealing with noneconomic issues, the pricing system can be expanded to provide a mechanism whereby multiple gainers pay into a
financial pool that compensates losers and funds programs to mitigate adverse impacts.

Some of the specific issues identified by the committee include establishment, acceptance, and funding of impartial management institutions; establishment of equitable and efficient guidelines and
procedures for evaluating and mitigating third-party effects; removal
of legal and institutional barriers to transfers; providing for re-
versibility of transfers; and providing means for public funding of appropriate social and economic impacts.

At this writing there are many unanswered questions related to the
water pricing issue, and I anticipate that future committees within
the American Society of Civil Engineers will grapple with the same
issues.

Questions

1. Explain the differences between water "requirements," "needs," and "demands."

2. How has the meaning of the term "water conservation" evolved over the twentieth century?

3. What is the approximate per capita water use in the United States? If strict water conservation were applied everywhere, how much could this figure be reduced, in your opinion? What would be the relationship between this figure and industrial water use?

4. It is widely reported that agricultural water use wastes a lot of water and that urban needs in dry areas could be met by reallocating water from agriculture to cities. What is your opinion about this conclusion?

5. Pricing is one tool to promote conservation. What is a "conservation rate" for water use? Do you believe that it would be effective?

6. What are "dual water systems," and how do they relate to conservation? Do you know of any examples? See Chap. 21.

7. In your opinion can direct reuse of water be implemented on a widespread basis? What are the pros and cons?

References

Bruvold, William H., *Municipal Water Conservation,* California Water Resources Center, Berkeley, September 1988.

Clark, Jim, and Dennis Bode, Water Conservation Annual Report, City of Ft. Collins, January 1993.

Council for Agricultural Science and Technology, Effective Use of Water in Irrigated Agriculture, Report 113, Ames, IA, 1988.

Davis, Laura L., Ft. Collins Water Metering Case Study, Paper for Class CE 544, Water Resources Planning, Colorado State University, April 1988.

Ft. Collins City Council, Agenda Item Summary, No. 29, November 3, 1987.

Ft. Collins Coloradoan, Council Inaction on Water Meters Will Haunt City, editorial, December 6, 1987.

Grigg, Neil S., *Water Utility Operating Data, 1984,* American Water Works Association, Denver, 1988.

Keller, Andrew A., and Jack Keller, Effective Efficiency: A Water Use Efficiency Concept for Allocating Freshwater Resources, *USCID Newsletter,* October 1994–January 1995.

South Platte Team, South Platte River System in Colorado: Hydrology, Development and Management Issues, Working Paper, Colorado Water Resources Research Institute, Ft. Collins, January 1990.

Task Committee on Water Pricing Policy, Final Report, to Executive Committee of the Water Resources Planning & Management Division, American Society of Civil Engineers, September 30, 1991.

Western Governors' Association, Water Conservation and Western Water Resource Management, Denver, May 1984.

18

Groundwater Management

Introduction

Groundwater is the source of much of the nation's water supplies, and protecting it against contamination must be a major policy objective. However, the tasks of planning, managing, and coordinating groundwater resources are, in many ways, more complex than similar tasks for surface water. Both groundwater quantity and quality must be managed, but groundwater is more difficult to detect and monitor, and does not lend itself to the same management as surface water.

Groundwater problems range from simple cases of managing individual wells to problems covering vast regions, such as the Ogallala Aquifer, which is described later in the chapter. Alluvial aquifers generally follow river basin lines and can be managed along with surface water, but nontributary aquifers may follow completely different paths. They may not follow political boundaries at all, and create special problems of management.

The goal of this chapter is to illustrate approaches to groundwater policy and strategy, groundwater control programs, and practical problems with groundwater management issues. Three case studies are presented: one on regional groundwater management, one on saltwater intrusion control, and one on the formulation of groundwater policy and strategy.

Nature of Groundwater Problems

Groundwater storage and flow paths make up one of the key elements of natural water systems. The details about groundwater hydrology are beyond the scope of this discussion, but an overview will be presented. The nature and occurrence of underground water and the movement of

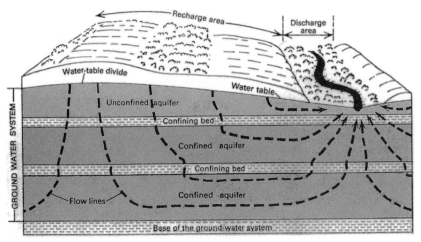

Figure 18.1   Movement of water in groundwater systems. (*Source: Heath, 1984.*)

water through groundwater systems is shown in Fig. 18.1 (Heath, 1984.)

Although groundwater is the source of drinking water for most people, it is often ignored in planning programs and taken for granted. The problem was expressed this way by the U.S. Water Resources Council (1980):

> The role of groundwater in water supply planning often has been slighted in the past, one reason being the belief that groundwater could not be adequately evaluated in terms of availability, chemical quality, and economics, or in conjunctive supply with surface water resources. However, substantial progress in groundwater analytical capability in recent years has made that resource more amenable to rational planning and management operations.

The Water Resources Council listed five attributes of groundwater that make it important to water resources planning: widespread occurrence over much of the nation, general capacity to transmit water over great distances, generally large storage capacity, unusually good chemical and bacteriological quality of the groundwater, and manageability. Most groundwater withdrawals in the United States are for irrigation, largely in California and the Southwest. Groundwater is of enormous importance as a source of drinking water in all parts of the country, as well as in other parts of the world.

Planning problems occurring with groundwater fall into several categories. These include groundwater assessment, development of groundwater supplies, and remediation of groundwater contamination. The case studies in this chapter cover several scenarios. In my own experience with state government, I have encountered the following types of groundwater problems: preparation of county-level information reports, assessment of hazards from injection sources, interstate negotiations over groundwater pumping, regulation of large pumping sources, and assessment of pollution from landfills. These categories of problems illustrate both supply and contamination concerns.

Role of Law in Management

In surface water, the law covers allocation and withdrawals, water quality, and various environmental issues. With groundwater law, these same general issues arise, but groundwater also involves issues of property rights.

According to Bowman (1990), groundwater law has been adjusted to reflect the desire and need to prevent resource depletion and conflict among users. Courts today view groundwater as a shared public resource subject to management and regulation, more than as private property with rights of unlimited use. The private property view was a holdover from the past, when not much was known about the extent or rate of movement of groundwater. In that situation, the property owner would consider water under the property as belonging with the property and not be aware that it flows from place to place. Also, according to Bowman, state legislatures are passing groundwater management statutes to try to settle groundwater disputes and prevent so much conflict in the courts. This is referred to as a "management doctrine" for groundwater, and is meant to become a new type of groundwater law. One specific example of implementation is to designate groundwater management areas for regulation of withdrawals. Bowman made a nationwide survey to study groundwater management area programs and found that 27 states allow for the formation of special management areas to address groundwater quantity issues. Some of these also are concerned with groundwater quality. There is variation in that some states leave the management responsibility with local groundwater users and others give it to a central state agency. States with local management include the heavily irrigated states of Colorado, Kansas, Nebraska, and Texas.

Bowman traced the history of groundwater law in the United States. He stated that American courts and legislatures have experimented for about 100 years. Early procedures viewed groundwater as private property with rights of unlimited use, which was not adequate when the intensity of use increased and conflicts rose, so the shared view of groundwater arose. Evolution from the property-base rules of capture, the rules requiring sharing, gave rise to the concept of the management doctrine, which recognizes interdependence, acknowledges groundwater as a shared public resource, and allows flexibility to regulate withdrawals suitable for a particular aquifer.

Groundwater quantity mechanisms used in states include use permit requirements, water use monitoring and reporting, well spacing requirements, well construction standards, prioritized allocations, restricted usage in times of shortage, and other similar provisions.

Groundwater Strategy

Whipple (1990) developed principles for a state groundwater strategy. According to Whipple, a groundwater strategy would be a policy document that outlined how a state should manage its groundwater, consistent with budgetary limitations and provisions of law. He reported on the strategy developed in New Jersey, which covered both water supply and water quality. This strategy had two directions critical to success: a state program to control nonpoint sources of groundwater pollution and a strategy to correct the lack of coordination among state programs regulating groundwater contamination.

In the western states, groundwater is heavily relied upon for irrigation. Surprisingly, California and Texas have no state-wide groundwater regulation. A professor at the University of California at Los Angeles, John Dracup, was quoted as stating that "groundwater in most of California is still managed like a can of soda with a lot of straws" (Brickson, 1991).

Groundwater management has not been a burning issue in California because problems of groundwater depletion do not coincide with political boundaries and often require regional coordination. Also, they usually do not provoke a crisis and there is a general lack of public knowledge and regionally specific scientific data on groundwater, according to Brickson. Management options which have been mentioned include reducing or eliminating overdraft through groundwater regulation, pump taxes, more groundwater recharge, and other measures.

Case Study: Ogallala Aquifer

The Ogallala Aquifer covers a six-state region from South Dakota to Texas that depends mainly on irrigation for crop production. The aquifer system has three parts: Southern High Plains, Central High Plains, and Northern High Plains (U.S. Geological Survey, 1983). Other regional aquifer systems include California's Central Valley, the Southeastern Carbonate Aquifers, the Atlantic Coastal Plain, and the Basin and Range Lowlands.

Growth in irrigation began in the Ogallala Aquifer in the 1930s, and by the 1950s some 3.5 million acres were irrigated, with annual use being about 7 million acre-feet. By 1980 the acreage had grown to 15 million, and the water use to 21 million acre-feet. Groundwater tables were under severe decline in a number of areas. Congress authorized a study which began in 1978 and was completed in 1982 (Six-State Study, 1982).

The outcome of the study was the development of six alternative management strategies. These were a baseline strategy; a strategy to stimulate voluntary action through research, education, demonstration programs, and incentives; a strategy which combined education with demand reduction through a regulatory program; a strategy to add local water supply augmentation; a strategy which would include intrastate surface water transfers; and a strategy that would involve interstate surface water transfers.

Now, some 12 years after the completion of the study, it appears that no action has been taken. However, local water districts and farmers appear to have taken matters into their own hands, leading to improved water management in the region. Like other large-scale interstate studies, it is apparent that difficulty in taking joint action may overwhelm the parties involved.

Case Study: Saltwater Intrusion

This case study involves freshwater development and saltwater encroachment in the Amsterdam, Netherlands, Dune Water Catchment area. The case study was originally contributed by A. J. Roebert and was published as Case History No. 2 in a UNESCO volume about groundwater problems in coastal areas (Custodio, 1987).

According to the case study, the Dune Water Catchment Area of the Amsterdam Water Supply Board is located along the Dutch North Sea Coast, south of the city of Haarlem. Water withdrawal in this area of some 36 km^2 started about 1853 with open canals which drained the dune area. This enabled exploitation of the upper phreatic aquifer. Later, a large stock of water was discovered deeper in the subsurface. Fresh water is limited to a freshwater lens under the dunes, and since 1903 water has been extracted from the lower subsurface by a system of wells. The rate of extraction of water has increased gradually.

By 1970 the upper aquifer was receiving a total of 52 million cubic meters of artificially recharged river water per year. Otherwise, the natural source of replenishment is precipitation, which amounts to about 13 million cubic meters annually. The system has been overdrawn for more than 25 years, and that is why the upper aquifer is now artificially recharged.

With the annual recharge the capacity has reached 83 million cubic meters. Water extraction from the lower aquifer has been virtually stopped for a number of reasons, one being contamination of wells with saltwater.

Throughout the catchment area, fresh water is found to a depth of about 60 m. The wells are screened between 25 m and 35 m and there is at least 25 m of fresh water between the screens and the underlying saltwater zone. However, a vertical flow accounts for most of the saltwater intrusion in the wells. Saltwater encroachment by 1956 happened after a long period of intensive extraction from the lower aquifer.

Saltwater encroachment in the lower aquifer occurs from two difference sources: the rise of the freshwater/saltwater interface and its extension into a dispersion zone, and local upcoming of brackish water under individual wells.

The Amsterdam Water Supply Board uses the lower aquifer as a reserve for periods when the quality of Rhine River water is particularly poor for artificial recharge. During such periods, good-quality drinking water may be obtained by a process of mixing infiltrated river water with water from the lower aquifer.

The above information is generally extracted from the case study provided by Mr. Roebert, an employee of the City of Amsterdam Water Supply Board. Obviously, the extraction of dune water is an important element in the Amsterdam water supply, and it is hoped that more can be learned about this case in future years.

Case Study: National Groundwater Strategy

In the 1980s the U.S. Environmental Protection Agency began to seek a national groundwater strategy. It began with the Carter administration and continued during the Reagan administration.

At the time, the EPA saw its strategy as structured around four main needs: building and enhancing institutions at the state level; expending controls over major, inadequately addressed sources of contamination, such as leaking storage tanks, surface impoundments, and landfills; issuing guidelines for EPA decisions affecting groundwater protection and cleanup; and strengthening the EPA's organization for groundwater management at the headquarters and regional levels.

The strategy was printed and disseminated in draft form, and there were hearings and regional forum discussions, but in the end it became apparent that adopting a broad strategy on a problem with so many different facets and local conditions was not wise policy.

It is hard to say when a potential policy idea dies, but at this time it seems clear that the EPA has chosen not to pursue the strategy as conceived in the 1980s, and the general elements of the strategy have simply become part of the EPA's overall management approach including legislation and program rules. Action on the groundwater front remains with the states.

Questions

1. How are courts inclined to view groundwater—as a shared public resource subject to management and regulation, or as private property?

2. In the past, some states have tried to stay away from passing groundwater legislation, leaving the issue to the courts. What is the trend today?

3. Name several management tools used by states to regulate groundwater use.

4. What are some of the pro's and con's of groundwater recharge?

5. Saltwater intrusion is sometimes difficult to measure. Why? Can you suggest how a monitoring program should be organized to detect and measure it?

6. As a state, Colorado has integrated its surface and groundwater rights systems. What are some of the technical problems to be expected in such an approach?

7. The Ogallala Aquifer is clearly a shared, multistate problem. What management measures do you suggest to handle it? Should a federal–interstate compact be organized? Should the states be left alone to handle it? Should a laissez-faire approach be followed?

References

Bowman, Jean A., Groundwater Management Areas in the United States, *Journal of Water Resources Planning and Management,* Vol. 116, No. 4, July–August 1990, pp. 485–502.

Brickson, Betty, California's Groundwater Resources after Five Years of Drought, *Western Water,* November–December 1991.

Custodio, E. (ed.), Case History No. 2, Fresh Water Extraction and Salt Water Encroachment in the Amsterdam Dune Water Catchment Area, in *Groundwater Problems in Coastal Areas, A Contribution to the International Hydrologic Program,* UNESCO, Paris, 1987.

Heath, Ralph C., Ground-Water Regions of the United States, U.S. Geological Survey Water-Supply Paper 2242, Washington, DC, 1984.

Six-State High Plains Ogallala Aquifer Regional Resources Study: Summary, High Plains Associates, Boston, July 1982.

U.S. Geological Survey, *National Water Summary,* Washington, DC, 1983.

U.S. Water Resources Council, *Essentials of Ground-Water Hydrology Pertinent to Water-Resources Planning,* Bulletin 16, rev., 1980.

Whipple, William, Jr., and Daniel J. Van Abs, Principles of a Groundwater Strategy, *Journal of Water Resources Planning and Management,* Vol. 116, No. 4, July–August 1960.

19

River Basin Planning and Coordination

Introduction

Today, river basin problems are on the front burner because of policies that decisions should be made on the basis of "eco-regions," which, for water resources, are watersheds and river basins. Chapter 16 discussed this in terms of watershed management. On a larger scale, the watershed becomes the river basin, where water issues are often focused because of interdependencies between land and water uses.

As a result, river basins are the natural accounting units for water management. However, there is a dilemma: How can water managers coordinate decisions in river basins when political decisions are made by cities, districts, states, and federal agencies using jurisdictional boundaries that do not coincide with the river basins?

This fundamental problem, which arises because political units do not coincide with river basins, is a main reason for conflict and difficulty in coordinating water resources management. To deal with it, John Wesley Powell, the explorer of the Colorado River Basin, is reported to have advocated state boundaries in the West that coincided with river basins—so "Colorado" might have gone from the Rockies to the Gulf of California.

Unfortunately, river basins usually lack political unity. Abel Wolman said: "basin approaches come into criticism by some on the score that basins are essentially non-economic or social units. Viewed by themselves, they represent artificial spheres of action irrelevant to societies' needs. The engineer-planner finds them convenient, because he sees them as continuous hydrologic worlds" (Wolman, 1980).

Wolman gave a series of examples of issues on great river systems where he had personal experience: the Danube, Rhine, Great Lakes, Colorado, Rio Grande, Lower Mekong, Niger, Nile, Indus, Senegal, Lake Chad, Tennessee, and Ohio. He concluded that "river planning must be viewed as a continuum of process, and not as a finished and inviolate set of projects." He offered two simple conclusions: "While nature's laws treat the river and its tributaries as the arteries which make a river basin a single living unit, man's laws frequently treat those same water courses as the boundaries of separate sovereignties, as divisions and barriers, not as connecting links"; and "This review of existing schemes and the nature of the administrative problems related to international and regional water resources development should illuminate one basic conclusion—that there is no best organizational arrangement to accomplish the planning, construction and operation of an integrated river basin development. This follows from the fundamental fact that no two rivers are alike, and that the social, economic, and political environments within each river basin impose different demands on organization for development" (Wolman, 1980).

The basic goal of river basin management actions is to achieve the best integrated results for the human and ecological systems in the basin. Problems faced involve both quantity and quality of water. Water withdrawals, consumptive uses, and return flows form the focus of the water quantity issues. Water withdrawals are controlled by water allocation law (see Chaps. 6 and 15), and both the volume and timing of withdrawals require coordination and control. Consumptive use is important because water that is consumed does not return to the stream. Return flows affect physical, chemical, and biological systems. The issues involve conflicts over industrial operations, land use, environmental management, recreation, and urban development, and include planning, construction, and operational issues involving new and existing facilities.

Thus, river basins offer opportunities to apply the principles of integration to water resources management, but they present political challenges. Kenney and Lord (1994) state: "Every major river basin in the United States is either international, interstate, and/or substate; no basin conforms exactly to the contours of a state boundary. As a consequence, water resources administration in the United States has been characterized by multijurisdictional conflicts from the first days of the Republic. In fact, the calling of the Constitutional Convention was, in large part, prompted by concerns over how navigation policies affected interstate commerce."

Another political challenge in river basins is to empower disenfranchised populations in the decision-making process. Robert Hunt, an anthropologist at Brandeis University, observed this at the 1995 convention of the American Society of Civil Engineers (ASCE) Water

Resources Planning and Management Division. Responding to the question, "What is the main diversity issue in water management?" he stated that the critical issue faced by the water industry is empowering minorities and underrepresented groups in river basin management, because if they are brought to the decision-making table, river basin management will improve and a better use of resources will enable economic and social conditions to improve.

Thus, the challenge in river basin management is to achieve the lofty ideals of geographic and ecologic integration in the messy context of river basin politics. Future water managers must work together on this challenge in river basins, in spite of political barriers. Past river basin plans reflected good intentions, but they often stayed on the shelf. Special interests collide over management and regulatory actions in river basins, and solutions require focused political, legal, and financial measures which must be well justified by water managers and their consultants.

Most of the chapter focuses on institutional and management measures in river basins, but techniques for studying water quantity and quality actions are also needed. Systems analysis techniques, such as the models and data management systems described in Chap. 5, are the basic building blocks. Any problem involving the allocation of water needs a river basin model to account for the withdrawals, uses, and returns. The accounting model is the basis for evaluation of impacts and the examination of alternatives. These quantitative techniques are illustrated in this chapter's case studies.

Physical Setting of River Basins

The reason that river basins are natural accounting units is that they are natural geographic units. They are also, to some extent, ecological units because their terrestrial and aquatic communities are interdependent. Ecological systems are more closely related in small basins than in large basins, where several ecological systems may be present.

Figure 19.1 illustrates a river basin formed by the ridge lines of the hilltops defining its contributory watersheds. The figure shows an interesting aspect of how river basins are shown. Compare it to Fig. 3.1, which illustrates a river basin development as seen for the 1950 report of the President's Commission. Figure 19.1 has essentially the same features, located in about the same places, with few changes. Perhaps if a similar figure was developed today it would show more environmental features and relatively fewer structural measures.

Studies of river basin morphology show that rivers begin with steep slopes and flatten as they progress toward the sea. Tributary drainage areas increase and river bed sediment sizes diminish as the

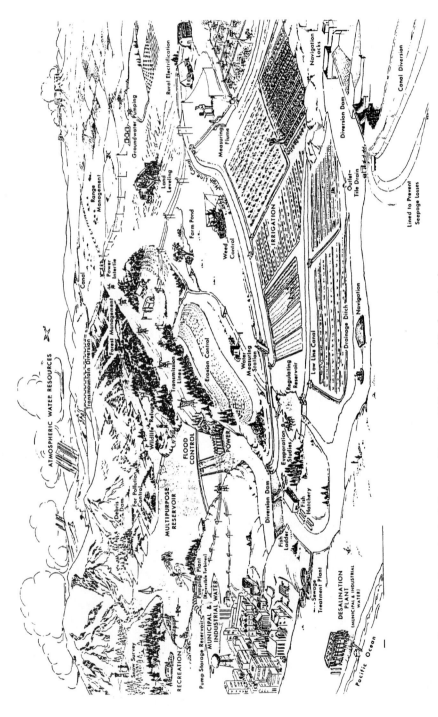

Figure 19.1 Comprehensive watershed diagram. (*Source: U.S. Bureau of Reclamation.*)

rivers flow downstream. Streams can be classified according to *stream order,* and quantitative relationships are available to show these changes (see, for example, Petersen, 1986). Some rivers, such as the Mississippi, begin in alpine regions and flow to warmer zones, as in the Mississippi Delta region, where the stream ecology changes completely. Often, major population centers are located near the estuaries, like New Orleans on the Mississippi.

The earth's major rivers are well known for their beauty and large scale. Think of the incredible diversity of the regions that these rivers traverse—for example, the Nile, flowing 4000 miles across the equator, beginning in mountain regions and ending with a whimper in the Nile Delta as it waters small plots of farmland; or, in contrast, the Amazon, not quite as long as the Nile, but with an annual discharge 61 times as great (National Geographic Society, 1984).

Institutional Setting of River Basins

The institutional settings of river basins cause complexity in the river basin management process. A problem might involve, for example, substate regional issues; or a basin might involve interstate and state-wide issues. A river might also be international in scope as, for example, the Nile, which involves several African countries; or the Mekong, which involves several countries in Southeast Asia.

The main players in river basin management are the government units that the river touches, i.e., cities, counties, special districts, states, and national government units and agencies; water users; and interest groups, including industries, environmental groups, farm groups, and others. Motivating the players to work together is difficult. Special interests may decide to seek their goals by focused, single-interest actions, such as political action or lawsuits. Gaining cooperation is impossible in the absence of a central authority, which is usually missing, or an effective coordinator. This is why much of river basin management focuses on the search for coordination mechanisms.

A coordination mechanism for a river basin is the institutional arrangement to coordinate among agencies and interest groups operating in the basin. The mechanisms vary along with the institutional settings and the physical and economic conditions in the basins. For intrastate basins, arrangements for coordination range from informal networks among water managers to more formal commissions, authorities, and water management districts. River basin authorities and water management districts can also be organized by state statutes. Florida's water management districts are well known, as are Nebraska's natural resources districts, coordinated by the State Department of Natural Resources, and California's water management districts. River basin associations are also in use as volunteer regional management entities.

Coordination mechanisms for interstate streams can involve several different forms. Kenney and Lord (1994) summarize the historical development of these mechanisms, and identify seven different categories:

Interstate compact commission.   An appointed group of people charged to oversee an interstate water compact. The Pecos River Commission, organized by a compact approved by Congress in 1948, is discussed in Chap. 19. It broke down and was unable to prevent a 14-year lawsuit between Texas and New Mexico, followed by a Supreme Court decree.

Federal–interstate compact commissions.   An interstate compact commission that includes formal provisions for federal agency involvement. The Delaware River Commission, discussed later in this chapter, is an example of a federal–interstate compact commission. It allocates water among the big water users in the basin, including New York and Philadelphia, and is thought to be successful.

Interstate council.   An appointed group, usually less formal than an interstate compact commission.

Basin interagency committee.   A committee that usually consists of designated representatives from federal agencies operating in a basin.

Interagency–interstate commission.   A committee, usually legislatively authorized, that includes both state agency and state government representatives.

Federal regional agency.   A federal agency created specifically to operate in a regional area. There is only one example in the United States, the Tennessee Valley Authority.

Single federal administrator.   A single federal official with designated duties. The only one in operation in river basins is the Secretary of the Interior, who has designated authority in the Colorado River Basin.

Reviewing the performance of these different types of arrangements is beyond the scope of this discussion, but a few observations may help. Six river basin commissions (RBCs) were established by the Water Resources Planning Act of 1965 (Fig. 19.2).

After 15 years at $3 million per year, the commissions did not coordinate projects effectively or resolve many difficult water management issues. In reviewing the performance of the RBCs, the U.S. General Accounting Office (1981b) concluded that they had made meaningful

Figure 19.2 U.S. river basins. (*Source: U.S. Geological Survey.*)

contributions that included providing a forum for communications between states and other parties, coordinating river basin studies, and
providing guidance and assistance on other water issues, but in spite
of these benefits, the RBCs "did not become the principal coordinators
of water resources projects as they were intended, that is, to prepare
up-to-date, comprehensive, coordinated, joint water plans and meaningful long-range schedules of water resources priorities."

The General Accounting Office (GAO) noted that while the federal
government was spending some $200 million annually on water re-
sources planning including grants to states and to the RBCs, federal
and state representatives did not use their collective authority to
carry out commission objectives. Much of the water resources planning was done without input from the RBCs. Federal and state agencies did not view the commissions as having authority to establish
joint plans, establish priorities, or coordinate actions. The GAO noted
that success of the RBC concept depends on "cooperation of State and
Federal Members."

U.S. experience shows that well-intentioned, generalized designs
for river basin commissions have not worked well. Why? The basic
reason is lack of incentives for participants to work together. The benefits of the commissions—improved communication, citizen involvement, increased dissemination of information, and state involvement

in federal decision making—have been peripheral to the political process.

River basin management apparently works in some other countries. The Germans have operated successful river basin water quality authorities for a number of years. The French have successful basin-wide planning and financing of water management as a core feature of their national water law. The British organize all water services along river basin lines, and recently privatized the river basin authorities. See Chap. 8 for details.

Evolution of River Basin Planning in the United States

A look at the past will help us to understand what works and what does not in river basin planning. When attention to water projects bloomed in the early twentieth century, the river basin became the planning unit in the United States. The Rivers and Harbors Act of 1927 authorized the U.S. Army Corps of Engineers to study multipurpose river basin development. These studies were known as "308" reports. The New Deal's National Resources Planning Board focused on river basin studies.

During the New Deal, the nation launched its most famous experiment in river basin management: the Tennessee Valley Authority. The TVA was created in 1933 by federal law as an attempt to infuse an underdeveloped region with economic and social development. The TVA's roots were anything but smooth (Lowitt, 1983). In the 1920s, when the concepts for the TVA were planted, two key debates and struggles were under way in the nation. The first was a struggle over public versus private power. The second debate was about water development itself. Superimposed over these debates was the concept of "regional planning," an infatuation of President Franklin Roosevelt.

It was the demise of federal leadership in water that caused the difficulty in river basin planning, and the demise of the Water Resources Council programs ended the experiments of the Water Resources Planning Act that financed planning studies and much of the coordination. In the U.S. experiment with Level B river basin studies, the Water Resources Council's consultants concluded that river basin studies had been useful, but:

> Water resources planning, including the Level B study programs, operates alongside the political process by which local interests, Congressmen, and federal agencies make project-by-project water resource investment decisions. Level B studies are part of a continuing effort to rationalize that process by contributing to regional or basin-wide planning and decision making, particularly with regard to coordinating federal program activities within hydrologic regions or subregions (Field, 1981).

Roles in River Basin Management

The evolution of river basin management in the United States illustrates the importance of clearly defined roles. In fact, allocating roles and responsibilities is a central feature of the evolution of federalism in the United States—that is, the allocation of responsibilities among the federal, state, and local governments.

In the past, a major benefit of river basin planning was the coordination of federal program activities, but now that the federal presence in water resources *development* has diminished, a new allocation of roles is needed, one that focuses on coordination mechanisms.

Fundamentally, roles in river basins are the same as in other venues of water management: providing services, regulating, planning and coordinating, and providing support. The role of service providers in a river basin is to look after the interests of customers. The problem is in defining who are customers. Is the environment a customer? Are downstream interests also customers?

Realistically, service providers mainly serve their direct customers. They might try to consider the needs of others as well, but when supplies become tight, money becomes an issue, or politics take over, this view, along with voluntary cooperation, may break down, and regulatory measures are needed. This regulatory role is necessary because there is no natural regulating mechanism in place.

Thus, the two principal roles in river basin management are service provision and regulation. The role of planners and coordinators is to try to facilitate the service provision and to minimize the regulation, considering all customers and all socioeconomic and environmental objectives.

By working together to coordinate river actions, service providers can benefit each other. Regulators can participate in coordinated planning, hoping to achieve the desired results without imposing the final measure, regulation. Planning and coordination seek to avoid using sledgehammers to swat flies.

Water Administration in River Basins

Part of the regulatory mission is to control and administer the water uses. This function is described in the discussion of the Colorado system in Chap. 15. To illustrate, in Colorado's court-based system of water allocation, judges issue decrees about water adjudication, but the state engineer makes the decisions about how to manage the decrees on a daily basis. The state engineer operates seven river basin offices, each of which is headed by a division engineer who oversees a group of water commissioners who administer the water diversions and records. The system does not always work as neatly as it seems to on paper, and the local water commissioner is coordinator as well as administrator. Thus, this regulatory function takes on a lot of coordination at the grass-roots level where the water commissioner is actually the "river master."

In the case of the Pecos River (see Chap. 15), the state engineer's office of New Mexico employs a "water master" who oversees, reports, and controls. The interstate river master, appointed by the U.S. Supreme Court, carries out the same general functions, but is limited to overseeing the delivery of water at the state line.

It turns out that the functions of river master or water master, i.e., implementing limits, coordinating, and working out disputes on the ground, are critical regulatory and coordination roles in river basins, especially where water supplies are limited.

Case Studies

Scenarios for river basin management should be approached on a case-by-case basis. Differences are created by river size, location, institutional structure, ecology, development, and other natural and man-made characteristics. The players in the river basin must be identified, however, and their needs and preferences stated. Following this general sequence of steps, a number of river basin problem scenarios can unfold.

In *Water Resources Planning* (Grigg, 1985), I described four river basin case studies. Two were intrastate, regional basins. The first one, an urbanizing basin in the East (Upper Neuse), illustrates the competition for water in the Piedmont area of North Carolina and shows the problems that occur when a coordinating mechanism is missing. The next one, an urban-agricultural basin in the West (South Platte River, Colorado) shows that a voluntary, cooperative approach is the best way to improve water management in the difficult arena of western water management, as described elsewhere in this book. This basin includes the site for the Two Forks project, which is described in more detail in Chap. 10. Also, this case study is expanded in this chapter to illustrate interstate aspects and endangered species issues.

The first of two interstate river basins described in the earlier book was the Level B study of the Yadkin-Pee Dee River, which flows from Virginia to South Carolina. This was one of the few Level B studies run by state governments under the authority and funding of the Water Resources Planning Act. The study generated a great deal of conflict, and the final report was comprehensive, as it was intended to be. No projects have resulted from the study so far, but the ultimate benefits from the study will be its data, its baseline studies, and its role as a historical document about river conditions and management. In fact, conflicts in the basin seem to have increased. In North Carolina, a furor has developed over plans for interbasin transfer from the basin (see Chap. 15).

In the second interstate basin, the Potomac River, a Washington Metropolitan Water Supply Task Force was formed. Models were created through the Interstate Commission on the Potomac River Basin, which had a staff member with modeling expertise and an interest in collaborative problem solving. The planning effort was successful, and the number of new reservoirs planned was reduced to one. McGarry (1983) attributed the success of the effort to five factors: Local leaders realized that they could not depend on the federal government for a solution; elected officials were involved at the decision-making level; citizen leaders were effectively involved; traditional planning concepts were replaced by a regional perspective; and several individuals took personal responsibility to persevere despite obstacles.

To illustrate additional principles of river basin management, I will present four more cases: an expanded description of the Platte River system, including Colorado, Wyoming, and Nebraska; the Colorado River Basin; the Apalachicola-Chattahoochee-Flint river system in Georgia and Florida; and the Delaware River Basin. Another river basin case, the Albemarle-Pamlico system in Virginia and North Carolina, deals with estuary water quality management, and is presented in Chap. 22. Many of the same water management elements are present in this case, and it also involves a conflict over interbasin transfer (Virginia Beach water supply case, Chap. 15).

Platte River system

Three states are involved in the Platte River Basin: Colorado, Wyoming, and Nebraska. In addition to the interstate issues, there are many intrastate issues, including basin-wide management within Colorado, Nebraska, and Wyoming. Figure 19.3 shows the three-state basin upstream of the confluence of the South Platte and North Platte Rivers. Further downstream is the Big Bend region of the river, where the whooping cranes and sandhill cranes stop over on their migratory routes. Further still is the confluence of the Platte with the Missouri River.

The most visible conflict in the basin is in Nebraska, where environmental issues focusing on the survival of the whooping crane have resulted in years of studies, court actions, and most recently, political statements about developing a species "recovery plan." Basically, ecological interests are calling for more water to flow through Nebraska and for revised schedules of water flow, including periodic flood flows to "flush" the sediments and new seedlings from the river. As several

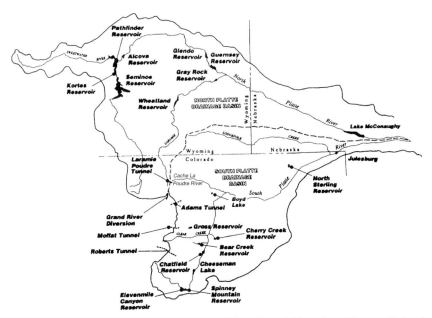

Figure 19.3 Platte River Basin projects in Colorado and Wyoming. (*Source: Federal Energy Regulatory Commission, 1994.*)

ecological species are involved, the amount and schedules of the flows demanded would be quite complex.

When Nebraska's water projects were in their planning stages, mostly in the 1930s, water concerns focused on power, irrigation, and flood control. For example, a 1931 history of the Central Platte region contains a diagram showing a "comprehensive plan" for Platte River development which includes power, irrigation, and flood control (Hamaker, 1964). The Flood Control Act of 1944, the Pick-Sloan plan for coordinated development of the Missouri River Basin, also authorized projects for irrigation, power, and flood control, and was the "comprehensive plan" of the time. The Water Resources Planning Act provided funds for the Missouri River Basin Commission and for Level B studies. In Nebraska this resulted in a state water framework study (Level A plan) and a Level B study of the Platte River Basin in Nebraska. These were preceded by work done by the Missouri River Basin Interagency Committee (MRBIAC), which completed a Missouri Basin Framework Study. This framework study was conducted during 1964–1969 with involvement of 10 basin member states and all affected federal agencies. Data from the study were later incorporated into the Level B study and the Nebraska State Water Plan (Missouri River Basin Commission, 1976, and working papers, 1975).

The Level B study, prepared by the Missouri River Basin

Commission, was completed in 1976. At that time, it was expected that the Level B study would be updated and kept current by the Missouri River Basin Commission through the "comprehensive coordinated joint planning process" (CCJP), but that did not occur because the commission was abolished in the early 1980s and its files transferred to the Corps of Engineers office in Omaha, Nebraska.

As part of the State Water Planning and Review Process, the Nebraska Natural Resources Commission undertook a Platte River "Forum for the Future" project. The purposes of the forum were "to provide a vehicle to develop and improve the general understanding of the Platte River," and "to provide a means for developing a consensus among those responsible for decisions concerning use of the Platte River waters" (Nebraska Natural Resources Commission, 1985). This process apparently did not serve to provide the consensus needed, since the report concluded that "the Forum process has not yet resolved any conflicts." In the 1980s and 1990s, the focus in Nebraska shifted to relicensing of hydropower projects by the Federal Energy Regulatory Commission (FERC), a long-term process. This is still under way at this writing (1994).

The failures and frustrations of these planning processes are made evident by an editorial cartoon in *Water Current,* a publication of the University of Nebraska Water Center (Fig. 19.4).

Colorado developed its South Platte water early in the state's history.

Figure 19.4 Grave of Nebraska Water Management Board. (*Source: Water Current, University of Nebraska Water Center, September 1991.*)

Main reservoirs of the Denver Water Board were completed in the early part of the twentieth century, including those to capture transmountain diversion water. Squabbling over the North Platte's waters in the 1930s led to the approval in the New Deal period of water development in Wyoming. This was one of the factors that led Colorado to pursue water from the Colorado Basin, and turn away from dependence on North Platte River water (see Colorado Big Thompson case study in Chap. 12, and Tyler, 1992).

Within Colorado, a study on the South Platte River concluded that a voluntary, cooperative approach was the best water management approach. By the early 1990s, however, it had become clear that cooperation would not emerge, that court battles, agency decisions, and adversarial processes would continue to rule the day. The adversarial processes center around several conflicts, three of which I will describe briefly here.

The Two Forks proposal was the most visible conflict (see Chap. 10). Briefly, the Denver region proposed a new, mainstem reservoir to meet water supply needs into the twenty-first century. After a $40 million environmental impact review process and endorsements by Colorado's governor, the Corps of Engineers' district engineer, and the regional administrator of the EPA, the EPA's newly appointed national administrator, William Reilly, vetoed the permit application.

The second conflict was the "reserved rights" case. Here, the U.S. Forest Service claimed reserved water rights under the organic enabling act of the Forest Service for the use of water for natural purposes, such as sediment flushing. After a long and expensive battle in state courts, the petition was denied in 1991.

The third issue was the city of Thornton's secret acquisition of a northern Colorado ditch company's shares. Here, the city planned a secret campaign to buy up land and water in northern Colorado, and divert the water from the farms to Denver to meet the suburb's growing needs. Again, an expensive court battle ensued.

These three conflicts probably cost $50–$75 million in legal and engineering fees within a span of about five years, in addition to paralyzing other, more productive uses of the time and energy of countless officials. Many hard feelings were generated, and the climate for cooperation and coordination was soured considerably.

Wyoming is a more rural state than Colorado, but it has its share of water issues. Like Colorado, one category of issue will involve resolving water pressures brought by federal agencies trying to implement the recovery plan in Nebraska.

It is apparent that, due to interstate compact issues and other legal and political matters, joint planning for management of the Platte River Basin's waters has enjoyed only limited success, and the prospects for the future are for more conflict. No basin-wide decision group exists to implement any "comprehensive plans," should they be developed, but the "recovery plan" may be the venue for the next stages of comprehensive planning.

Colorado River system

In the western United States, water history is dominated by the development and operation of the Colorado River Compact. *Time* magazine devoted a cover story to the Colorado River in 1992, and said that its conflicts were probably the most visible in the nation at that time (Gray, 1991). The Colorado Basin (Fig. 19.5) provides water for 20 mil-

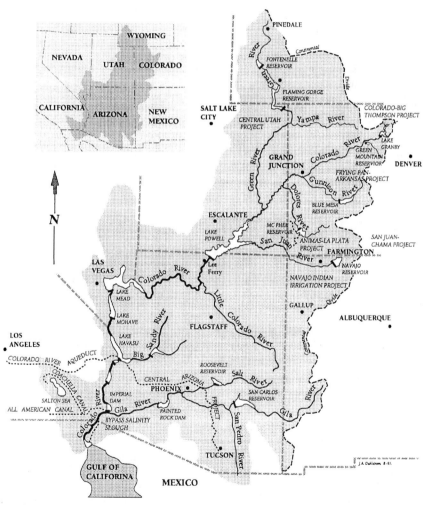

Figure 19.5   Colorado River Basin. (*Source: Northern Colorado Water Conservancy District.*)

lion people in seven states and for 2 million acres of farmland. Seventy percent of San Diego's water comes from the river. Arizona takes water for Phoenix and Tucson. The Imperial Valley, a major agricultural region, depends on the river. A prolonged California drought reached its sixth year in 1992, and cities started to desalt water.

The development of the Colorado River Compact makes fascinating reading in the politics of water, and many books and papers have been written about it. The "institutional history" chapter of one such book occupies 42 pages, and is just a summary (Hundley, 1986). The 1922 Compact is based on overestimated water supplies: It counts on 16.5 million acre-feet, but the river produced only 9 million acre-feet during the drought. The lower three states, California, Arizona, and Nevada, have gotten more than their share; now the upper four states worry that this overuse may become institutionalized.

The institutional history begins, as do most histories of the development of the Colorado River, with the explorations of John Wesley Powell, the famous Civil War veteran who explored and mapped the basin in the 1870s. The earlier history of the basin, while interesting, does not involve water development measures. In 1902, Powell's nephew, Arthur Powell Davis, was one of the early advocates of comprehensive development through large storage reservoirs. Settlers in California's Imperial Valley, a fertile area of 600,000 acres near the Mexican border, wanted a diversion canal from the river. This became the All-American Canal, so-called because it was all in the United States, whereas a previous diversion scheme had been a collaborative effort between the United States and Mexico. The U.S.-Mexican canal meant that U.S. farmers had no control over increased uses in Mexico.

In 1924 Los Angeles leaders worked with other city officials to create the Metropolitan Water District of Southern California (MWD). The MWD became a major player in Colorado River development, and was allied with Imperial Valley farmers. Their goal became the Boulder Dam and the All-American Canal.

A compact was needed, and under the leadership of Delph Carpenter, a Greeley, Colorado, lawyer, the compact was developed, but only after a long and difficult quest, including the personal leadership of Herbert Hoover, later President of the United States. A tribute to Carpenter's leadership on the Colorado River Compact has been published by the Colorado Water Resources Research Institute (Colorado Water Resources Research Institute, 1991).

The Colorado River Basin story is still being written. Several major dams have been completed, including Glen Canyon Dam. Both the upper basin states (Utah, Colorado, Wyoming, and New Mexico) and the lower basin states (California, Arizona, and Nevada) have separate and joint interests. Relationships with Mexico are a factor. There is a separate compact for the upper basin states. Southern California's explosive growth of the 1980s, exacerbated by the California drought, have alarmed the upper basin states, who now figure that, because of environmental issues, they may not be able to use their full entitlements.

The Colorado River Basin history illustrates the physical, legal, political, and environmental complexities of managing a large, stressed river basin in an arid region. During 1992, the state of Colorado and the U.S. Bureau of Reclamation began to work on a special decision support system for the river basin.

The Colorado River Decision Support System

During the early 1990s, Colorado's state water agencies decided that a computer-based decision support system would aid them to work with the Bureau of Reclamation and other states to manage the Colorado River. The resulting Colorado River System Decision Support System (CRDSS) illustrates principles of systems analysis and planning (see Chap. 5).

Lochhead (1993) identifies the following as changes that are apparent in the Colorado River Basin. The basin states and 10 Colorado River Indian tribes have, during 1991 and 1992, discussed their mutual concerns, with the central theme being a recognition that initiatives for change should originate by the interests most affected, and that change need not be revolutionary, but can, on the other hand, be worked out within the framework of existing law, without the intervention of outside parties.

Within this context, major river issues can be addressed. The problems are interrelated, and illustrate the need for a decision support system. Politicians and managers are increasingly aware that flexibility in river operation is going to be required to resolve problems like this. The good news is that, by working together, many things can be accomplished.

A needs analysis and feasibility study was completed in 1992, and it identified three general purposes of the CRDSS: interstate compact policy analysis including evaluation of alternative operating strategies, determination of available water remaining for development, and maximization of Colorado's compact apportion (Dames and Moore, and CADSWES, 1993).

The CRDSS project has part of its roots in conflicts between two state water agencies, the state engineer's office and the Colorado Water Conservation Board. As a result of this, and other factors, the state developed an organizational and coordination structure around Colorado River matters.

The plans for the CRDSS involve a large number of water users, modelers, and managers, some with extensive experience in water resource modeling and databases. The earliest concepts for the CRDSS were developed by local water managers, staff members of the state engineer's office, and researchers. They conceived of a configuration that would include spatial and relational databases (tabulations of numerical data), a toolbox with models and utilities, and a graphical user interface (GUI).

The feasibility report listed the following characteristics of decision support systems which enable them to "provide a dynamic, efficient, and effective information base for understanding and solving the problems at hand":

- Unambiguous treatment of data and information
- Efficient and effective visualization of information
- Data analysis techniques allowing full use of information
- Formal techniques for plan evaluation
- Rapid evaluations of policy alternatives
- Dynamic modeling of system state and updating of information

I believe that modeling and database development have been oversold in water resources management for a long time, so a conservative approach is necessary. In the case of the CRDSS, the first issue is to make sure that the "big river model" runs—and runs identically for all parties—to provide for coordination in problem solving. This will also require a coordinated approach to data collection and management in the states.

In the long run, there is a built-in "bottom-up" dependence of the CRDSS on first data, then models of the Colorado River System in Colorado, then the "big river model." If the approach to CRDSS took a "top-down" approach, it would fail, in my opinion. This is an approach that is 100 percent dependent on the success of the combined models-data package, rather than one where the models and data are disaggregated so that if only a partial approach is available, it can be used.

Apalachicola-Chattahoochee-Flint River system

The Apalachicola-Chattahoochee-Flint (ACF) is a 400-mile-long river system beginning above Atlanta, Georgia, and flowing to the Gulf of Mexico (Fig. 19.6). The basin drains a total of about 19,800 mi^2. Florida contains only 2500 mi^2 of basin area, but is the location of the sensitive Apalachicola Bay at the mouth of the river system. Alabama, as a riparian state, has about 2800 mi^2 of the drainage

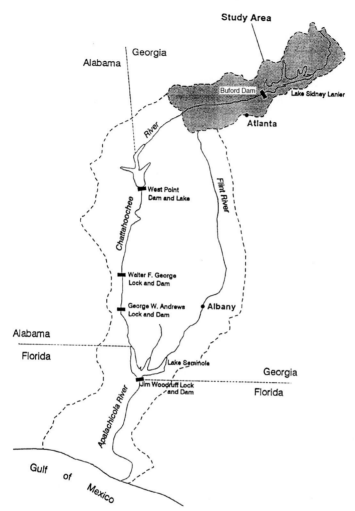

Figure 19.6   Apalachicola-Chattahoochee-Flint River Basin. (*Source: U.S. Army Corps of Engineers.*)

area. There are five major Corps of Engineers dams on the river and some 10 authorized nonfederal dams.

Development of the ACF began about 1828 with federal work on the Chattahoochee River. By 1874 navigation work began on the Chattahoochee to develop a 4-ft by 100-ft navigation channel upstream to Columbus, Georgia. In 1935 an emergency flood control project was authorized in the vicinity of West Point, Georgia. The Comprehensive Development Plan for the ACF was approved in 1945–1946, providing for a 9-ft navigation channel from the Gulf Intracoastal Waterway, and a program of dam construction for navigation, flood protection, hydropower, and streamflow reregulation. Dam construction began in 1947 with Jim Woodruff Dam, completed in 1957. This was followed by Buford Dam and Lake Lanier, completed in 1957, Walter F. George Dam, completed in 1963, George Andrews Dam, completed in 1963, and West Point Dam, completed in 1970.

To provide for navigation needs, the federal projects on the ACF Inland Waterway and River System are to be operated as a system. Storage at Lake Lanier is about 65 percent of the total system storage, so operation of this facility is a critical element of the overall system operation. Downstream activities with a stake in the system operation include urban water supply, including Atlanta; industrial and agricultural users; in-stream fisheries; recreation; and navigation. Water quality is a large concern on the system, as is water adequacy for Apalachicola Bay.

Conflicts on the waterway have been growing. The drought of 1986 exacerbated conditions, especially as they relate to navigation on the lower reaches, and to recreation on Lake Lanier. Environmental interests have been pushing for a management plan for water to Apalachicola Bay. In 1990 the state of Alabama filed a lawsuit against the Corps of Engineers to enjoin the Corps from reallocating any Lake Lanier storage from in-stream uses to water supply. This lawsuit led to approval of a comprehensive study by the Corps and the states. The purpose of the comprehensive study is "to describe the water resource demands of the basins, determine the capabilities of the water resources and to evaluate alternatives which best utilize the water resources to benefit all user groups within the basins (U.S. Army Corps of Engineers, 1991).

Here we have a basin in an area of the nation with abundant water resources, with rainfall averaging over 50 in per year. The growth of the megacity, Atlanta, the environmental concerns, fragmented local-state-federal interests, regional competition, interest-group advocacy, and political involvement all create conditions for conflict rather than cooperation in organizing the management system for the river basin.

Figure 19.7 illustrates how some neighbors perceive Atlanta's growing water demand. The Coosa River Basin is adjacent to the ACF, northwest of Atlanta, and flows into Alabama.

Delaware River

The Delaware River Basin includes four states: New York, Pennsylvania, Delaware, and New Jersey (Fig. 19.8). Conflicts that led to the federal–interstate compact began with New York City's 1920s plans to divert water supply from its headwaters (Hansler, 1980). This amounts to an interbasin transfer because New York City

Figure 19.7 An editorial view of Atlanta water demand. (*Source: Montgomery Advertiser.*)

is located in the Hudson River Basin. The conflict led to Supreme Court decrees of 1931 and 1954 in which the court granted authority to divert water subject to New York City's agreement to augment low flows with releases (U.S. General Accounting Office, 1981a).

Believing that Supreme Court decrees are not the way to engage in comprehensive river basin planning, the states began to negotiate with the federal government, and by 1961 the Delaware River Basin Compact (DRC) had been approved, and the Delaware River Basin Commission (DRBC) had been organized.

This was the first federal–interstate water compact, and it was not developed without testing some intergovernmental issues. Seven federal agencies opposed the compact, believing that it might be unconstitutional and involve conflict of interests for federal agencies. However, the states agreed to give Congress the right to amend the compact and the President a veto over elements of the comprehensive plan that he believed to be against federal interests, and these agreements cleared the way for passage of Public Law 87-328, which created the commission.

Figure 19.9 is a photo of a DRC meeting of March 27, 1991, at which New Jersey Governor Jim Florio voiced support for expanding storage capacity in the Francis E. Walter Reservoir. The photo illustrates a typical setting for a commission meeting, with the commissioners seated at a table, joined by the executive director and with support staff also in evidence.

The key management element of the Delaware River Basin

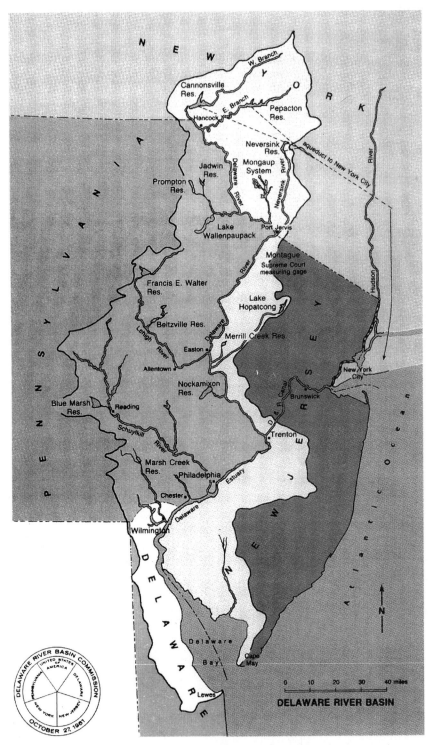

Figure 19.8 Delaware River Basin. (*Source: Delaware River Basin Commission.*)

Figure 19.9 Delaware River Basin Commission meeting showing New Jersey Governor Jim Florio. (*Source: Delaware River Basin Commission Annual Report, 1991.*)

Compact is the comprehensive plan, which has been evolving since about 1962. Gerald Hansler, Executive Director of the DRBC, wrote that "the Comprehensive Plan is the yardstick against which all water resources programs and projects with the basin are measured. Some 1500 proposed projects have been reviewed since 1962 and many modified to conform with the Plan. Appropriate programs of Federal and State water resource management agencies are integrated into DRBC's Plan" (Hansler, 1980).

The kinds of issues faced by the commission are drought water management, estuary water quality, toxic substances control, groundwater overdraft, and flood losses. Obviously, facing these issues involves considerable conflict on the commission.

One of the features of the DRBC operations is the use of a River Master to supervise diversions from and releases to the river. The River Master is an employee of the U.S. Geological Survey, and makes an annual report to the Supreme Court; see, for example, Sauer (1987).

New York State, being farthest upstream, questions the benefits of the commission and has indicated that it will limit its funding. Holding the states together in the compact arrangement, funding the commission's staff, and making the commission's work effective in planning, coordinating, and managing the waters of the Delaware will be continuing challenges.

Conclusions about River Basin Planning and Management

As shown by the case studies, river basin management is hampered by political and institutional problems. As Abel Wolman (1980) explained: "Nature's laws treat the river basin as a single unit, man's laws treat them as separate sovereignties; and there is no best organizational arrangement for integrated river basin development." A summary of experience might be: "geographic and ecological integration prevented by political disintegration"!

Efforts in the United States to make river basin management work effectively have not been entirely successful. Now, the role of the federal government in river basin management has diminished, and a new state–federal–local approach is needed.

Lessons from case studies

The case studies reviewed in this chapter provided different lessons.

- The Upper Neuse case study shows problems when a coordinating authority is missing, and illustrates the need for a state government water supply regulatory authority.

- The Level B study of the Yadkin-Pee Dee River generated conflict, and no management actions seem to have resulted so far. One lesson seems to be that top-down basin studies are not effective.

- On the Potomac River, a task force used simulation models to reduce the number of new reservoirs needed. Success was attributed to local, not federal initiative; involvement of elected officials; citizen leadership; regional perspective rather than traditional planning; and individual perseverance in the face of obstacles. The use of the model, coupled with negotiation, seems to be a key factor as well.

- The Albemarle-Pamlico system illustrates the need for broad-based and comprehensive water quality management planning, using a combination of regulatory and voluntary authorities.

- The South Platte River shows that a voluntary, cooperative approach may be needed in the arena of Western water management, but this is still a theory until proved to work. In Colorado, three disputes have cost millions in professional fees, and the climate for cooperation and coordination has soured.

- The interstate Platte River system illustrates the limits of joint planning without a central authority to enforce decisions. The Missouri River Basin Commission has been dismantled, and no decision maker exists to implement any "comprehensive plans."

- In Nebraska, the Level B Study of the 1970s is defunct and not followed. A state-led Platte River Forum did not provide consensus, and by the late 1980s the focus had shifted to relicensing of hydropower projects, an expensive and long-term process. The case illustrates the limits to cooperation when financial and environmental values are involved.

- Conflicts on the Delaware River led to the nation's first federal–interstate water compact. The key element of the compact is the comprehensive plan. The compact uses a river master to supervise diversions and releases. Future effectiveness of the compact will be a continuing challenge, but the institutional approach seems to have worked.

- The Apalachicola-Chattahoochee-Flint river system is in the process of conflict resolution at the time of the writing. Conflicts on the waterway have been growing since the 1970s, and the drought of 1986 uncovered many issues. Three states and the Corps of Engineers are working to find long-term solutions through a new institutional mechanism. A solution such as on the Delaware may be the key.

- The Colorado River's conflicts involve growth, drought, and stress over water in the West, and may illustrate the world's most complex transboundary water system.

Large-scale, comprehensive river basin plans

Top-down basin studies are not effective for large-scale river basin problems. Local initiative is required, and the preference will be for the use of voluntary, cooperative approaches.

A regional perspective rather than traditional local planning is needed. This creates a difficult intergovernmental problem which may need intervention by a central authority to solve.

For complex water quality problems such as estuaries, there is a need for water quality management planning with both regulatory and voluntary authorities.

For large-scale interstate problems, a compact may be only way to deal with persistent conflicts.

While comprehensive river basin planning is tricky and may not be helpful, it may help when a legally binding compact is available, and if actions are keyed to the plan.

While commendable in their goals, large-scale river basin studies seem unlikely to provide immediate benefits. However, the use of the comprehensive plan in conjunction with a compact on the Delaware River may illustrate effective application of master river basin planning. However, when the commission lacks any authority, as in the defunct commissions established under the Water Resources Planning Act, there is nothing to make them successful.

Ultimate benefits from the Level B studies may be their data and their roles as historical documents about river management. Rather than expensive efforts to prepare a comprehensive, coordinated plan, water service providers and regulators might work with regional organizations to keep historical documents and maps in a single location as a database. A planning or research organization could then host a periodic conference to review the status of basin development and management.

Effective leadership combined with technical tools

As shown by the Potomac River effort, success is possible in river basin management when effective leadership is combined with technical tools to focus on a specific problem. The involvement of elected officials, coupled with citizen leadership and individuals who persevere in the face of obstacles, are key requisites to success. The use of effective models, coupled with negotiation, may be necessary for complex problems.

River basin management in arid regions

In Western water situations with insufficient water and where both intrastate and interstate water conflicts are involved, comprehensive and coordinated joint planning will remain extremely difficult. Any planning or management effort will have to be based on legal principles and instruments. Recognizing this difficulty from the beginning will be essential. Political leadership can offset the tendency toward high cost and divisive solutions in these situations. In large-scale river basins in arid regions, such as the Colorado River, planning must be based on high-level political negotiation and established through legally binding contracts.

River basin management in humid regions

Even in humid regions, an interstate compact may be the only avenue for effective management for interstate streams in high-density, urbanizing regions. Even where there is abundant water, basins with growth in demands, coupled with environmental concerns, create conditions for conflict rather than cooperation. The Delaware River Basin Commission is a model that provides several decades of experience, and the ACF is a situation that is beginning to emerge. In any of these cases, making the management arrangement work in the face of state disincentives will be the main challenge.

Estuary water quality management

Estuaries are a special case of river basin management. Solutions through management planning must be multijurisdictional, multifaceted, and interdisciplinary. Perhaps the recent U.S. experience in management planning for estuaries can provide lessons for river basin management in the context of working out issues rather than coordinating projects.

Staffing for river basin planning efforts

On basins of all scales, emerging problems can be anticipated by a planning staff, but without regulatory authority, a coordination effort might not be successful. Maintaining a planning staff is essential, but participants will be reluctant to pay for it. This is a key problem in coordination of water management in all situations. Drawing staff from water service providers on a temporary basis will probably not work in the absence of cooperative attitudes. Maybe the best approach is to hold periodic basin conferences, with arrangements to publish proceedings. A research organization could analyze possibilities for coordination, publish suggestions, and then water service providers and regulators could implement them under encouragement from political leaders.

Benefits

Consultants to the WRC found two classes of benefits: rationalizing the decision-making process, and contributing to regional planning. Rationalizing decision making involves improved communication between agencies, citizen involvement in resource issues, increased dissemination of information, and state involvement in federal decision making. These are "good government" principles, and indicate the close relationship between river basin planning and the political process. Benefits to regional planning include comprehensive planning, analyses of issues, development of planning methodologies and analytical tools, data collection, and implementation review.

Final word

Unless an effective coordinating authority is available, there will be a need for a regulatory authority; otherwise there will be deadlock. This was proved by the Pecos River interstate compact, in which the 1948 compact specified a cooperative two-state commission, but there was no tie-breaking vote. A lawsuit took 14 years to settle and generated considerable conflict for the states.

Realistically, we must recognize the limits of large-scale joint planning without a central authority to enforce decisions. We must also recognize the limits to cooperation when financial and environmental values are involved. In other words, what I wrote earlier may be the controlling factor: What we face in river basin management is how to achieve the lofty ideals of geographic and ecologic integration in the messy context of river basin politics.

Questions

1. How can water managers improve coordination of decisions in river basins when political decisions are made by cities, districts, states, and federal agencies using jurisdictional boundaries that do not coincide with the river basins?

2. The main players in river basin management are the government units the river touches, water users, and interest groups. Discuss the roles of each in reaching agreement on river management strategies.

3. What is a "coordination mechanism for a river basin," and what is it used for?

4. Can river basin management be carried out without a central authority or "czar" to make decisions? If so, how?

5. What contributions can a decision support system for a river basin make to problem solving? Who should have responsibility to develop and manage it?

6. In the Pecos River dispute between New Mexico and Texas, why did the Supreme Court appoint a River Master? What is the main duty of the River Master?

7. For the ACF basin, give your opinion about the proper roles of the federal, state, and local governments in water coordination, and suggest who might fulfill the data collection, reporting, and enforcement roles in a coordinating mechanism for the basin.

References

Colorado Water Resources Research Institute, Delph Carpenter, Father of Colorado River Treaties, Text of Governor Ralph L. Carr's 1943 Salute to Delph Carpenter, Ft. Colllins, September, 1991.

Dames and Moore, and CADSWES, Feasibility Study, Colorado River Decision Support System, Denver, January 8, 1993.

Field, Ralph M., and Associates, Regional and River Basin Level B Studies: A Summary Report, U.S. Water Resources Council, Washington, DC, 1981.

Gray, Paul, The Colorado: The West's Lifeline Is Now America's Most Endangered River, *Time,* July 22, 1991, pp. 20–26.

Grigg, Neil S., *Water Resources Planning,* McGraw-Hill, New York, 1985.

Hamaker, Gene E., *Irrigation Pioneers: A History of the Tri-County Project to 1935,* Central Nebraska Public Power and Irrigation District, Holdrege, NB, 1964.

Hansler, Gerald M., The Delaware River Basin, in *Unified River Basin Management,* American Water Resources Association, May 4–7, 1980.

Hundley, Norris, Jr., The West against Itself: The Colorado River—An Institutional History, in *New Courses for the Colorado River, Major Issues for the Next Century* (G. D. Weatherford and F. Lee Brown, eds.), University of New Mexico Press, Albuquerque, 1986.

Kenney, Douglas S., and William B. Lord, Coordination Mechanisms for the Control of Interstate Water Resources: A Synthesis and Review of the Literature, Report for the ACF-ACT Comprehensive Study, U.S. Army Corps of Engineers, Mobile District, July 1994.

Lochhead, James, S., Colorado's Role in Emerging Water Policy on the Colorado River, *Colorado Water Rights,* Vol. 12, No. 2, Denver, Summer 1993.

Lowitt, Richard, The TVA, 1933–1945, in *TVA: Fifty Years of Grass-Roots Bureaucracy* (Edwin C. Hargrove and Paul K. Conkin, eds.), University of Illinois Press, Urbana, 1983.

McGarry, Robert M., Potomac River Basin Cooperation: A Success Story, in *Cooperation in Urban Water Management,* National Academy Press, Washington, DC, 1983.

Missouri River Basin Commission, Platte River Basin—Nebraska—Level B, Missouri River Basin, June 1976, and working papers, including Fish and Wildlife, dated July 1975.

National Geographic Society, *Great Rivers of the World,* Washington, DC, 1984.

Nebraska Natural Resources Commission, 1985 Nebraska Natural Resources Commission, State Water Planning and Review Process, Platte River Forum for the Future, January 1985.

Petersen, Margaret S., *River Engineering,* Prentice-Hall, Englewood Cliffs, NJ, 1986.

Sauer, Stanley P., William E. Harkness, and Bruce E. Krejmas, Report of the River Master of the Delaware River for the Period December 1, 1985–November 30, 1986, U.S. Geological Survey Open File Report 87-250, Reston, VA, 1987.

Tyler, Dan, *Last Water Hole in the West,* University of Colorado Press, Boulder, 1992.

U.S. Army Corps of Engineers, Mobile District, Plan of Study, Alabama-Coosa-Tallapoosa and Apalachicola-Chattahoochee-Flint River Basins, draft, April 1991.

U.S. General Accounting Office, *Federal-Interstate Compact Commissions: Useful Mechanisms for Planning and Managing River Basins,* Washington, DC, February 20, 1981a.

U.S. General Accounting Office, *River Basin Commissions Have Been Helpful, But Changes Are Needed,* Washington, DC, May 28, 1981b.

Wolman, Abel, Some Reflections on River Basin Management, in *New Development in River Basin Management,* Proceedings, IAWPR Specialized Conference, Cincinnati, June 29–July 3, 1980.

20

Drought and Water Supply Management

Introduction

Drought is one of the most pervasive and worrisome problems faced by water resources managers. It is the most complex of hydrology phenomena, and embodies issues related to climate, land use, and water use norms, as well as management issues such as preparedness. Moreover, droughts are "creeping disasters," and easy to ignore until it is too late. Although severe droughts are rare, some level of drought occurs in the United States every year, and in other countries drought disasters bring famine and suffering on a regular basis.

Drought is, for the most part, a complex management problem with many actors. Drought preparation requires individual and collective action to provide for secure water supplies and to make advance plans to share and conserve when supplies are short. Drought response requires plans to allocate scarce supplies and to take care of those most in need.

This chapter defines drought, explains concepts of risk and water security, and explains the roles of local, state, and federal government agencies that relate to drought actions. Role definition, coordination, and implementation are critical elements. Getting the actors organized to prepare for drought is the first part; the second part is to get them to work together in a coordinated manner.

Local government water managers, those "on the firing line," may use the chapter to help select levels of risk for water supply planning and to prepare drought response plans. State water officials may see their special roles in drought water management. Federal officials have critical roles, especially in managing federal reservoirs, and in data collection, analysis, and management.

Importance of Water Supply to Safeguard against Drought

Failures in water supply can have serious consequences to cities, industries, and other water users such as irrigation, hydropower, recreation, and wildlife. While it is not possible to prevent water shortages entirely, effective water resources management can minimize problems. Surprisingly, textbooks and technical reports dealing with water resources planning offer little guidance about what to do about drought, and the research literature deals mainly with responding to drought after it occurs.

Drought is a normal climatic feature. That it is also a serious and continuing worldwide water issue is due more to lack of effective water planning and management than to climate. No region is immune to drought problems. In the United States there were serious droughts in the 1950s, 1960s, 1970s, and 1980s, and the 1990s brought new drought problems. Each drought has been adapted to and the nation has survived, but the effects have been serious, though often underreported.

Polls have shown that people's greatest concerns about water are about two issues: shortages and contamination. The consequences of failure in water supply can be dramatic, and the risk is increasing with the interdependence and vulnerability of water systems. Environmental stakes have also risen, and drought has been shown to have dramatic effects on ecosystems, sometimes effects caused by people rather than nature. The threat to agriculture from drought threatens food supplies and farm income, especially in nations where food supplies are marginal. All of these reasons illustrate why drought is so important politically.

The solution to drought is to work together to find ways to prepare for it and to mitigate its adverse consequences. Preparing for drought in water resources management must be a *process,* not a project (or a prescription) to be started and finished when a drought is over. There is a need to institutionalize this process of improving overall water resources management with the lessons learned from droughts.

Understanding Drought

Not only is drought difficult to prepare for and respond to, it is difficult to understand due to its complexity. The first category of complexity is scientific/management complexity. This begins with the fact that there are definitional problems with drought. There are differences between what are called meteorological, agricultural, hydrologic, and economic droughts; each is a different phenomenon, and there is no simple way to explain the concept of drought as one phenomenon. Drought is multifaceted.

The complexity of drought makes it difficult to forecast drought onset or duration, and this factor is responsible for drought being called a "creeping disaster," as opposed to other disasters that hit suddenly and which can be responded to with single, comprehensive efforts. The water manager never knows when the drought begins or when it ends, and policy makers do not know when the disaster hits (although they certainly know when they are in the middle of it). Drought scope, timing, and intensity are difficult to describe. Since drought varies geographically, complex maps and statistics are required to characterize it fully. Drought is difficult to analyze hydrologically; our research has shown that not much is known about low-flow hydrology. There are disagreements about the ecological effects of drought. All of these complexities make drought difficult for policy makers to understand.

Economics often drives the politics of drought response, and drought is complex economically. The impact of drought on different sectors of the economy is hard to quantify, and drought damages have never really been adequately known. Drought effects involve different water management purposes/objectives, and some of these are intangible; all involve considerable social accounting. Normally, nonagricultural drought damages seem low; but they affect food security and stabilization programs for rural economies. As a result, drought policy involves much of the complexity of agricultural and rural development policy.

The legal aspects of drought are complex. One of the techniques to combat drought, to introduce flexibility into water management and transfer systems, encounters legal obstacles and complexity that give way to political difficulties.

As with much of water resources management, the complexities of drought cause political conflict. Because drought is an intergovernmental problem with vertical and horizontal dimensions, it requires extraordinary levels of coordination. It also involves conflict among different interest groups (economic slices). Drought involves large stakes and regional conflicts, exacerbating the potential for conflict. In sum, drought water management is a political minefield.

If drought is complex politically, it goes without saying that it is complex for management agencies. There is normally no single drought preparation and management agency. The principle of unity of command is not observed in drought water management. Each sectoral agency has a stake in drought, and this impedes coordination.

Complexity of drought impedes understanding, and leads to impasse. The political complexity of drought impedes bureaucratic solutions.

Defining and measuring drought

I use the term "drought water management" to mean to prepare water supplies and to manage water during drought. Other terms appear in the drought research literature, such as "drought management," "drought response," and "drought mitigation," but drought water management is different from these in that it relates only to water resources management activities, not to the full spectrum of disaster response.

Following the definition that water resources management is the application of management programs and water control facilities to control natural and man-made water resources systems, *drought water management* means all activities that provide for planning, design, implementation, regulation, and operations management in response to the drought hazard.

The simple concept "drought" causes much confusion in water management discussions. One source of confusion is the definition of drought. To begin with, drought has two dictionary definitions: (1) a period without enough rain, or (2) a period of shortage. The first kind—a period without enough rain—is called a "meteorological drought." This meteorological drought leads to the second kind of drought—shortages—which are called by various names, such as "hydrologic," "agricultural," and "socioeconomic" drought.

Palmer (1965), who developed a widely used drought index, formulated this definition, which is normally regarded as being a definition of meteorological drought: "an interval of time, generally of the order of months or years in duration, during which the actual moisture supply at a given place rather consistently falls short of the climatically appropriate moisture supply."

Changnon (1987) states that drought is difficult to define because it is not a distinct event with a recognizable start or end and it is the result of many complex factors. He presents the diagrams shown as Figs. 20.1 and 20.2 to explain drought. Figure 20.1, essentially a diagram of the hydrologic balance, shows that precipitation deficits will show up later in the hydrologic cycle. Figure 20.2 shows that precipitation deficits will show up sequentially in runoff, soil moisture, streamflow, and groundwater.

Changnon (1987) defines agricultural drought as "a period when soil moisture is inadequate to meet evapotranspirative demands so as to initiate and sustain crop growth." He defines hydrologic drought as "periods of below-normal streamflow and/or depleted reservoir storage." To this should be added depletion of groundwater storage.

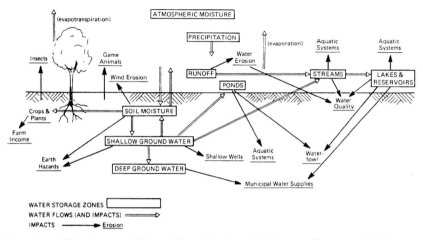

Figure 20.1 Hydrologic conditions affected by drought. (*Source: Changnon, 1987.*)

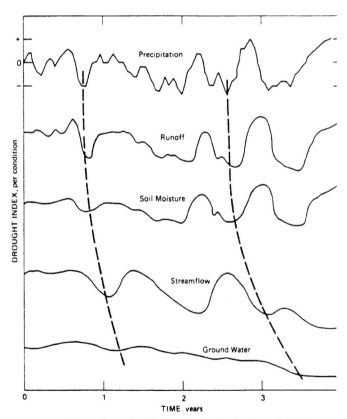

Figure 20.2 How drought affects the hydrologic cycle. (*Source: Changnon, 1987.*)

Changnon states that economic drought is "a result of physical processes but concerns the areas of human activity affected by drought." This can be interpreted to mean shortages in water supplies necessary for social and economic activity. Although in meteorological terms drought is caused by an extended period of dryness, economic drought means not having sufficient water to meet demands due to hydrologic shortfalls below expected levels of supply. The "expected level of supply" is a socioeconomic part of the definition of drought.

These two aspects of drought problems, supply and demand, are a source of confusion, but they have been handled quite nicely by a task force in Portland, Oregon, which separated the definitions of "drought" and "drought water shortage" as two separate concepts (City of Portland, 1988). "A drought in the Portland Water Bureau service area is a period of abnormally dry weather sufficiently prolonged to cause lack of water." A drought water shortage is a "lack of water caused by a drought which creates an immediate threat to the public health, safety and welfare." Also, "Lack of water includes effects of increased demand and reduced supply, both due to abnormally dry weather. The health factor includes water quality impacts due to low water storage in the reservoirs."

Another factor—expectation—also complicates the concept of drought. Expectation determines whether or not a drought is a hazard or a disaster. For example, a prepared water manager may state, "We expected an eventual drought—we didn't know when it would occur, but we planned for it." Meanwhile, the unprepared person may state, "What a disaster—we were totally surprised by this drought and now we're running out of water."

Drought indices

As with other complex phenomena, scientists seek to wrap several parameters into one to create an "index." If a successful drought index could be found, it would go a long way toward describing the complexities in the different kinds of drought. Unfortunately, the indices that have been developed fall short of this goal because they are linked to the definition of drought and involve the same ambiguity and complexity.

In principle, drought indices provide a measure of the difference between needed and available water resources and can be part of the decision support systems relating to drought. A local water utility might use a drought index to trigger water use restrictions and to inform the public about the availability of water supplies. A river basin authority might use an index to inform about and coordinate the use of water throughout a basin. A state might use an index to measure the availability of water resources statewide. At each of these levels, indices can be used for reporting, research, or management actions. Different users of indices will have different decision support requirements.

In general, water managers need indices to measure climatic and hydrologic trends and fluctuations. An international study team reported that a numerical index is needed to characterize the "intensity" of drought events (UNESCO/WMO, 1985). The simplest index mentioned by them is to compare the depth of precipitation and/or runoff for a given duration with a long-term mean.

W. C. Palmer developed the United States' most widely used general index, the Palmer Drought Severity Index (PDSI). This is referred to as one of the "more complex climate based indicators" by the UNESCO/WMO panel. The panel cites literature where many indices are identified. Since the PDSI is cited so much, it will be briefly described.

Management use of drought indices

To water managers, drought means problems in meeting demand. In that sense, drought means not having sufficient water to meet demands because supplies fall below expected levels. The "expected levels" are socioeconomic, because expectations can be adjusted.

Because of this link with socioeconomics, a drought index that will be useful to management must incorporate aspects of demand—that is, how adequate are supplies to meet demand?

To design an index for a particular situation, the following approach might be used:

$$\text{Index} = \frac{\text{available water supplies}}{\text{expected or mean water supplies}}$$

Available water supplies might include surface water, stored water, groundwater, and soil moisture. The definition given earlier from the City of Portland (1988) leads to the "Portland water supply index." This index provides guidelines for the city to assess the adequacy of its water supplies.

At the river basin level an example is the index used for the Apalachicola-Chattahoochee-Flint (ACF) river basin during the 1986 drought (Davis et al., 1987). Because of the dominance of Lake Lanier in this basin, the following index was used:

$$I = R + \text{Sum}(D_j) \qquad \text{with Sum taken from } j = 0 \text{ to } 3$$

where I is the monthly index value; R is the difference between the current lake elevation and the long-term monthly mean; and D is the difference between a given month's mean rainfall and the long-term mean. In this index, values of 0 to -2 are normal; -3 to -4 are below normal; -5 to -6 are mild to moderate shortage; -7 to -8 are moderate to severe shortage; and -9 to -10 are extremely severe shortage. This index was developed for the U.S. Army Corps of Engineers Mobile District, by the University of Alabama.

At the state level, numerous indices have been used or identified. In Colorado the State Engineer's Office (SEO) publishes a surface water supply index. It was designed in 1981 by the SEO and the U.S. Soil Conservation Service to link the effects of snowpack, streamflow, precipitation, and reservoir storage using a weighted probability formula. Values of the surface water supply index are published monthly by the SEO (Colorado Water Resources Research Institute, 1990).

As the reader might expect, because of the complexity of drought, drought indices have not advanced very far. Water managers must prepare their own indices to reflect local conditions. Research is needed to provide guidance on how to do that.

Assessment of Risk: Safe Yields of Water Supply

Any drought management or water supply planning methodology requires that the relationship between expected supply and demand be known to peg the risk of failure of water supply. Two terms frame this issue—risk of failure and security of water supply. Risk of failure is just what it says—the risk of running short of water—and security means the degree of insurance against failure that is provided.

The droughts of the 1970s and 1980s helped U.S. water authorities to realize the need for careful risk-based planning to prevent shortages. However, textbooks in hydrology or water supply do not provide much guidance about water supply security, especially for complex systems involving multiple sources. More has been written about how to react to drought after its onset, with emergency programs of allocation, restrictions, and interim supply arrangements, but the need for these might be prevented with proper preparation.

The information needed to assess the adequacy of water supplies is the probability that the raw water supply system will fail, that is, run out of water. This is usually presented in terms of the return period of the drought planned for or the annual probability of running short.

Risk and security are two sides of the same coin: What is the risk of failure, or what is the security of our supplies? Return period is a measure of the *security* of raw water supplies. Probability of failure, P, is a measure of the *risk* of failure of water supplies. Security is thus measured by risk; the riskier the supplies, the less security there is, and vice versa.

The concepts of return period and failure probability are well known, but for droughts they are complicated by the fact that drought *duration* must be considered as well as system yield and return period for a fixed time period, such as one year.

Water supply yield

The concepts of water supply yield and safe yield are useful for the measurement of risk and security. The concept of safe yield has a parallel concept in electricity production, that is, *firm power*. Several other terms, such as dependable yield, firm yield, and reliable yield are also used.

The term "safe yield" was apparently coined originally for groundwater, but it has been extended to surface water without formal definition. The concept is completely different for surface water and groundwater systems. You will find it discussed in groundwater texts, but you will not find much discussion in standard hydrology texts; for examples, see Maidment (1992), McCuen (1989), and Bras (1990). I anticipate that this problem will be corrected as concepts of risk analysis become more advanced as they relate to water supply reliability.

In the case of a groundwater source, yield is the amount of water that can be withdrawn safely without impairing the aquifer either through overdraft, mining, contamination, or other means. The term "safe yield" is used to denote this level of yield. The American Water Works Association (1984) defines the safe yield of a well as "the amount of water which can be withdrawn annually from a groundwater basin without producing undesired results such as permanent lowering of the water table."

In the case of surface water, other considerations come into play. Consider three situations: withdrawal from a single stream diversion point, withdrawal from a single reservoir, and withdrawal from a water supply system consisting of several streams and reservoirs.

For the single-stream diversion point, if all of the flow can be diverted, safe yield will be the lowest flow for the period under consideration. If an in-stream flow requirement is in place, safe yield will be the increment remaining after the in-stream flow requirement is met.

For a single reservoir, safe yield for a period under consideration is the amount that can be withdrawn by utilizing the storage available. In their handbook on drought management, the American Water Works Association (1984) defined safe yield for a reservoir as "the amount of water that can be withdrawn from storage in a specified interval of time (usually during a dry period or drought).

For a complex system of streams and reservoirs, safe yield is the integrated product of the system, but little has been written about this practical problem faced by utility planners. A system simulation is necessary to determine safe yield of systems. For example, in the Ft. Collins drought study described by Frick, Bode, and Salas (1990), the hydrologic yields of the different water rights and ditch systems had to be simulated for various years to see how the combinations of rights produced total yields.

Demand as a factor in safe yield

Risk of failure can consider demand as either static and fixed, or as a variable. If demand management is considered as an element of supply, as it increasingly is, then evaluating the risk of failure becomes even more complex.

Both supply and demand are implicit in a general understanding of drought, where the term is taken to mean not having sufficient water to meet demands due to hydrologic shortfalls below expected levels of supply (a drought water shortage). The "expected levels of supply" form the socioeconomic part of the definition. Both supply capacity and demand vary randomly from day to day, with demand following trend lines over time, and with capacity changing from time to time as a result of new projects and depletions of various kinds.

The fact that demand is normally quantified in terms of averages, or *average annual demand,* hides considerable information about variation, distribution, and elasticity of demand. Hiding all of this information within one statistic is an impediment to assessment of risk and a source of confusion in water supply planning. Comparing supply and demand should consider the fact that demand depends on climate and other factors, and that demand can be managed. However, most water supply plans consider demand to be static, equal to the average level, and do not consider demand variation.

The concept for comparing supply and demand is simple, but difficult to implement in practice. A water supply plan will forecast the supply and the demand; if all storage facilities never quite reach empty during a period of analysis, according to the forecast, then the supply is said to be adequate. The period of accounting, or *planning horizon,* must be considered, since if there is enough water for a multi-year period but an individual year is extremely dry, then the supply is inadequate.

Considering the variations of both supply and demand reveals the complexity of drought and shows that the question of whether the raw water supply is adequate is really a complex question dealing with hydrology (what is the yield of the supplies), statistics (what is the risk that the supplies will fail), economics (what will be the demand management strategies) and politics (will the people accept the demand management).

The essence of planning procedures is to assess the expected ratio of supply to demand and to make a judgment as to whether the factor of safety, however it is expressed, is adequate. Thus the normal sequence of all water supply studies is: Select a planning period, such as 25 years into the future; assess the demand; forecast the supply for the critical period; judge whether the excess of supply over demand is satisfactory. The challenge is, of course, to properly evaluate the demand and the supply.

An example of safe yield in Ft. Collins

There are variations in use of the term "safe yield." For example, Ft. Collins, Colorado, used this definition of safe yield in an unpublished planning study: "Safe yield is defined as the maximum average annual demand level for which demands can be satisfied each year of a representative hydrologic period during which a constant demand condition is maintained."

Four concepts are embodied in this definition: average annual demand level, representative hydrologic period, constant demand condition, and satisfaction of demands. In this sense Ft. Collins uses safe yield to compare supply and demand. The average annual demand level is an average, but it includes a stochastic time series component of variation. Satisfaction of demands is a dynamic concept that is normally taken as static. This incorporates both level of demand, which is stochastic and can be managed, and timing of demand, which is stochastic but cannot be managed. A constant demand condition means that demand is in general constant, that growth trends have not been incorporated.

This definition of safe yield is embodied in Fig. 20.3, which was used by the city of Ft. Collins to determine its water supply needs.

A proposed definition

I recommend a simple definition of safe yield: "Safe yield is the minimum yield (over a period such as a day, month, or year) statistically expected from a water supply system in a specified planning period (such as 50 to 100 years)." With this definition must go the caveat that in assessing whether this supply yield is adequate, the variability of demand must be considered.

Practical measures

Practical measures and rules of thumb will inevitably be practiced. The American Water Works Association states that in the Northeast, common practice is to design for a drought of probability .05 and to add a reserve of 25 percent to the storage volume. This provides secu-

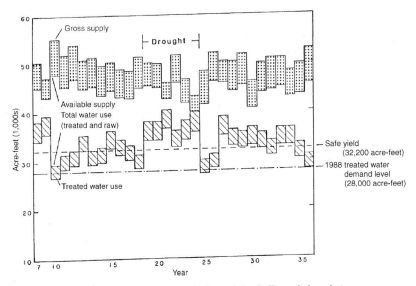

Figure 20.3 Safe yield analysis. (*Source: City of Ft. Collins, Colorado.*)

rity against about a .01 drought, and for droughts more rare than this, the approach is to use demand management. Guidelines for computing safe yields in the Northeast are given in an article in the *Journal of the New England Water Works Association* (Progress Report, 1969).

Added security from independent sources

Just as one can add security to one's income by diversifying, one can ensure water supplies by gaining access to independent sources. This can be shown theoretically through a statistical analysis that illustrates that the variability of the sum of independent sources is less than the variability of each one separately.

Roles in Drought Water Management

Defining and assuming roles of local, state, and federal government agencies to prepare for drought is the critical missing element in drought preparation and mitigation. What is meant by "role definition"? It means to consider the overall problem of drought preparation and response and to allocate tasks. It means to be sure that when plans and initiatives are needed, they will be done by responsible parties. It means that no important functions are left undone and that effective coordination is carried out to provide water management organizations and the public with confidence that their interests are being looked after.

Local government responsibilities

Local government water managers are "on the firing line" in the sense that they are responsible for providing the supply during dry as well as wet periods. The California Department of Water Resources has produced a guidebook for utilities on how to manage drought (California Department of Water Resources, 1988). The steps are: Forecast supply situation in relation to demand; assess drought mitigation options; establish triggering levels; develop drought demand reduction program; adopt the drought plan. The American Water Works Association (1984) has also published a handbook. This handbook, originally developed by the New England River Basins Commission, lists the following steps: Identify supply situation in relation to drought; assess supply augmentation options; assess demand reduction options; develop plan for sequential emergency measures; and select water saving hardware/software. Another comprehensive guidebook was prepared for the Corps of Engineers Institute for Water Resources (Dziegielewski et al., 1983). It recommends drought alert procedures; assessment methods for expected water supply deficit; models for forecasting water consumption; estimation methods for the costs of emergency supplies; procedures to estimate potential demand reduction and supply conservation; and methods to determine monetary losses from supply cutbacks.

Moreau and Little conducted two surveys of municipal utilities to cover the Southeast drought of 1986 and the nationwide drought of 1988 (Moreau and Little, 1989; Moreau and Lawler, 1988). They found that comprehensive approaches are mostly lacking.

A survey I conducted found the following: In general, the concept of balancing raw water demands and supplies and providing for a margin of safety or security is accepted in the profession, but there is little uniformity in procedures for determining adequacy of supplies; the concept of safe yield is not used consistently to state the reliability of water supply systems, and definitions of safe yield are not uniform; for the most part, planning for raw water does not take into account the fluctuation of demand, as most utilities plan for average conditions; statistical approaches to describing drought in terms of return period and duration are not in general use; where utilities rely on either groundwater or a large river for a supply, there is little inclination to be concerned with drought return periods; and there is little discernible difference in planning procedures in the different regions of the United States, whether humid or semiarid in climate.

State government roles in drought water management

State government has logical roles which fall into four categories: coordination, mainly through task forces; providing data and technical assistance; providing emergency aid to local governments and agriculture; and regulatory actions, mainly restricting water use. "Coordination" involves taking leadership at critical times, both to prepare for drought through water resources planning and policy development and to respond effectively during drought.

Wilhite (1990) concludes that state government "typically has played a passive role in drought monitoring, impact assessment and response." He states that in recent years states have made impressive strides, but "many states in drought prone regions have not yet developed plans and recent contact with states that have plans has disclosed varying degrees of dissatisfaction with the existing . . . procedures."

The Western States Water Council conducted an evaluation of state response capabilities and prepared a matrix of capabilities (Willardson, 1990). The categories of the assessment are state authority, governor's emergency powers, state water law, state drought response capacity, problem areas identified, and special state concerns.

Role of the federal government

The federal government has important roles in operation of federal reservoirs, in coordination, and in data management. The Corps of Engineers assessed the purposes of their reservoirs and their susceptibility to drought and concluded that both the types of uses and the quantities of water used have increased during the past 50 years (U.S. Army Corps of Engineers, 1990b). As the principal federal water management agency in the Southeast, the Corps assessed lessons learned from the 1986 drought (U.S. Army Corps of Engineers, 1988). Three principal findings emerged: the Corps' guidance document on drought contingency planning needs to be reevaluated (this document spells out management tasks ranging from establishment of a drought management committee to the evaluation of reservoir operating rules); Corps authorities and those of other federal agencies need to be more clearly described for the time when the drought moves from the concern level to disaster; a Corps workshop on preparation of drought contingency plans would be helpful to transfer 1986 experience to other regions. Some of the specific lessons learned were: the need for a drought contingency plan; the importance of the drought management committee; the value of water supply and water use data; the need for up-to-date water control manuals and reservoir rule curves for low-flow operation; the use of a simulation model for assessing impacts;

the need for open communication and public information; the need for memoranda of cooperation; the need to have a drought monitoring and response plan; and the value of division and district coordination.

Making Drought Water Management Programs Work Better

Since the 1980s a number of studies have recommended improved management measures. The Corps of Engineers (1990a) undertook a National Study of Water Management during Drought and reached a number of general conclusions. The Corps found that although much had been done in the United States to reduce vulnerability to drought since the Dust Bowl days of the 1930s, there remains a moving target and future effects of drought are likely to be more severe than the 1988 drought, especially if global warming occurs. No consensus exists about national strategy, but the conclusion of experts is that better planning, better data, better study techniques, and more coordination, collaboration, and communication will improve the situation.

In an analysis of California's needs, the Association of California Water Agencies (1989) identified several other needs: better interconnections, increased conjunctive use, increased efficiency of water use (in the Delta), protection of existing supplies, better financing of measures, and relaxation of constraints on water transfers.

The U.S. General Accounting Office (1993) evaluated how federal agencies monitor and coordinate the government's response to drought, and concluded that collecting and reporting data is a "collaborative, multilevel effort led by the federal government," and that users are generally satisfied with the data provided by federal agencies. No single federal agency has the lead responsibility, but when drought has been severe, temporary committees have been set up to coordinate; because drought has had more significant effects in recent years, these temporary expedients may no longer be adequate.

As a result of a research project of the late 1980s, Evan Vlachos and I reached a number of conclusions about needed drought roles and policies (Grigg and Vlachos, 1993). In general, water supply agencies must become as self-reliant as possible, but self-reliance must be balanced with collective security of water supplies. This means to ensure that individual and area-wide water supplies are adequate to deal with drought risk, including maintaining adequate reserves. Water supply agencies must band together to develop and test regional drought contingency plans. There may be roles here for regional and state governments and for federal agencies.

More water storage and management capability may be needed. This can be provided by storage projects, reallocation of storage, provisions for water banking and conjunctive use, and by demand management programs.

The crucial role of state governments in drought water management must be recognized by both executive and legislative branches. Each state will require a unique response. There is an important role for the state's governor here, to make sure that all agencies are coordinated. State governments must assess their roles in helping local water agencies and other economic sectors to prepare for drought and to mitigate its effects.

Each state should designate a drought management authority and formulate advance agreement on actions to be taken. Coordinating authorities with modeling and simulation capabilities should explore their potential contributions and test their roles. Each state should periodically evaluate its plans and performance in drought water management.

Education of the public and officials about the future problems of drought needs continuing attention. How to do this needs attention from state water agencies.

In general, we need to make drought water management a *process,* not a project (or a prescription) to be started and finished during and after a drought. The nation is still looking for the correct allocation of roles. Recent legislation and agency planning efforts illustrate this. Drought researchers should unravel the complexity of drought and present usable explanations and drought indices. Ways must be sought to increase the flexibility of water transfers to aid in drought mitigation, both intrastate and interstate. Drought water managers need to improve role coordination for the various situations represented by environmentally unique water districts in different states and regions. Again, if we commit to long-term risk management, we can optimally allocate limited resources and implement effective, timely, and coordinated responses that address broad water issues.

Management Strategies

Like floods, management strategies for drought can be summarized as structural and nonstructural, hardware and behavior. Figure 20.4 illustrates the range of responses identified by Grigg and Vlachos (1990).

Examples of Drought Problems

During the 1980s and 1990s the United States and other nations were hit with a number of droughts, some quite severe. These are generally described in Grigg and Vlachos (1993), but the droughts

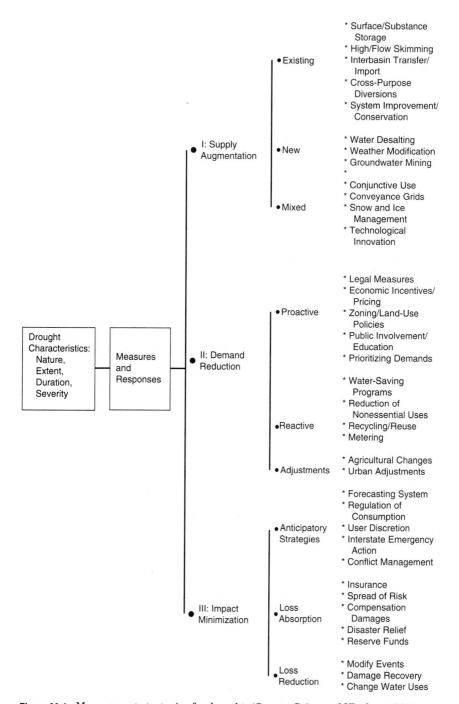

Figure 20.4 Management strategies for drought. (*Source: Grigg and Vlachos, 1990.*)

were so widespread and the effects and measures so variable that it is impossible to describe all aspects. Two examples will serve to illustrate the range of problems.

Western drought of 1980s

Climatic effects of the 1988 drought were severe in the West, and Fig. 20.5 illustrates the effect on the Mississippi River. The figure illustrates a year of Mississippi River flows at Vicksburg and shows highest and lowest average monthly discharges. The flows for the year 1988 are thus seen to fall below the low points, beginning in late May and continuing into July.

Drought in Athens, Greece

To illustrate how drought affects cities in other nations, a case of Athens, Greece, will be cited (Karavitis, 1992). Water resources of the greater Athens area are unevenly distributed. In this regard, the context of the metropolitan Athens area may be summarized in the following components: the area is semiarid and subject to shocks due to the water scarcity; there has been a rapid and high concentration of population and activities with marginal and decaying infrastructure; and the management experience of complex systems seems to be lim-

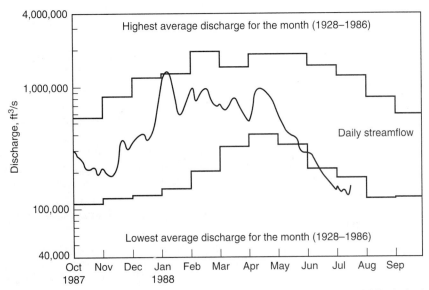

Figure 20.5 Mississippi River flows during summer 1988 at Vicksburg, Mississippi. (*Source: U.S. Geological Survey.*)

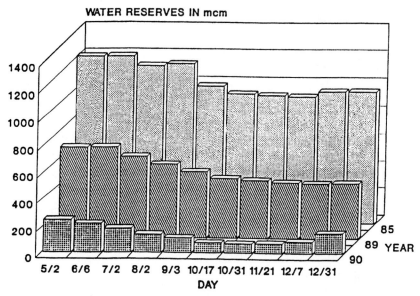

Figure 20.6 Drought impacts on the city of Athens, Greece. (*Source: Karavitis, 1992.*)

ited. Hence, Athens is under stress with very little resilience to natural hazards.

Karavitis found that for the 1990–1992 droughts in Athens (the 1990 drought was the worst on record), the databases were neither complete nor reliable and the water allocation plan was at best haphazard. To illustrate the severity of the drought, the water reserves of Athens are presented in Fig. 20.6. To give an indication of the magnitude of the problem, in October 1990 the Athens metropolitan area had water reserves of only 56 days.

Karavitis found that in Athens, drought impact assessment, alert mechanisms, and contingency planning were generally not satisfactory; responses were piecemeal, with an emphasis on short-range actions. Measures were applied well after the onset of the event, with little consideration to the crucial timing factor. Moreover, response since the drought has not moved to solve the underlying problems of supply and preparedness, according to Karavitis (personal communication, 1994).

Questions

1. What is meant by the statement, "drought is a 'creeping disaster,'" and what then would be a "sudden disaster"?

2. Why is drought water management a process, not a task to be completed at a point in time?

3. Define drought. What is the difference between a "meteorological drought" and a "hydrologic drought"?

4. Explain how you might use a *drought index* for coordination of scarce supplies using a permit system in a river basin.

5. Explain the concept of safe yield of water sources for both groundwater and surface water. Define safe yield of a system made up of different water supply sources.

6. Explain why drought is the "most complex of hydrologic phenomena."

7. The text states that one can increase security to water supplies by gaining access to independent sources, and that one can show this theoretically through a statistical analysis to illustrate that the variability of the sum of independent sources is less than the variability of each one separately. Can you show this with a statistical analysis?

8. What, in your opinion, are the most important roles of local, state, and federal governments in drought water management?

References

American Water Works Association, *Before the Well Runs Dry: Volume II, A Handbook on Drought Management,* Denver, 1984.

Association of California Water Agencies, *Coping with Future Water Shortages: Lessons from California's Drought,* Sacramento, 1989.

Bras, Rafael L., *Hydrology: An Introduction to Hydrologic Science,* Addison-Wesley, Reading, MA, 1990.

California Department of Water Resources, Office of Water Conservation, *Urban Drought Guidebook,* March 1988.

Changnon, Stanley A., Jr., *Detecting Drought Conditions in Illinois,* Illinois State Water Survey, Champaign, 1987.

City of Portland, Oregon, Bureau of Water Works, Drought/Water Shortage Plan, April 1988.

Colorado Water Resources Research Institute, *Colorado's Water: Climate, Supply and Drought,* Colorado State University, June 1990.

Davis, C. Patrick, and Albert G. Holler, Jr., Southeastern Drought of 1986—Lessons Learned, *Engineering Hydrology,* 1987.

Dziegielewski, B., Duane D. Baumann, and John J. Boland, Evaluation of Drought Management Measures for Municipal and Industrial Water Supply, Report 83-C-3, Institute for Water Resources, Ft. Belvoir, VA, December 1983.

Frick, David M., Dennis Bode, and Jose D. Salas, Effect of Drought on Urban Water Supplies. I: Drought Analysis and II: Water Supply Analysis, *Journal of Hydraulic Engineering,* Vol. 116, No. 6, June 1990.

Grigg, Neil S., and Evan C. Vlachos, Drought and Water-Supply Management: Roles and Responsibilities, *Journal of Water Resources Planning and Management,* Vol. 119, No. 5, September/October, 1993.

Grigg, Neil S., and Evan C. Vlachos (eds.), Proceedings, November 1988 Workshop on Drought Water Management, Washington DC, February 1990.

Karavitis, Christos A., Drought Management Strategies for Urban Water Supplies: The Case of Metropolitan Athens, Ph.D. dissertation, Colorado State University, 1992.

Maidment, David R. (ed.-in-chief), *Handbook of Hydrology,* McGraw-Hill, New York, 1992.

McCuen, Richard H., Hydrologic Analysis and Design, 1989.

Moreau, David H., and Andrew J. Lawler, The Southeast Drought of 1986: Impact and Preparedness, paper presented at the NSF-sponsored Drought Water Management Workshop, Washington DC, November 1988.

Moreau, David H., and Keith W. Little, Managing Public Water Supplies during Droughts: Experiences in the United States in 1986 and 1988, Water Resources Research Institute Report 250, September 1989.

Palmer, W. C., Meteorological Drought, Research Paper No. 45, U.S. Dept. of Commerce, Weather Bureau, Washington, DC, 1965.

Progress Report of Committee on Rainfall and Yield of Drainage Areas, *Journal of the New England Water Works Association,* June 1969.

UNESCO/WMO, M. A. Beran, and J. A. Rodier, rapporteurs, *Hydrological Aspects of Drought,* Paris, 1985.

U.S. Army Corps of Engineers, The National Study of Water Management During Drought, The Drought Preparedness Studies, Ft. Belvoir, VA, November 1990a.

U.S. Army Corps of Engineers, Hydrologic Engineering Center, Lessons Learned from the 1986 Drought, Report for the Institute for Water Resources, IWR Policy Study 88-PS-1, Ft. Belvoir, VA, June 1988.

U.S. Army Corps of Engineers, Hydrologic Engineering Center, A Preliminary Assessment of Corps of Engineers' Reservoirs, Their Purposes and Susceptibility to Drought, Davis, CA, December 1990b.

U.S. General Accounting Office, Federal Efforts to Monitor and Coordinate Responses to Drought, GAO/RCED-93-117, June 1993.

Wilhite, Donald A., Planning for Drought: A Process for State Government, International Drought Information Center, University of Nebraska, 1990.

Willardson, Tony, State Drought Response Capability, presented at State/Federal Water Related Drought Response Workshop, Houston, TX, April 16–17, 1990.

Regionalization in Water Management

Introduction

Regional cooperation offers a powerful tool to water managers, but there are formidable barriers to carrying it out. This chapter illustrates some of the approaches available for regional cooperation. First, theory and general observations about regional integration are presented; then case studies are presented to illustrate regional water supply development, regional investments with economies of scale, and a regional environmental issue.

The chapter begins with a summary of a paper on regionalization (Grigg, 1989). I consider this to be the "theory" of the issue, and although the theory is helpful, anyone on the front lines of human and political struggles over local government, special districts, and local–state–federal relations knows that the reality depends on the facts on the ground. In other words, finding regional solutions is a political issue, not a technical problem.

Theory of Regionalization

Definition and types of regionalization

Regionalization in water management is integration or cooperation on a regional basis, or any plan that integrates management actions in a river basin, metropolitan area, or other geographical region. Going beyond the concept, agreement on a definition of regionalization in the water industry seems to have been difficult. A committee of the American Water Works Association (AWWA) struggled and finally presented this complex definition: Regionalization is "a creation of an appropriate management or contractual administrative organization or a coordinated physical system plan of two or more community water systems in a geographical area for the purpose of utilizing common resources and facilities to their optimum advantage" (American Water Works Association, 1981).

There has been no shortage of policy studies that advocated regionalization to ameliorate problems in the water industry. Advocates of regionalization point out that it can provide benefits in three areas: economics, service, and water quality. In economics, regionalization offers economies of scale in capital facilities and operational costs. As pointed out by Clark (1979, 1983), this does not mean that large central systems always offer scale economies, but there may be other, more subtle ways in which regionalization will pay off, often in cooperation in operating programs. Service can improve by extension through regionalization of high-quality and reliable water supplies to rate payers in outlying or needy areas.

Water quality benefits can result from centralized management. For example, the writer's utility provides high-level laboratory service to outlying utilities, and this service will aid them in meeting safe drinking water requirements. Consolidation of utilities and water quality standards might mean that all areas rise to the standards of the highest level. Regionalization can also benefit raw water quality through source protection programs.

One might think of the benefits as the "economics" of regionalization, and the barriers as the "politics" of the issue. On a practical basis, the barriers are formidable. One listing is by Thackston and others (1983), who evaluated water policy in Tennessee: loss of community control of income, need for legislation, public indifference, lack of regional cooperation, distrust and provincialism, regional inefficiencies and bureaucracy, added complexity, inequities in financing, physical design problems, personnel difficulties, public–private compatibility problems, lack of fiscal credibility, and redistribution of financial assets. They concluded that institutional issues are the greatest impediment to regionalization, and that success in regionalization is based on "70 percent politics, 20 percent engineering and 10 percent luck."

Political and institutional barriers are familiar, but there may also be sound engineering reasons to be skeptical of some regionalization proposals. Clark's studies (1979) show that regionalization does not always provide economies of scale in capital investments for water supplies.

Examples of Regionalization

Appropriate applications of regionalization, according to the AWWA, include "an urban complex of water systems that might be more effectively operated under a single management structure; or an urban complex of independently owned or operated water systems that could have a master coordination plan...or rural or suburban water systems, remotely located that could obtain economies of scale under a single management structure."

After reviewing a number of case studies, I compiled the following list of typical arrangements:

- Regional management authority
- Consolidation of systems
- Central system acting as raw water wholesaler
- Joint financing of facilities
- Coordination of service areas
- Interconnections for emergencies
- Sharing of any management or service responsibility

Regional water supply authorities are a good example of the possibilities and problems. They can coordinate supplies for the benefit of all in a region. The cities of St. Petersburg and Tampa and the counties of Hillborough, Pinellas, and Pasco in Florida created a new raw water authority in 1974 through an interlocal agreement under Florida legislation that encourages regional water supply development (Hesse, 1980). Adams (1994) describes how member governments contract with the authority to develop water, and then buy the water from the authority. The authority has no taxing power but can issue revenue bonds. In analyzing the authority's effectiveness, Adams observes that from its creation until 1990, the authority succeeded in developing additional water supplies and serving its member governments. However, with additional conflicts in the last few years over source development, the authority's effectiveness has diminished.

Cooperation can definitely reduce the need for infrastructure development. In the Washington, DC, metropolitan area, water organizations cooperated to reduce future storage requirements. At issue were problems that went back 20 years. They were resolved through a task force approach among federal, state, and local agencies, with local leaders involved at every step (McGarry, 1983).

Raw water is a good tool for cooperation, as in the California Bay area. Some have called for regionalization and consolidation of systems, but progress has been through cooperation in raw water supply efforts without changing the management of distribution systems by the existing utilities (Gilbert, 1983).

Private water companies can spread costs among customers to achieve equity in rates. In West Virginia, 12 districts of the West Virginia Water Company, a subsidiary of the American Water Works Service Company, Inc., went to single-tariff pricing. This spread the costs to help the smallest and weakest areas. The approach was upheld by the Public Service Commission of West Virginia (Limbach, 1984). This spreading of cost is one of the barriers to regionalization in other areas.

In other countries we see examples of regionalization that work well. In England, regionalization, followed by privatization, became national policy. In Japan, regionalization is practiced almost without question. See Chap. 8 for details.

Regional authorities are also attractive in developing countries where adequate water service can be a matter of life and death in rural areas. An effort reported by Donaldson (1984) was started in 1938 and since 1961 has involved some $7 billion in international loans and national matching funds. See Chap. 25 on water and sanitation in developing countries for more details on this subject. Also in that chapter, the recent experience of Venezuela in regionalizing services is reviewed.

Introduction to Case Studies

Coordination is often required to balance regional water supply, either in a river basin or a metropolitan region. The first case, regional aspects of water supply in the Denver, Colorado, region, illustrates that coordination and cooperation might solve the problem and avoid the need for expensive and heavy handed regulation, but achieving coordination is often a very difficult and expensive task. Regionalization can help in investment in regional infrastructure systems by providing for economies of scale. The second case study illustrates how a water district led a regional study of water supply needs in a rural region of northern Colorado. Environmental issues need attention at the regional scale, surpassing the ability of single local governments to solve them. Issues such as lake level control or remediating past environmental problems are examples. The third case study illustrates the Kissimmee–Lake Okeechobee–Everglades ecosystem (KLOE) and shows the role of a regional water management district in resolving problems that transcend a single government agency.

Case Study: Denver Metropolitan Water Supply

The Denver metropolitan water supply case illustrates the struggle between an older, established central city water provider and newer suburbs. This is a common scenario and can be seen, among other places, in the Virginia Beach case study (Chap. 15) and the West Coast Regional Water Supply Authority cited earlier in this chapter. The Denver case also set the stage for the Two Forks permit battle (see Chap. 10).

Denver's growth in a semiarid region was made possible by its water supply. The water supply system began about 1859 when the city was a mining camp (Cox, 1967). From 1859 to 1872 residents relied on individual supplies from private wells or streams. A private Denver City Water Company served surface supplies from 1872 to 1878, but by the 1880s several private water companies were competing, and they were all consolidated into the Denver Union Water Company in 1894, the predecessor of the present Denver Water Board (DWB). It built Cheesman Dam in 1905, still part of the system. The next 40 years saw tremendous growth in Denver's water system. In 1936 the Moffat Tunnel was completed, bringing into reality the dream of bringing West Slope water to Denver.

Chapter 10, which described Denver Water's Two Forks project, illustrated some aspects of Denver's water system. Figure 21.1 shows the Front Range of Colorado and the major water facilities. Figure 21.2 presents conceptual details in the form of a flow diagram (Smith, 1972). Although the figure is somewhat dated, the major entities are still in operation and the figure would not be changed greatly if it were updated to current conditions.

A proliferation of suburban water agencies began in 1948 when the DWB raised rates. The 1950s drought tested the systems of Denver and the suburbs, and by the early 1960s Denver had completed Dillon Reservoir, which holds 254,000 acre-feet. In the 1960s several projects were launched, just before environmental activism began to increase. By the 1970s, environmental opposition to DWB policies had forced a conservation program, agreements to release in-stream flows, and a citizen's advisory committee.

Part of the environmental concerns of the 1970s was that Denver sought a new surge of growth. This concern led to the Two Forks controversy, probably the most significant water supply battle in the United States during the 1980s (Milliken, 1989). See Chap. 10 for a discussion of Two Forks.

The Two Forks controversy has national implications; it is not just a local water skirmish. After Two Forks there are new organizational

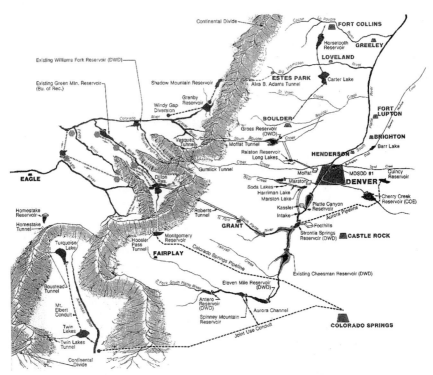

Figure 21.1   Front Range water supply system. (*Source: U.S. Army Corps of Engineers, Omaha District, October 1986.*)

initiatives. A Front Range Water Authority and a Metropolitan Water Authority have been organized, the city of Thornton announced a "City-Farm Program," there were meetings of the "Group of 10," a metro cooperation group that included water supply in its aims, and there are new, state-wide initiatives. Private developers announced numerous proposals for new projects. In 1994, the main action was the Front Range Water Forum, coordinated by the state's department of natural resources.

The Denver Water Department (DWD) decided not to file suit over the permit veto, and DWD engineers have stated that they learned several things from the affair. One lesson was that they could not think just about their problems; they also had to consider their neighbors. Another lesson is that in water supply planning, they had to study the impacts.

On the wastewater side, Denver's systems grew gradually. Denver built its first sewer in 1881. By the late 1950s, 45 different agencies in the region were collecting and disposing of wastewater, and there were

METROPOLITAN DENVER WATER SYSTEM

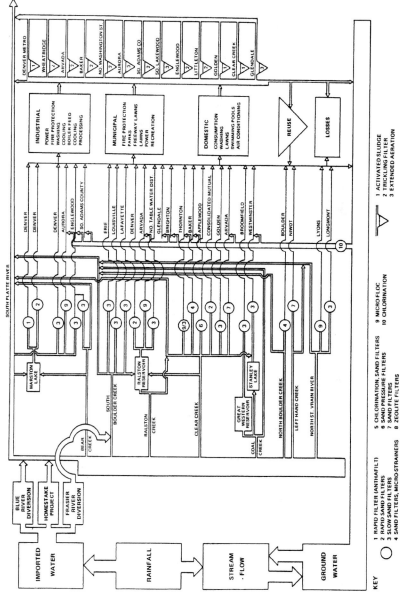

Figure 21.2 Metropolitan Denver water system. (*Source: Smith, 1972.*)

447

21 treatment plants, mostly small and overloaded. In 1960, enabling state legislation was passed, and in 1961, 13 communities joined to create the Metropolitan Denver Sewage Disposal District No. 1. The district's plant has evolved since it began to operate in 1966, and the district was renamed the Metro Wastewater Reclamation District in 1990. Today, it treats about 150 million gallons of wastewater daily and serves about 1.2 million people in the region. The district has 20 member municipalities and lists another 24 special-connection customers, mostly special districts.

During the past decade, the major water issue in the Denver area has been the struggle over a new supply. In addition, the region faces significant questions of water policy. Wastewater issues have generally been less controversial, but the city faces regulatory burdens and cost increases. Denver's experiences are not unique, as other large U.S. municipalities face similar problems.

Northern Colorado Pipeline

In 1989, the Northern Colorado Water Conservancy District authorized a study of regional water supply needs along Colorado's Front Range north of Denver. With cooperation and partial financing from a number of entities, the study was to identify alternatives to enhance water availability and water quality in the region at the lowest cost, and provide a basis for maintaining stable water supplies for the future of northeastern Colorado outside the Denver metro area. Figure 21.3 shows the study area (Northern Colorado Water Conservancy District, 1991).

The study divided water supply entities into three zones: within the NCWCD boundaries (40 entities), entities between the boundaries and county lines (11 entities), and entities south of the county lines on the northern tier of Denver (8 entities).

The study processes included municipal and industrial demand projections, assessment of water availability for municipal and industrial uses (including an assessment of potential gains from regional integration of supplies), development of alternatives for enhanced use of regional water supplies, and alternative institutional frameworks for regional water supply.

It was concluded that regional integration of municipal and industrial water supplies in the study area would offer substantial benefits. As a result of the study, the NCWCD has initiated a Southern Water Supply Project. Although the initial phases of the proposed pipeline project seem to offer substantial benefits and not to involve the development of new water supplies, the project has experienced substantial opposition.

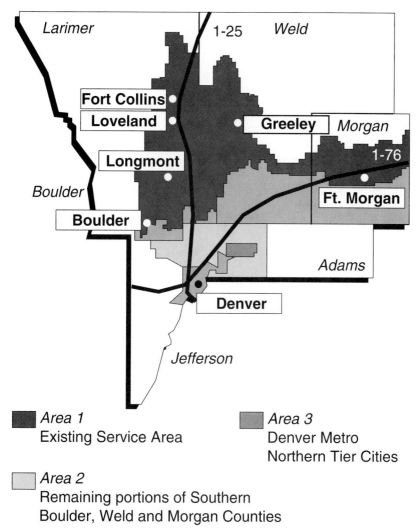

Area 1
Existing Service Area

Area 3
Denver Metro
Northern Tier Cities

Area 2
Remaining portions of Southern
Boulder, Weld and Morgan Counties

Figure 21.3 NCWCD regional pipeline project map. (*Source: Northern Colorado Water Conservancy District, 1991.*)

South Florida Water Environment

The general environmental condition of South Florida was described briefly in the case study comparing the wetlands of Florida's Everglades with South America's Pantanal (Wade et al., 1993). In this case study, the regional environmental management issues are highlighted.

The state of Florida was a pioneer in regional water management, dividing the state into water management districts. According to a brochure developed cooperatively by the districts:

The state's water management districts are regional agencies organized along hydrologic boundaries rather than political lines. Organized according to the 1972 Water Resources Act, their authority is delegated through the Florida Department of Environmental Regulation, as well as directly by the Florida Legislature. Regulatory programs, long-range planning, surface water restoration and water resource education are just some of the ways the districts protection Florida's natural water systems.

Figure 21.4 shows a map of the district boundaries.

In this case, one of the districts, the South Florida Water Management District (SFWMD), has major regional responsibilities. Much of the material that follows is from the SFWMD 40th Annual Report, which summarizes the history of the district (South Florida Water Management District, 1989).

Figure 21.4   Florida water management districts. (*Source: Florida Districts.*)

Some of the statistics of the SFWMD are as follows: It covers all or part of 16 counties with more than 5 million population; it has a total area of 17,930 mi^2; it operates about 1500 miles of canals and 215 primary water control structures; and it has a budget approaching $200 million, mostly from ad-valorem property taxes.

The SFWMD has two primary basins. The Okeechobee Basin is based on the Kissimmee-Okeechobee-Everglades (KOE) ecosystem, and the Big Cypress Basin includes the Big Cypress National Preserve and some 10,000 islands. Figure 21.5 illustrates the two complex ecosystems.

The story of the KOE's environmental problems began early in the twentieth century, before Florida's explosive growth occurred. Agricultural drainage was viewed as a way to create productive land from "swamp." From 1913 to 1927, 440 miles of drainage canals were dug along the lower east coast and south of Lake Okeechobee by the Everglades Drainage District. In 1926 and 1928, killer hurricanes hit South Florida and killed close to 3000 people living in farming areas. Damage was in the billions of dollars. The federal government responded with a levee around the lake. Then, in the period 1931–1945, several drought years brought low water supplies, peat fires, and saltwater intrusion. This was followed by a disastrous wet year in 1947, when over 100 in of rain, twice the normal amount, hit the east coast area. Two hurricanes also hit that year, and flooding was catastrophic. The response was the authorization of a Corps of Engineers flood control project and the creation of the Central and Southern Florida Flood Control District, which later became the SFWMD.

The first decade of the district's work focused on flood control, and enormous facilities were constructed, including dikes, pumping stations, canals, and levees. The second decade, from 1959 to 1969, was a time of tuning up, testing the system, and continuing construction. Heavy rainfall in 1960 produced record lake levels in Lake Okeechobee, with a water level rise of 2 ft. The project proved that it could move large quantities of water and serve its purpose. Also in this period, providing water supplies for growing urban populations in South Florida became a priority, especially as the region faced the drought of 1961–1965. Thus the district's "water supply era" began.

One of the largest projects of the 1960s was the channelization of the Kissimmee River. Over nearly a 10-year period, the river was deepened and straightened by the Corps of Engineers. A river that had been 103 miles long was shortened to 56 miles with a loss of about 35,000 acres of wetland habitat and a 90 percent reduction in migratory waterfowl. By the mid-1980s, the district had studied alternatives for restoring the river, and the restoration project began in 1994. It is expected to take about 15 years to complete at a cost of

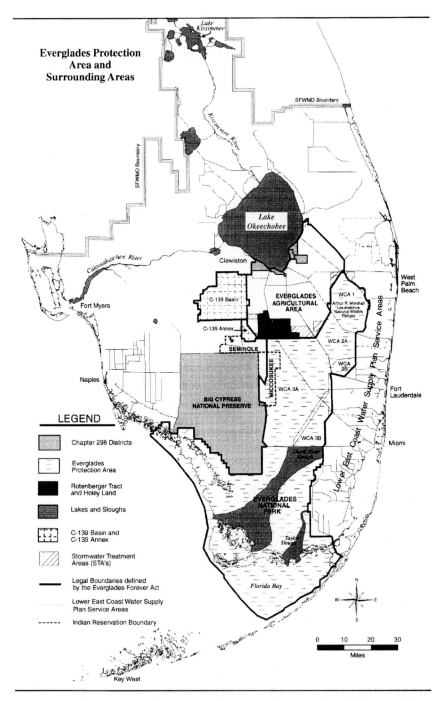

Figure 21.5 Everglades and surrounding area. (*Source: South Florida Water Management District, 1989.*)

some \$372 million in 1992 dollars. Figure 21.6 illustrates the main features of the restoration. Figure 21.7 is a photograph of the pre-restoration river.

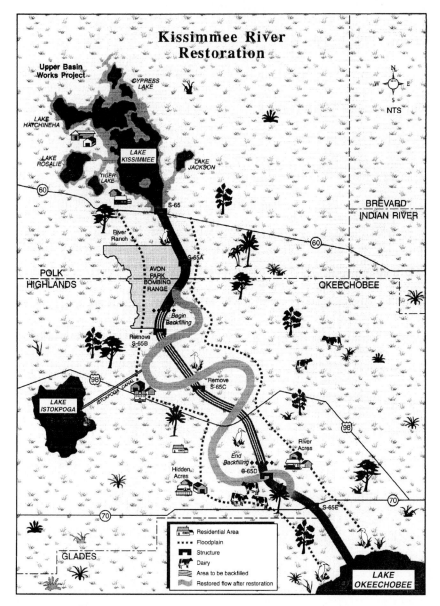

Figure 21.6 Kissimmee River restoration project. (*Source: South Florida Water Management District, 1989.*)

Figure 21.7 Kissimmee River before restoration. (*Source: South Florida Water Management District, 1989.*)

The district's water supply responsibilities revealed the delicate balancing needed to reconcile flood control, water supply, aquifer replenishment, and ecological systems management. The year 1988–1989 brought a record-breaking drought to the region, perhaps a 1-in-50 year drought. Water restrictions were imposed in 9 out of 12 months. Lake Okeechobee dropped 3 ft below normal levels. The district had to back-pump water into the lake, at the risk of harming aquifers.

During the 1970s and 1980s Florida was one of the fastest-growing states in the nation, with a combination of immigrants seeking opportunity and senior citizens fleeing cold weather. Growth and environmental issues began to achieve center stage, and earlier water management programs in South Florida came under sharp criticism (Florida Water Management, 1983). Problems included overdrainage, saltwater intrusion, polluted wells, overpumping, and hazards to species. Even the Biscayne Aquifer's recharge system was threatened. This aquifer, extending nearly 3300 miles from the southeast to the Gulf of Mexico, is quite porous and fragile. Contamination of the aquifer has been recognized as one of the most serious environmental problems facing the state. The Everglades National Park system, including the KOE ecosystem, has become perhaps a more visible environmental issue.

About 1970, at the same time as the nation became more aware of environmental issues, the SFWMD entered an era of intense ecological concern. In 1971, the U.S. Geological Survey issued a report saying that Lake Okeechobee was in danger, being overburdened with too many nutrients and aging too rapidly. In 1972, the Florida legislature passed the Water Resources Act, the Florida Comprehensive Planning Act, and the Environmental Land and Water Management Act, and created five regional water management districts, including the SFWMD with its present name. The district's mission also became more comprehensive, and its staff began to include biologists and other environmental scientists as well as computer specialists.

As a result of the 1970s emphasis on environment, projects began to be adapted to work with the natural KOE ecosystem. Still, ecological problems resulting from past practices continued to appear. In 1986, an intense algae bloom covering 120 mi^2 at its peak was recorded. A Lake Okeechobee Technical Advisory Committee (LOTAC) advised about measures such as phosphorous removal, aquatic weed management, expansion of best management practices, improved agricultural practices, and diversion and aquifer storage and recovery of nutrient-rich waters. Florida passed the Surface Water Improvement and Management Act (SWIM) in 1987. This led to the preparation of comprehensive plans for regional water improvement.

In 1993, a $465 million government-industry plan was announced that would distribute most of the cost to sugar growers, state government, and the South Florida Water Management District via a property tax (Gutfeld, 1993). In 1994, Florida passed the Everglades Forever Act, which is intended to be a fundamental plank in the preservation and restoration program for the Everglades. It provides for construction projects, research, and regulation to implement the program. One of the key features of the program will be the intense coordination needed to implement it. The coordination needed for this regional environmental effort is similar to that needed for estuary restoration projects (see Chap. 22).

One of the projects is the Everglades Nutrient Removal Project, located in Palm Beach County. Figure 21.8 shows a demonstration project using wetlands for phosphorous removal.

The SFWMD is also involved in water conservation by implementing an innovative dual distribution system whereby homeowners irrigate their yards with a combination of treated water and canal water. The concept is being tested in Cape Coral, in southwest Florida, with 12,000 homeowners participating. Figure 21.9 illustrates the piping system, including groundwater withdrawal.

Figure 21.8 Demonstration wetland water treatment area. (*Source: South Florida Water Management District.*)

Conclusions

Regionalization in water management tests the abilities of governments to work together. The Denver metro case study illustrates the difficulty of finding a mode for regional cooperation in water supply. The Two Forks veto shows the greater strength of environmental values over economic values. Central city–suburb conflicts are clearly illustrated in the case.

The northern Colorado pipeline illustrates the theory of finding regional infrastructure solutions that make economic sense. Also, the key role of a regional authority in integrating the plans of local water utilities is evident.

The South Florida case study contains many environmental lessons, but from the standpoint of regionalization, it illustrates how a regional authority can evolve from a single purpose (flood control) to become a comprehensive water management district, and provide a meeting place for conflicting interests to work out multiobjective strategies that include ecological goals.

The water supply industry will face economic and structural changes in the future. If it is to find new supplies, meet quality standards, and hold costs down, regionalization may help.

While regionalization has advocates, it has received little sustained policy attention from the federal or state governments. The concept is complex and difficult to understand or analyze. It is controversial

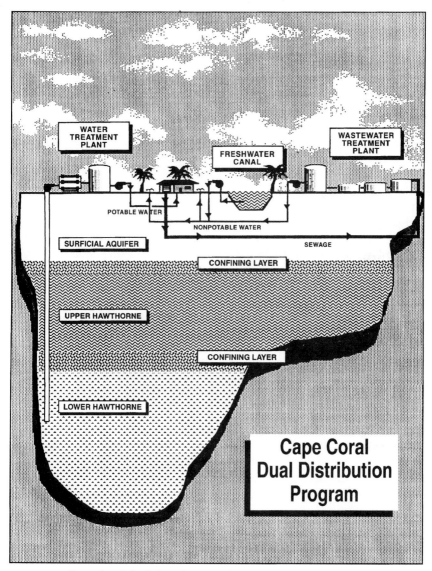

Figure 21.9 Cape Coral dual distribution schematic. (*Source: South Florida Water Management District.*)

among professionals and a hot political topic at the local level. Barriers are formidable.

What is needed to encourage regionalization in the face of formidable obstacles? Progressive action can be taken at local levels to explore possibilities. An example of the benefits available can be seen in the case of the metropolitan Washington, D.C., water supply example case that was cited. Counties or regional planning councils can authorize cooperative studies of the best ways to cooperate or integrate water utility functions.

Two important incentives are regulatory and financial. Regional water or wastewater planning and system development clearly work best with appropriate financial incentives. Public utility commissions, as well as state government regulatory authorities, can force coordination.

The state level is the most logical place for sustained policy attention to regionalization. Daniel Okun, a long-time advocate of regionalization, points this out in his article about needed state initiatives (Okun, 1981). He states that action at the state level is needed because federal intervention will not be favorably received and local action is unlikely except in a few major population centers. Humphrey and Walker (1985) reach the same conclusion. They see the state roles as facilitator, broker, and regulator to resolve institutional issues that evade local solution because of a desire to maintain local autonomy. They see as promising the scenario where states control the major water supply sources, thus enabling pricing and allocation to be used to promote regional solutions. They see the success of state action to be dependent on tailoring plans to individual local needs at times when action is clearly needed, such as during drought.

Actions at the local and state levels can be encouraged at the federal level by studies which evaluate the effectiveness of regionalization and establish improved databases. Without the federal presence, a vacuum in policy analysis and research sometimes occurs.

There is a need for improved management databases to serve the water industry. These would provide data on financial, facility condition, and performance parameters. They could aid in overcoming some of the management problems that stem from the lack of both market incentives and economic regulation in the water industry.

Questions

1. Define "regionalization in water management." Explain how the terms "integration" and "cooperation" relate to the term.

2. Give two practical examples of types of regionalization in water management.

3. What would be the main theoretical benefits of regionalization?

4. What would be some barriers to success in regionalization?

5. Regarding types of regional arrangements, do you believe a regional management authority or a central system acting as raw water wholesaler will be the best method for a large urban area?

6. What are the advantages of joint financing of facilities?

7. What are the similarities between regional problems and river basin management problems?

8. In the final analysis, a regional problem is a political problem. There is an old saying, "All politics are local." How does this relate to a typical regional services problem in the water industry? Can you give an example from your own experience?

References

Adams, Alison, West Coast Regional Water Supply Authority, Case History, Department of Civil Engineering, Colorado State University, unpublished, November 18, 1994.

American Water Works Association, Regionalization: Why and How, *Journal of the American Water Works Association,* May 1981.

Clark, Robert M., Water Supply Regionalization: A Critical Evaluation, *Journal of the Water Resources Planning and Management Division, ASCE,* September 1979.

Clark, Robert M., *Economics of Regionalization: An Overview,* American Society of Civil Engineers, New York, 1983.

Cox, James L., *Metropolitan Water Supply: the Denver Experience,* Bureau of Governmental Research and Service, University of Colorado, Boulder, 1967.

Donaldson, David, Regional authorities support small water systems in the Americas, *Journal of the American Water Works Association,* June 1984.

Florida Water Management: A Shift Away from Structural Planning, Water Information News Service, May 31, 1983.

Gilbert, Jerome B., Coordination of Major Independent Systems, in *Cooperation in Urban Water Management,* National Academy of Sciences Press, Washington, DC, 1983.

Grigg, Neil S., Regionalization in the Water Supply Industry: Status and Needs, *Journal of the Water Resources Planning and Management Division, ASCE,* May 1989.

Gutfeld, Rose, Agreement Is Reached on Framework of $465 Million Plan to Save Everglades, *The Wall Street Journal,* July 14, 1993.

Hesse, Richard J., A Regional Approach to Public Water Supply, in *Energy and Water Use Forecasting,* An AWWA Management Resource Book, Denver, 1980.

Humphrey, Nancy, and Christopher Walker, *Innovative State Approaches to Community Water Supply Problems,* Urban Institute Press, Washington, DC, 1985.

Limbach, Edward W., Single Tariff Pricing, *Journal of the American Water Works Association,* September 1984.

McGarry, Robert S., Potomac River Basin Cooperation: A Success Story, in *Cooperation in Urban Water Management,* National Academy of Sciences Press, Washington, DC, 1983.

Milliken, J. Gordon, Water Management Issues in the Denver, Colorado, Urban Area, in *Water and Arid Lands of the Western United States,* Mohamed T. El-Ashry and Diana C. Gibbons (eds.), Cambridge University Press, Cambridge, 1989.

Northern Colorado Water Conservancy District, Regional Water Supply Study (draft), May 1991.

Okun, Daniel A., State Initiatives for Regionalization, *Journal of the American Water Works Association,* May 1981.

Smith, Francis L., Jr., The Urban Water System: A Comprehensive Analysis, M.S. thesis, University of Denver, 1972.

South Florida Water Management District, 40th Annual Report, 1988–1989, West Palm Beach, FL, 1989.

Thackston, Edward L., Frank L. Parker, Michael S. Minor, James D. Bowen, and William S. Goodwin, *Water Policy in Tennessee: Issues and Alternatives,* Vanderbilt University, April 1, 1983.

Wade, Jeffry S., John C. Tucker, and Richard G. Hamann, Comparative Analysis of the Florida Everglades and the South American Pantanal, Interamerican Dialogue on Water Management, Miami, October 27–30, 1993.

22

Water Management in Estuaries and Coastal Waters

Introduction

A large fraction of the world's population lives very near the coastlines, but one of the toughest jobs in comprehensive water management is managing water quality and ecological issues in estuaries and coastal waters such as Chesapeake Bay, in large water bodies such as the Great Lakes, and in small seas such as the Mediterranean Sea and the North Sea. These valuable water bodies harbor diverse fish and wildlife species and are deeply appreciated for their beauty. However, on the one hand, large population and industrial regions are located near them, and on the other, their ecologies are fragile and their biological productivity is vital to fish and wildlife.

If the ecological balances in coastal and estuarine areas are to be preserved, water managers must play key roles. However, estuaries are the interfaces between river basins and the ocean, so the effectiveness of river basin management determines the health of the estuary. Chapter 19 outlined the difficulty of river basin management; estuaries involve another set of multiple entities, and coordination of water management actions is even more difficult.

In this chapter we discuss the nature of estuaries, the causes of their problems, and the management issues. Several case studies show the actions necessary for estuary protection. The general approach is to implement comprehensive management plans. Interestingly, the same principles—action plans based on science, and commitments to coordinated approaches—also apply to river basin management (see Chap. 19) and to other complex issues such as water and sanitation in developing countries (see Chap. 25).

Nature of Estuaries

An estuary is the mass of water formed by the confluence of a freshwater channel and the sea. Several examples of estuaries are presented in this chapter, including Chesapeake Bay, Apalachicola Bay, the Albemarle-Pamlico Sound area and associated Chowan River, and the California Bay-Delta region.

Many of the world's major cities lie astride estuaries and related harbors, and much of the world's population lives close to and depends on estuaries for income and food. There are about 850 estuaries in the United States alone (National Academy of Science, 1983). Estuaries are among the most biologically productive of all water systems. The annual benefits provided by estuaries were estimated at $82,000 per acre in 1971 (Macfarland and Weinstein, 1979). At least two-thirds of the commercial and sport fish landed in the United States are dependent on the estuarine environment during some portion of their life cycle. With both commercial and sport fisheries under great pressure, the value of the estuarine habitat has increased.

Estuarine complexities make it difficult to model water flows and quality changes. Hydrodynamics and material transport in estuaries are determined by tides, river flow, density differences, winds, and short-period waves.

Figure 22.1 illustrates the concept of an estuary's ecosystem, which

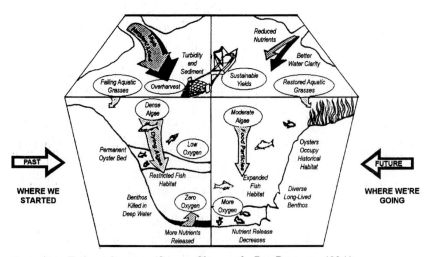

Figure 22.1 Estuary diagram. (*Source: Chesapeake Bay Program, 1994.*)

includes nutrient balances, freshwater inflows, grass and submerged aquatic vegetation, fisheries, benthos, and various food chain species.

The U.S. Environmental Protection Agency has a National Estuary Program which was established under the Water Quality Act of 1987. The EPA experiences will be described in a later section (U.S. Environmental Protection Agency, 1988a). The National Oceanic and Atmospheric Administration (NOAA) also has an active program of estuary studies, and an initiative to improve estuarine and coastal ocean science (National Oceanic and Atmospheric Administration, 1988).

Estuaries are sometimes associated with deltas, as in the case of the Bay-Delta region in California (one of the chapter's case studies). In its simplest form, a delta is a triangular deposit of sediment at the mouth of a river or a tidal inlet, but in more complex forms delta regions are lands formed by ancient deposits and making up large, flat coastal regions such as the Bay-Delta system, much of the state of Louisiana, the Nile Delta, and much of Holland.

Causes of Estuary Problems

In spite of considerable research, little is known on a global scale about the characteristics of estuaries or the loads they bring to the oceans. UNESCO compiled a world register of rivers discharging into the oceans in 1977 (UNESCO, 1977). The register showed that of 260 major rivers discharging into the oceans, only 25 percent were regularly monitored for water quality (see Chap. 14 for a discussion of water quality monitoring programs). The most polluted estuaries were in the industrialized countries, where the most data were collected.

Specific water problems in coastal areas fit a pattern and include nutrient enrichment, eutrophication and nuisance algae; threats to dissolved oxygen levels; shellfish bed closures; lost and altered wetlands; disappearance of submerged aquatic vegetation; threats to living resources from toxics; diseased fish and shifts of fish species; salinity intrusion; and groundwater problems (Davies, 1985, 1988).

Business Week (1987) devoted a cover story to "Troubled Waters" on October 12, 1987. After reviewing the situation, the editors gave their opinion: "coastal waters, breeding ground for all types of marine life, have largely been neglected in the cleanup effort. The result? Ever-increasing incidents of beach closures, fish kills and hepatitis from eating contaminated shellfish. And the problem is destined to get worse: Three out of every four Americans are expected to live within 50 miles of the sea by the year 1990."

This report, discussing the oceans but really focused on coastal waters, is echoed worldwide. In 1988 the director of the Oceans and Coastal Areas Programme Activity Centre of the U.N. Environment

Programme (UNEP) stated: "There's no mystery to marine pollution. The worst problem today is the huge quantity of raw sewage and industrial effluent spewed into the sea, with no thought to consequences, from coastal cities all over the world" (U.N. Environment Programme, 1988).

The U.S. Congress held hearings on coastal waters in 1988 and stated: "The continuing damage to coastal resources from pollution, development, and natural forces raises serious doubts about the ability of our estuaries, bays, and near coastal waters to survive these stresses. If we fail to act and if current trends continue unabated, what is now a serious, widespread collection of problems may coalesce into a national crisis by early in the next century (U.S. Congress, 1989).

After reviewing reports and holding hearings around the country, two subcommittees of Congress concluded that coastal waters are in poor shape; a proper response will take a cooperative, national effort; basic programs such as the Clean Water Act must be maintained but coastal waters should get more priority within them; and pollution control efforts should emphasize the coastal ecosystems approach (U.S. Congress, 1987).

Management Responses and Issues

The threats to estuaries have spurred action in some countries, such as the United States, which has passed numerous environmental laws. However, more is needed. Congress's observation that "a proper response will take a cooperative, national effort" is the main challenge, because of the difficulty of mounting coordinated and cooperative efforts using partnerships.

As a result of the National Estuary Program, which was established under the Water Quality Act of 1987, the EPA prepared an Estuary Program Primer (U.S. Environmental Protection Agency, 1988a). The Primer summarizes lessons that were learned primarily from the EPA's extensive Chesapeake Bay and Great Lakes programs. It explains the EPA's recommended management approach, based on "collaborative, problem-solving approaches to balance conflicting uses while restoring or maintaining the estuary's environmental quality." At the same time, it advocates a basic process consisting of problem identification, characterization, and phased management approaches. Thus, it combines the straightforward planning process with a politically realistic one. It is summarized briefly by the EPA as follows: "the program is woven together by two themes: progressive phases for identifying and solving problems and collaborative decision making."

Figure 22.2, from the EPA Primer, shows the overall process as including the Governor's nomination; the convening of a "management

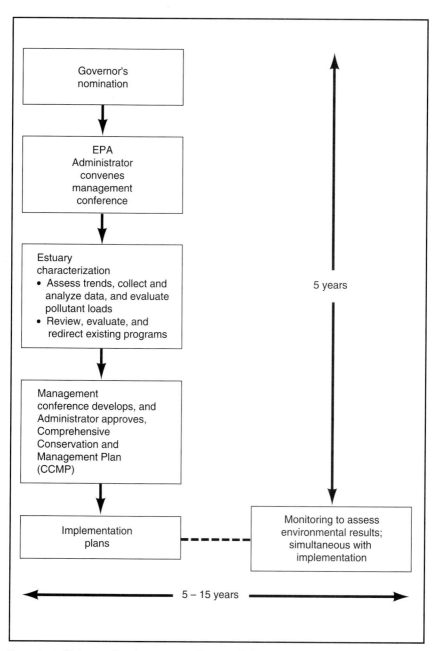

Figure 22.2 Estuary planning process. (*Source: U.S. Environmental Protection Agency, 1988a.*)

conference" by the EPA administrator; estuary characterization; a "management conference" which produces a comprehensive conservation and management plan (CCMP); implementation plans; and continuous monitoring. Thus, there are four phases: planning initiative, or building a management framework; characterization and problem definition; creation of a CCMP; and implementation. The committee structure of the conference targets four constituent groups: elected and appointed policy-making officials from all governmental levels; environmental managers from federal, state, and local agencies; local scientific and academic communities; and private citizens and representatives from public and user interest groups. The latter can include businesses, industries, and community and environmental organizations.

Case Studies

Introduction

The major restoration programs studied by the EPA were Chesapeake Bay and the Great Lakes, but quite a few other estuary and lake restoration programs provide experience about water management principles. In this section I present brief discussions of a few of them, with emphasis on a program where I can report on the basis of personal experience, the Albemarle-Pamlico program. First, I will distill from the reports of others some of the main points and case histories.

Chesapeake Bay Program

Chesapeake Bay (see Fig. 22.3) represents perhaps the largest, most complex estuary management program in the United States. If restoration efforts are ultimately successful—a formidable task—they will prove that a combination of regulatory and voluntary efforts can succeed in complex, large scale, interjurisdictional water management.

Started in 1976, the Chesapeake Bay Program is the oldest of the U.S. estuary recovery efforts. The bay's statistics are staggering: about 20 million recreational activity days per year; over one-fourth of the national catch of oysters; more blue crab catch than all other areas of the United States combined; more than half of the nation's soft-shell clam catch; wetlands that are a major stop for waterfowl in the Atlantic Flyway; spawning grounds for 90 percent of the striped bass along the East Coast; home to 200 species of finfish; and more (U.S. Environmental Protection Agency, 1980).

The basic problem in Chesapeake Bay is that the bay's rich resources have been in serious decline due to the usual culprits: population and overuse. Indications include an overall seafood harvest that is only a "shadow" of nineteenth-century catches; oyster catches that

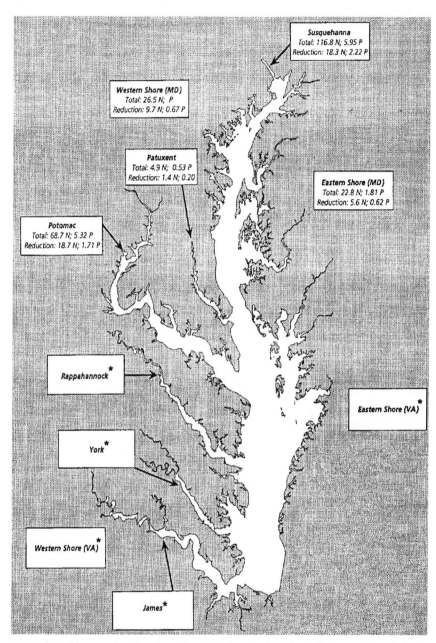

Figure 22.3 Chesapeake Bay: target nutrient reductions by tributary and region, in millions of pounds. (*Bay-wide modeling has shown that these Virginia rivers have less effect on the bay than rivers to the north. Nutrient reductions will improve local conditions and additional tributary-specific modeling will be conducted to develop reduction targets for these rivers. Between now and 1997, when reduction targets are expected to be developed, Virginia will implement an interim 40 percent reduction strategy.) (*Source: Chesapeake Bay Program, 1993.*)

are about a third of 1960 levels; stunted offspring of finfish species such as shad and river herring; eutrophication; loss of submerged aquatic vegetation; and general impacts from the waste coming from 5000 pollution sources that include factories, farms, municipal treatment plants, and city streets (*US News and World Report*, 1986).

There has been success on the Chesapeake Bay Program, although the jury will remain out for years as the region continues to develop and its politics change. According to the U.S. Office of Technology Assessment (1987), the success so far can be attributed to four factors: preliminary research, adequate funding (over $30 million in 7 years); a long-term effort; and strong public participation.

The 1983 Chesapeake Bay Agreement was a historic document, and its signing was accompanied by great enthusiasm and optimism. However, one attendee at the ceremony, Jacques Costeau, warned the celebrants that there were enormous political disincentives to collective action, and the real difficulties, maintaining momentum and interjurisdictional cooperation, were still in the future (Chinchill, 1988). Another Chesapeake Bay Agreement was signed in 1987, this time with more detail than the 1983 version. Some were beginning to see that Costeau's admonition had been correct: The difficulties lay ahead. Some began to believe that the only way to achieve management of the bay would be to establish an interstate commission with the authority to set bay-wide policies and standards. The problem is to coordinate the policies and actions of all of the independent parties. However, Chinchill believes that the concept of a coordinated, ecosystem approach to restoring the bay is strong, and that there is room for optimism.

Great Lakes

As with Chesapeake Bay, the Great Lakes program is a complex, "large water body" management program, and if restoration efforts succeed, it will be a tribute to the coordinated, cooperative approach to water management.

The Great Lakes (see Fig. 22.4) were the other major water bodies where the EPA formulated theories about managing estuaries and large lakes. A task committee of the American Society of Civil Engineers (1992) characterized the techniques as those for management of "large water bodies," intending to show the similarities of the Great Lakes to other systems.

Like the Chesapeake, statistics on the Great Lakes are also staggering: 37 million people in two countries living in the watershed; 95,000 mi^2 of surface area; 20 percent of the world's freshwater (including the St. Lawrence River); and 10,000 miles of shoreline.

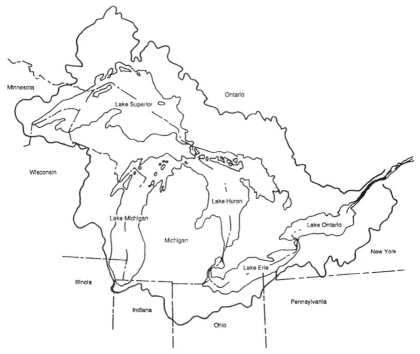

Figure 22.4 Great Lakes.

Problems are also massive: extinction of species; overfishing impacts, with 95 percent of the sturgeon gone; increased erosion; wastewater from 11,000 industries and 550 municipalities; and bioaccumulation of toxic chemicals (International Joint Commission, undated).

Institutional agreements on the Great Lakes trace back to 1909, when the United States and Canada signed the Boundary Waters Treaty that established the International Joint Commission (IJC). In 1972, the first international Great Lakes Water Quality Agreement was signed with a goal of restoring the chemical, physical, and biological integrity of the waters of the Great Lakes. Revisions were completed in 1978, 1983, and 1987. The 1987 revisions provided for "remedial action plans" (RAPs) in areas of concern, where restoration objectives have not been met, with adverse consequences to the lake system. The IJC identified 42 of these areas, 25 in the United States, 12 in Canada, and 5 in shared areas. In addition, "lakewide management plans" (LMPs) were called for in areas of open waters. Also, the 1987 agreement contained 16 annexes that dealt with specific issues such as airborne toxics, contaminated sediments, and control of phosphorus (U.S. General Accounting Office, 1990).

In 1978 the EPA established the Great Lakes National Program Office (GLNPO) as a focal point to plan, coordinate, and oversee restoration efforts. Coordination was initially unsatisfactory, so the 1987 Water Quality Act formally required the program office to identify problems, coordinate activities of organizations that could help in solutions, and report to Congress on progress.

The agreements and efforts that have been made to restore the ecosystem have been impressive, but results have lagged. Even the development of the RAPs is behind schedule. The General Accounting Office concluded that: "Even the difficult challenge of developing Remedial Action Plans and Lakewide Management Plans, which will involve substantial commitments of time and resources by many organizations, are just initial steps in planning the cleanup. Carrying the plans out will take decades and will require more effective pollution control programs by EPA and both the public and private sectors."

The 1987 Clean Water Act also required the EPA program office to develop a "Great Lakes initiative," which was finished in 1993. The program office also issued a "Great Lakes Water Quality Guidance," which is intended to provide procedures for calculating water quality criteria to protect aquatic life, human health, and wildlife (Whitaker, 1993). The Guidance, which is a 308-page document and mandates rules for 138 pollutants, is said to be "the most comprehensive reexamination of water quality regulation in decades" (*ENR*, 1993).

California's Bay-Delta region

Perhaps the most significant estuary on the West Coast of the United States is San Francisco Bay and its associated Bay-Delta region, which is made up of the delta of the Sacramento and San Joaquin Rivers (Fig. 22.5). Many of the management issues discussed in this book are at play in this complex system.

Like other deltas, the Bay-Delta region has its own geological history. According to the California Department of Water Resources' *Sacramento-San Joaquin Delta Atlas* (1993), "for millions of years, river flows and tidal action deposited sediment in the Delta, the low point of the Central Valley," and "These organic soils, up to 60 feet deep in some areas, were first farmed in the mid-1800's."

Human perceptions of delta development mirror what we find in other wetland regions. According to the Water Education Foundation's *Layperson's Guide to the Delta:*

> The Delta, as we know it, is largely a human invention. Early explorers found a vast mosquito-infested tidal marshland covered with bullrushes called tules. Later, trappers took advantage of the abundant wildlife. They were followed by farmers, some of them unsuccessful gold-seekers, who discovered in the Delta wealth of another sort: fertile soil. Over a century ago, these farmers, using Chinese laborers, began building a

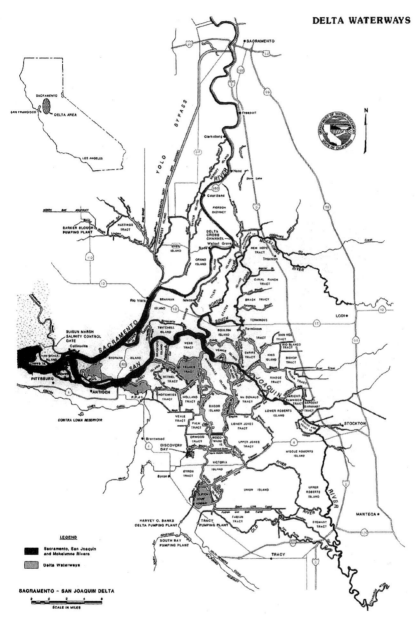

Figure 22.5 California Bay-Delta region. (*Source: California Department of Water Resources, 1993.*)

network of levees to drain and "reclaim" this fertile soil. Progressively higher levees were built to keep the surrounding waters out, lands were pumped dry, and what once was uncontrolled marshland was transformed into productive farmland. By 1930 more than 1,000 miles of levees surrounded close to 500,000 acres of farmland.

Progress is being made on agreements for water use in the delta. It has been a difficult struggle because the stakes are high. Some of the issues are striking a balance among urban, agricultural, and environmental water uses; negotiating correct water quality and salinity standards for delta water; providing water for species such as striped bass and salmon (note: striped bass were imported from the East Coast in 1879); preserving water quality for the 19 million Californians whose drinking water comes from the delta; maintaining flood protection programs and protecting levees from earthquakes; and managing harmful flows in the delta such as reverse flows from pumping actions (Brickson and Sudman, 1990).

One of the most contentious issues is balancing water uses because water rights are at stake. More than 5000 holders of water appropriation licenses or permits are at risk of losing water. According to Potter (1993): "As a result of the ESA [Endangered Species Act], State Water Project and Central Valley Project deliveries to state and federal water contractors were collectively reduced by about 250,000 acre-feet in 1992 to provide water for environmental purposes. This figure could easily reach 1 million acre-feet in 1993."

In concluding, Potter writes: "The major challenge for California today is to define a process for keeping the state's efforts and the federal efforts working in concert in order to achieve a "coordination and integration of water policy and management."

Albemarle-Pamlico Sounds and Chowan River

Albemarle and Pamlico Sounds are located in North Carolina just south of Norfolk, Virginia, and the Chesapeake Bay. They are major recreational and sports fisheries and important sources of commercial fishing revenues. The region has an important tourist industry. There have been water quality problems in the sounds and their tributaries, especially the Chowan River. Figure 22.6 shows the region, and Fig. 22.7 is a photograph of the Chowan River crossing at Edenton, which illustrates the general appearance of the river and sound.

In the early 1970s the state of North Carolina initiated studies of how to restore the Chowan River. The Chowan River Restoration Project (CHORE) was developed in 1979, and since then, a series of other actions have occurred (Grigg, 1979, 1981a).

Seasonal algal blooms occur naturally in the Chowan River, but usually they do not last long enough to cause nuisance conditions. In 1972, however, the seasonal blooms lasted from May until October, with disastrous results. Studies undertaken included primary productivity, nitrogen recycling, phytoplankton response to changes in water

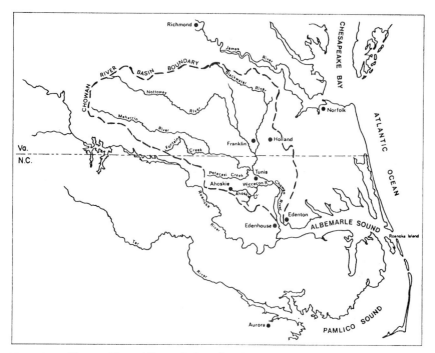

Figure 22.6 Chowan River–Albemarle Sound region.

Figure 22.7 Chowan River at Edenton. (*Source: North Carolina Department of Natural Resources and Community Development.*)

quality, and flow and water quality management modeling of the river system. There were no further serious algal blooms through 1977. As in many cases, this created a false atmosphere of security that the problem was transitory. However, a devastating algal bloom hit the river again in 1978.

These new blooms and their impact rapidly gained the personal attention of the governor, and the river was elevated to first priority in his environmental program. He directed that an action plan to clean up the river be developed by July 15, 1979, and the Chowan River Restoration Project was the response of the State Department of Natural Resources.

The project had long-term and short-term components. The long-term components were aimed at a permanent water quality management plan for the basin. The short-term plan intended to achieve cleanup as far as possible given economic conditions and knowledge about the problem. It had five components: reduction of industrial nutrient discharges; reduction of municipal nutrient discharges; control of farm and forest runoff; control of dry-weather raw water withdrawals by industries and municipalities; and consideration of innovative management approaches.

One stumbling block to rapid solution of the river's problems was lack of agreement among the two states, the local governments, and industries about the solutions needed. Because of the disagreements, Governor James Hunt of North Carolina asked the U.S. EPA to assign a top-level scientific team to audit the restoration project. The EPA responded by organizing a team which undertook their work during the summer of 1980. Their charge included the following: (1) to determine the technical adequacy of the existing database and study results relative to the best approach for solving the algal bloom problem in the tidal portion of the Chowan River; (2) to review the existing bi-state institutional arrangements and develop recommendations for modifications that encourage closer coordination; and (3) to make specific recommendations for any additional scientific evaluation deemed necessary to support regulatory controls.

The team's report was delivered in August 1980 (Chowan Review Committee, 1980). They generally concurred with the approach North Carolina had been taking to clean up the river, and they made recommendations for further study. In effect, the EPA report concurred with the feelings of state officials that the restoration of the river represented a major challenge to political and environmental officials and would require a long period of time for solution.

Now it is realized in North Carolina that the Chowan River difficulty is part of a larger water quality problem in the Albemarle region, and the state of North Carolina has sought to expand the Chowan River

Restoration Project to the entire region. The result has been the Albemarle-Pamlico Estuarine Study (1988). The study is a joint effort of the state, the federal government, and local interests, and includes a comprehensive report on status and trends in water quality and living resources and a comprehensive conservation and management plan.

Other examples

Problems such as with the Chesapeake and Great Lakes are found in a number of places, and management planning and action have been initiated, with varying degrees of success. For example, the Mediterranean Action Plan (MAP) is a complex, multinational program (U.N Environment Programme, 1985). The plan calls for a series of treaties, the creation of a pollution monitoring and research network, and a socioeconomic program to reconcile economic and environmental priorities. The North Sea is another example of an international water system needing management action (Rijswaterstaat, 1988). In the United States, a number of estuaries have received attention and management planning. For example, the Puget Sound Water Quality Authority (1987) published a comprehensive management plan. Another U.S. example is the Gulf of Mexico program launched in 1988 by the U.S. EPA. The action plan for the gulf will build on experiences in the Great Lakes and the Chesapeake (U.S. Environmental Protection Agency, 1988b). One of the gulf's estuaries is Apalachicola Bay, part of the ACF system described in Chap. 19.

Conclusions

Jacques Costeau's warning at the signing of the Chesapeake Agreement rings true: Development pressures and political disincentives throw up barriers to water management in the coastal zone where most people live.

The Albemarle-Pamlico region illustrates the problems:

- Trends in many places toward worse problems
- Need for comprehensive management programs
- Need for leadership by state governments
- Many unanswered technical questions
- Need to move ahead without all scientific answers
- Need for firm basis for partnership solutions

Partnership solutions can be based on management plans. To implement them, organizational structures are needed that include three levels of government: the scientific community, interest groups, and citizens. The EPA's process, which begins with collaboration between concerned parties in a management conference and then proceeds through phases, seems a good start, but perseverance is needed.

Management planning for estuaries requires a complex mixture of science and political action based on intergovernmental public–private cooperation. It applies knowledge of an estuary's complexities, which will never be fully understood, to solution strategies involving multiple players.

Like river basin planning, management plans for estuaries take many years to implement, and require sustained commitments of effort and scientific resources.

Questions

1. Compare the process of estuary management to that of river basin management. Who are the actors in each scenario? How do the ecological stakes differ?

2. How does the recommended management conference for estuary management compare to the coordination mechanism in river basin management? Would lessons from interstate river compacts apply to estuary management? Why or why not?

3. Is an estuary management plan a binding contract? Why or why not? How can one be enforced?

4. What is the state of the art of estuary water modeling? Can estuary models be used as decision support systems? Why or why not?

5. Compare the importance of nutrient modeling in estuaries to similar models in upland rivers and streams.

6. Should the federal government take a stronger hand in large-scale ecological problems such as the Great Lakes? What are the pros and cons of such an approach?

References

Albemarle-Pamlico Estuarine Study, *Albemarle-Pamlico Advocate,* Vol. 1, No. 1, Washington, NC, July 1988.

American Society of Civil Engineers, Task Committee on Water Resource Management of Large Water Bodies, *Management of Large Water Bodies,* New York, November 1992.

Brickson, Betty, and Ruth Schmidt Sudman, A Briefing on California Water Issues, *Western Water,* September/October 1990.

Business Week, Troubled Waters, October 12, 1987.

California Department of Water Resources, *Sacramento–San Joaquin Delta Atlas,* Sacramento, 1993.

Chesapeake Bay Program, *Progress at the Chesapeake Bay Program,* 1992–1993, Annapolis, MD, 1993.

Chesapeake Bay Program, *A Work in Progress,* Annapolis, MD, 1994.

Chinchill, Jolene E., Chesapeake Bay Restoration Program: Is an Integrated Approach Possible?, in *Water Policy Issues Related to the Chesapeake Bay,* William R. Walker (ed.), Virginia Water Resources Center, Blacksburg, VA, 1988.

Chowan River Review Committee (U.S. Environmental Protection Agency), *An Assessment of Algal Bloom and Related Problems of the Chowan and Recommendations Toward Its Recovery,* August 1980.

Davies, Tudor, Management Principles for Estuaries, U.S. Environmental Protection Agency, unpublished, 1985.

Davies, Tudor, Institutional Structures to Deal with Regional Water Problems: The Chesapeake Bay Example, 22d Water for Texas Conference, Houston, 1988.

ENR, Grand Plan for the Great Lakes, April 12, 1993.

Grigg, Neil S., Action Plan for the Chowan River Restoration Project, North Carolina Department of Natural Resources and Community Development, 1979.

Grigg, Neil S., The Chowan River Restoration Project, 1981 National Conference on Environmental Engineering, ASCE, Atlanta, July 1981a.

International Joint Commission, *The Great Lakes—St. Lawrence, Our Fragile Ecosystem,* undated.

Macfarland, J. W., and R. W. Weinstein, The Natural Estuarine Sanctuary Program, *Coastal Zone Management Journal,* Vol. 6, No. 1, pp. 89–97, 1979.

National Academy of Science, *Fundamental Research on Estuaries: The Importance of an Interdisciplinary Approach,* National Academy Press, Washington, DC, 1983.

National Oceanic and Atmospheric Administration, *NOAA Estuarine and Coastal Ocean Science Framework, Summary,* Washington, DC, January, 1988.

Potter, Robert G., deputy director of California Department of Water Resources, personal communication, October 1, 1993.

Puget Sound Water Quality Authority, 1987 Puget Sound Water Quality Management Plan, Seattle, 1987.

Rijswaterstaat, North Sea Directorate, *Management Analysis North Sea, Holland,* February 1988.

U.N. Environment Programme, The State of the Marine Environment 1988, *UNEP News,* Nairobi, April 1988.

U.N. Environment Programme, Mediterranean Coordinating Unit, *Mediterranean Action Plan,* Nairobi, Kenya, September 1985.

UNESCO, *World Register of Rivers Discharging into the Oceans,* Paris, 1977.

U.S. Congress, Committee on Merchant Marine and Fisheries, *Coastal Waters in Jeopardy: Reversing the Decline and Protecting America's Coastal Resources,* U.S. Government Printing Office, Serial No. 100-E, Washington, DC, 1989.

U.S. Congress, Office of Technology Assessment, *Wastes in Marine Environments,* OTA-O-334, U.S. Government Printing Office, Washington, DC, April 1987.

U.S. Environmental Protection Agency, *Chesapeake Bay, Research Summary,* EPA-800/8-80-019, Washington, DC, May 1980.

U.S. Environmental Protection Agency, *Estuary Program Primer,* Washington, DC, September 1988a.

U.S. Environmental Protection Agency, *The Gulf Initiative,* Washington, DC, 1988b.

U.S. General Accounting Office, *Improved Coordination Needed to Clean up the Great Lakes,* GAO/RCED-90-197, Washington, DC, September 1990.

US News and World Report, The Chesapeake Bay's Murky Future, October 20, 1986.

Water Education Foundation, *Layperson's Guide to the Delta,* Sacramento, CA, 1990.

Whitaker, James B., Launching the Great Lakes Initiative, *Water Environment and Technology,* June 1993.

23

Organization of
Water Agencies

Introduction

Chapter 8 described water industry structure, which is important to the institutional environment which determines how water problems are identified, organized, and solved. Organization is important because it drives mission statements and management roles. However, it is driven by programs that are funded as a result of legislation. Government agencies do what they are directed to do by their elected and appointed executives and enabling legislative bodies. This "command structure" aspect of government is responsible for some of the difficulties in coordination.

Lessons about organization of water agencies apply across the board. Agencies in Montana face issues similar to those in Alberta, and although the cultures are different, Florida faces similar problems to the island of Java, for example. The issues relate to the scale of government: Smaller nations face problems similar in scale to U.S. states, whereas large nations, such as China, India, or Russia, face national and regional problems similar to those of the entire United States. In all cases, organization of water agencies, particularly those dealing with planning and coordination, is a political issue.

In the sections that follow, I discuss experiences at the national level and in three state governments: one in the semiarid West, one in the humid Southeast, and one in the mid-Atlantic coastal region. In other parts of the book, other types of agencies are discussed.

The National Level: The United States

Types of agencies

At the national level, service delivery functions are done by "mission agencies," while regulatory functions are housed in agencies that administer laws related to particular industries or environmental and social concerns. The Defense Department houses the Army's Corps of Engineers, which manages a program of civil works that includes navigable waterways, reservoir construction and operation, assistance to urban areas, and comprehensive water planning. The Bureau of Reclamation emerged from the decision to settle the West by providing stable supplies of irrigation water. The U.S. Geological Survey emerged as the map maker, the geological investigation agency, and eventually as the national water survey. The Environmental Protection Agency was assembled incrementally from programs that began with public health and later expanded to the environmental arena. The U.S. Geological Survey has assumed the assessment function, an important part of planning, but as a data agency it stops short of actual planning and coordination (except for data coordination).

Planning and coordination

As described in Chap. 4, the search for a model for planning and coordination spans the period from the New Deal to the Reagan administration, about 50 years. The players have been the President, Congress and its committees, federal agencies, state governments, and diverse interest groups. By 1981, the concepts of planning and coordination as embodied in the Water Resources Planning Act (WRPA, Public Law 89-90, 42 U.S.C. 1962) were pretty much dead. What happened?

The Water Resources Council operated for around 15 years. However, it got into disputes about a number of issues, including the Principles and Standards, which were seen by some as impediments to project development. By 1981 the concepts of the Water Resources Planning Act were pretty much dead. The United States does not now have a formal mechanism for national water resources planning and coordination. There seems to be a consensus that the action for water planning and coordination should be at the state and local levels, with policy setting by national legislation.

State Water Agencies

Across the United States, 50 state governments face the same question: how to organize their water agencies for economic development, to meet social needs, and to protect the environment. In state governments, as with the federal government, service delivery is by "mission agencies," while regulatory functions are housed in agencies that mainly administer federal laws and their corresponding state statutes.

State water development agencies work toward the goals of resource *development,* as opposed to resource *protection.* As the United States has developed more stringent environmental laws, the protection goal has come to dominate in state governments, mostly since about 1970. Prior to that time, in both state and federal governments, the development agenda was dominant.

It is clear that the functions of state agencies divide along the lines of water industry organization: regulatory, service, support, and planning/coordination. Figure 23.1 shows some of the functions of these agencies as they appear in state governments. The service agency is shown as a water development authority, much as it would appear in a state where the state itself built and operated projects.

North Carolina's water agencies

The first state case is North Carolina's water agencies. I will focus on the state's Division of Water Resources (NCDWR), a planning and coordination agency housed within a larger department that also contains regulatory and resource development agencies. The organizational history of the NCDWR illustrates differences among regulatory, project development, and planning functions in a state water agency. My view was from 1979–1982, when I worked with the agency.

In February 1979, the water programs were being evaluated by their parent, the North Carolina Department of Natural Resources and Community Development (now the North Carolina Department of Environment, Health and Natural Resources, DEHNR). One of the issues was conflict between its regulatory and planning missions as they related to the water sector. The allegation was that the smaller programs were to some extent being neglected due to the bureaucratic dominance of the water regulatory programs.

The division director had developed a "functional" structure that featured three sections: environmental planning, environmental operations (permits and enforcement), and program support. Each had a section chief and a budget. To the division director, and to some others, the organization made functional and programmatic sense. To others, it mixed apples and oranges. The apples were regulatory programs and the oranges were project planning and assistance functions.

This reorganization, which did not last long, was only the latest in a long list of changes in the organization's development, which traced back to the 1920s (Howells, 1990). The dual nature of the concern with water resources, health and development, had been apparent for a long time. This duality still affects attempts to rationalize the control and

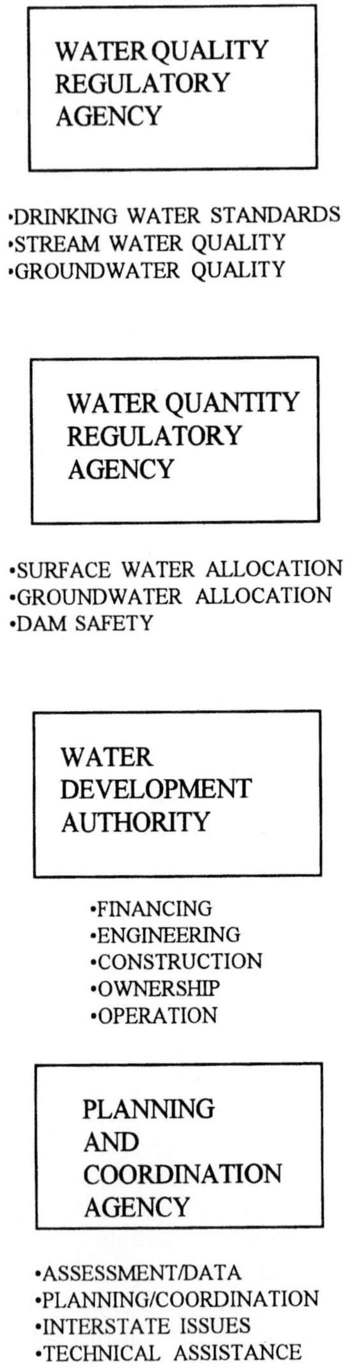

WATER QUALITY
REGULATORY
AGENCY

•DRINKING WATER STANDARDS
•STREAM WATER QUALITY
•GROUNDWATER QUALITY

WATER QUANTITY
REGULATORY
AGENCY

•SURFACE WATER ALLOCATION
•GROUNDWATER ALLOCATION
•DAM SAFETY

WATER
DEVELOPMENT
AUTHORITY

•FINANCING
•ENGINEERING
•CONSTRUCTION
•OWNERSHIP
•OPERATION

PLANNING
AND
COORDINATION
AGENCY

•ASSESSMENT/DATA
•PLANNING/COORDINATION
•INTERSTATE ISSUES
•TECHNICAL ASSISTANCE

Figure 23.1   Organization of state water agencies.

protection of water resources. In 1959 an act was passed establishing the Department of Water Resources and creating a Board of Water Resources. Later this department became the Department of Air and Water and Air Resources, and later still it was merged into the new Department of Natural and Economic Resources (1971), which became the Department of Natural Resources and Community Development (1977) and later the Department of Environment, Health, and Natural Resources (1989).

The organization in 1989 of the DEHNR completed the transfer of water-related offices and functions from the Health Department to the new, comprehensive department. It marked the evolution of a an initial concern centered on health to one that attempts to take a balanced look at health, environment, and development.

Alabama's water resources agencies

Like North Carolina, Alabama is located in the southeastern United States. The state, with a population of about 4 million, is heavily dependent on water-using industries. Its water resources programs also came under scrutiny, but the scrutiny was caused by drought problems in the 1980s. The governor established the Alabama Water Resources Study Commission (AWRSC) by executive order (Alabama Water Resources Study Commission, 1990). The work of the commission, and the subsequent legislative program, illustrate the evolution of a planning and coordination agency.

The duties of the AWRSC were to determine roles of the three levels of government and the private sector; to study the balance of water supply and demand and related problems; to evaluate Alabama's water laws and planning and coordination process; and to develop appropriate policies.

The commission went to work and involved a wide group of organizations and interest groups in its planning process, including a technical advisory committee with 10 federal and 15 state agencies, and 13 study committees that involved over 270 representatives of federal, state, and local governments, special-interest groups, and private businesses.

The list of state agencies that engaged in some aspect of water resources management illustrates the need for coordination:

- Regulatory agencies: Alabama Department of Environmental Management; Attorney General's Office; Alabama Department of Public Health; Alabama Public Service Commission
- Planning, development, or research agencies: Alabama Development Office; Alabama Department of Economic and Community Affairs;

Industrial Relations; Alabama Water Resources Research Institute; Geological Survey of Alabama

- Resource management agencies: Alabama Department of Agriculture and Industries; Alabama Forestry Commission; Alabama Department of Conservation and Natural Resources; Alabama Association of Conservation Districts; State Docks; Alabama Soil and Water Conservation Committee

Also, federal agencies involved in water resources participated in the study. They included the Corps of Engineers (Mobile District), the Environmental Protection Agency, the Farmer's Home Administration, the Fish and Wildlife Service, the U.S. Geological Survey, the National Weather Service, the Soil Conservation Service, and the Federal Energy Regulatory Commission.

Many industries are heavily involved in water issues in Alabama. The power companies, notably the Tennessee Valley Authority and the Alabama Power Company, are quite active. Both own and operate a series of mainstem reservoirs. Pulp and paper, chemical industries, coal development, and textiles are industries that are major players in Alabama water management circles. Also, the state's navigation systems, particularly the new Tennessee Tombigbee Waterway, are deeply involved in water issues.

The AWRSC found that Alabama has abundant water resources, but like many humid regions, it suffers from periodic maldistribution of water, contamination, and a group of issues that result from neglect of water resources management. Major issues identified by the commission were abandoned wells; citizen awareness; declining groundwater; droughts; environmental concerns; financing; floods; saltwater intrusion; septic tanks; surface water depletion and transfers; water quality; water quantity protection for future growth; and water resource management.

The overarching issue was coordination and policy setting. The commission stated (1990): "Alabama's lack of a state focal point and the virtual absence of emphasis placed on total water resources management has led to a wide gap in expertise. . . . It is urgent that Alabama close this gap in order to protect the state's share of interstate waters, effectively invest in projects of state significance, and assure adequacy and quality of water supply in the future."

Results of commission recommendations.   As a result of the commission's studies, the Alabama Water Resources Act was introduced into the legislature in 1991. It proposed an office of water resources and a water resources commission to launch a comprehensive assessment of the available water resources and to promote the coordination, development, conservation, enhancement, protection, management, and use of water resources. In effect, the agency would undertake most of the functions listed in the previous section.

The bill also provided for a regulatory program for water quantity which would begin to initiate restrictions on water withdrawals and uses in the state. Thus, it mixed planning and coordination with regulatory authority.

The bill failed in both 1991 and 1992. Reasons were opposition to the regulatory program, the financing program that would assess fees on water withdrawals, regional interests, and interagency competition.

The bill finally passed in 1993, and the Office of Water Resources was authorized on a tentative basis. The program to finance the agency by levies on water withdrawals was removed, and the regulatory program has not been authorized yet.

Colorado's water agencies

Organization of state water agencies in Colorado shows how conflicts in a water-short region can affect agency organization. Like other states, Colorado has a resource agency and a regulatory agency, and the planning and coordination function is not well organized. However, the regulation of water quantity in a water-short state takes on a much bigger focus than in humid states such as North Carolina and Alabama. The result is that in the resource agency, the Colorado Department of Natural Resources, water planning and the regulation of water quantity are divided into two agencies: the State Engineer's Office (Division of Water Resources) and the Colorado Water Conservation Board (CWCB).

The State Engineer's Office (SEO) is responsible for administering the surface and groundwater resources and for overseeing all diversions of water. It also administers the dam safety program. The Colorado Water Conservation Board is basically responsible for planning and coordination, but its functions illustrate problems inherent in water planning in environments where water is scarce. This is a complex subject, and a separate chapter is devoted to the general problems of Western water and the resulting conflicts.

Thus, Colorado has three powerful water agencies, the State Engineer's Office, the Colorado Water Conservation Board, and the Department of Health. Other agencies have roles as well, but these three administer the main programs described in this book (although another agency, the Colorado Water Resources and Power Development Authority, has important financing roles). In the next two sections I will provide more detail about the SEO and the CWCB.

Organization of the State Engineer's Office. Basically, the SEO office operates through seven division engineers located in river basins and a central staff in Denver. Each of the water divisions has a water court which is part of the state district court system. The water courts decide on changes in water rights, and the SEO administers them through the division engineers and their water commissioners, who oversee subdivisions of river basins.

Organization of the Colorado Water Conservation Board. As a state, Colorado has had difficulty in developing a policy on planning and coordination. The CWCB is the closest thing to a planning agency, but its members would deny that they engage in state-wide water planning. They might admit to some coordination, however.

The CWCB was organized in 1937 to serve as the state's agency to promote state-wide water development and to engage in project planning. It evolved over time into the agency we have today, which is the agency authorized to conduct state water planning under the Federal Water Resources Planning Act.

Actually, the CWCB made an attempt in the 1970s to develop a state water plan, but it was never published because of conflicts. Today, the term "water planning" is a sensitive one among water managers because of these conflicts. Nevertheless, planning and coordination do occur.

The CWCB staff carries out research on meeting federal legislative goals such as the Endangered Species Act, and it serves as a forum for water user groups to make recommendations on policy issues. With its statutory powers, most of the CWCB's activities are similar to those of other state planning and coordination agencies: to provide assistance to agencies and districts; to cooperate with the federal government; to prepare legislation; to study plans about interstate waters; to represent the state in interstate disputes; and to promote water conservation. Other authorities, such as to acquire land for flood protection or to acquire water rights, deal with specific goals of the state in water management.

Summary and Conclusions

From the experiences of the United States and three state governments, a few conclusions may start to explain the complex issues of water agency organization.

1. Water agencies are, for the most part, subject to the same political forces as other government agencies. This means that there will be frequent changes in philosophies and reorganizations, it is difficult to undertake "rational" approaches rather than "political" approaches, and there will be continuing conflict about means, ends, and priorities.

2. At the local water service level, it may be possible to isolate an agency as an "enterprise" or even to privatize a service and thereby escape the political buffeting, but at the state and national levels, agencies are where conflicts must be resolved, so they must be organized for coordination, for full disclosure of information, and to provide methods to work out differences over political objectives.

3. Government organization is driven by programs that are funded as a result of specific legislation. The legislative committees, interest groups, and agency staffers acquire expertise and interest in maintaining the programs, and turf wars develop. This makes the coordination task difficult.

4. Clear differences must be maintained among the functions of regulatory, resource management, and planning and coordination agencies. When these are mixed, political support for the agency programs may die and the agency may fail due to deauthorization or lack of funds.

5. Conflict is inherent in organizational structure, and matching the roles to the goals of the players is critical to government agency effectiveness. Unless roles are properly defined, confusion will dominate.

6. It is important to maintain a competent permanent agency staff and not rely too heavily on temporary political appointees. This will provide continuity in the "institutional memory." A good set of files and library in the agency is also essential. Of course, checks and balances between political and career employees are needed.

7. Although there are political differences between countries and regions, the generic lessons about organization of water agencies apply across the board.

8. With comprehensive water management being recognized as an intergovernmental responsibility, with increasing leadership required from state governments, the organization of state water agencies should receive greater scrutiny. Moreover, state governments ought to organize their water agencies with an eye to their own needs, not just to spend federal grant funds.

9. Regulatory agencies and resource management agencies will receive more attention than planning and coordination agencies due to the financial and political stakes involved in water management decisions. However, planning and coordination agencies (and data agencies) need more support and attention to make sure that water management decisions are coordinated.

10. It has historically been more difficult to carry out the functions of a water planning and coordination agency where water is short than where it is abundant. However, the emergence of environmental rules, interstate difficulties, and the need to regulate withdrawals are increasing the difficulty even in humid areas.

11. On a rational basis, it might be said that the provisions of the Water Resources Planning Act made up an ideal model for

federal–state water planning and coordination, but the model failed in the United States because of bureaucratic and political forces that affect water planning, coordination, development, and regulation. Now the nation must adapt to the failure and find a better model for planning and coordination.

12. It is a fact that some of the actors in water management will resist research, data collection, and coordination for political or self-interest reasons.

Questions

1. Activities of some water agencies, particularly at the state and federal levels, are driven by programs that are funded as a result of legislation. One of these agencies, the Corps of Engineers, has a working paper that identifies "stovepipe apartheid" as a problem in the agency. How does the funding by programs relate to stovepipe apartheid?

2. Of the problems caused by bureaucratic forces, budget competition, and political or personal conflicts, which in your opinion is the greatest cause of concern in the water industry and why? Can privatization help to cure any of these problems?

3. Throughout this book, the need for coordination in the water industry has been stressed. Why, in your opinion, did the Water Resources Council not provide the needed coordination at the national level?

4. What is the difference between a resource *development* agency and a resource *protection* agency? Does it matter if they are combined or separate? Why?

5. Can the enterprise principle be used for all water agencies? Why or why not?

6. What are the arguments for long-term staff in a water planning and research agency versus more frequent turnover and political appointments?

7. In 1995, some members of Congress proposed eliminating funding for the U.S. Geological Survey. Are water research and data agencies essential, or can these functions be handled by the private sector?

8. In the analysis of the Alabama Water Resources Agency, one question was whether it should be organized together with the Department of Environmental Management, which regulates water quality. Give an argument for putting the two agencies together and an argument for not putting them together.

References

Alabama Water Resources Study Commission, *Water for a Quality of Life,* Montgomery, October 1990.

Howells, David H., *Quest for Clean Streams in North Carolina: An Historical Account of Stream Pollution Control in North Carolina,* Water Resources Research Institute, Raleigh, November 1990.

24

Water Management in the Western United States

Introduction

Water management in the Western United States has become a battleground with a number of major policy issues for the nation, and it contains lessons that can be transferred to other regions. This chapter describes the nature and complexity of the issues, explains the national as well as regional aspects of the problems, analyzes the proposals for reform, and suggests some solutions. The goal is to present a balanced view of the situation, but I have to admit up front that no one involved in Western water will acknowledge that anyone else has a balanced view.

Water is said to be the life blood of the West, and in fact, this large and diverse region was only able to develop through a thorough exploitation of its water. The region depends on its water supply for everything from watering urban areas of southern California to producing cheap hydropower in the Northwest to preserving endangered whooping cranes in Nebraska. The limited water and the many interconnecting issues in the region explain the complexity of Western water and present special challenges.

As a result of its critical functions, Western water is high on the list of national policy issues, especially policies dealing with the environment. Much of the conflict surrounding Western water can be explained by the tension over development pressures and environmental values.

Simply stated, the environmental position is that water developers have raped the land and dammed too many streams to make sure that water-wasting agriculture gets all it wants. Water developers would say that the problem is caused by naive perceptions of new settlers who do not understand water development and storage and who do not appreciate where their food and water comes from. The truth is much more complex than either of these positions, and changes are inevitable. Indeed, water development in the West is, for all practical purposes, dead, and the environmentalists seem to have the initiative. Some, but not all, of the water issues are national, not just Western, and dealing with all of the these issues is an important national priority.

Topics related to Western water management are covered in other chapters. Chapter 10 covers the Two Forks case, which involves water users all the way from Omaha to Los Angeles. The Colorado River includes a number of case situations, from river basin management and the decision support system (Chap. 19) to the Colorado Big Thompson Project (Chap. 12). The Platte River case illustrates several issues, ranging from interstate compacts to preservation of endangered species (Chap. 19). The state engineer function in Western water management is described in Chap. 15, and Western water politics are described in the discussion of the Colorado water agencies (Chap. 23).

Western Water Issues

Western water issues revolve around conflicts between economic and environmental interest groups, water law, property rights, regional issues, and turf between agencies.

In the generally arid West there is heavy demand on limited water supplies, and streams have been heavily developed for storage and diversion. Much of the water, some 80 percent, is used by agriculture, and some say that it is wasted and improperly subsidized. Some allege that the region's water code, the appropriation doctrine, is out of date and ineffective. Farmers are feeling pressure from growing urban regions that may take their water and put them out of business. It is not easy to build new dams even when the water is available and needed due to growth and drought. The region has diverse environmental issues to deal with, including endangered species. Native Americans are claiming more of their water entitlements. Water marketing may begin to move water from one place to another, exacerbating some conflicts. The result of all of this is a spate of government and policy issues and much conflict.

The problems are predictable if we consider everything the region has been subjected to: transition from a frontier to a rural agricultural and mining region in the period 1850–1900; urbanization and intensification of agriculture from 1900 to 1945; exploding growth, especially in California, from 1945 to present; and now a postagriculture and mining period characterized by environmental activism.

The history of the West, including water development, is being rewritten. In earlier times the history told a tale of conquering the wilderness. Now emphasis is on the other sides of the story, with focus on Native American issues, environmental damage, and careless government policies. Some call the current histories "revisionist." Whatever your opinion, you must acknowledge that the wilderness-conquering paradigm is too simplistic.

Marc Reisner's (1987) book, *Cadillac Desert,* presents the story of water development, and generally assigns the "black hat" role to the water buffaloes such as William Mulholland, father of Los Angeles' water system, and to U.S. Bureau of Reclamation engineers. Reisner and Bates (1990) presented a sequel to *Cadillac Desert, Overtapped Oasis,* to put forward an analysis and proposals for reform. It presents the environmental side of Western water issues, and its main proposals will be summarized later.

The source of the problem is that there simply is not enough water in the West to supply all needs, from high-intensity agriculture, the biggest user, to giant urban centers such as Los Angeles and Phoenix, while still leaving enough in the streams to nourish fish and wetlands. The West is generally arid and has limited water supplies. Even the mighty Colorado, draining most of the West, is a midget stream compared to those in humid areas.

Economy, environment, and the appropriation doctrine

The issues between environmentalists and water developers explain much of the conflict. In general, the appropriation doctrine did not provide water for natural systems and wildlife; these were not even considered as beneficial uses under the property rights system through which ownership to water was assigned.

In the early days of development, most of the water was needed for agriculture and farmers staked out the water rights according to the appropriation doctrine, the West's water code, which gives first water right to the first appropriator: "first in time, first in right." Newer settlers allege that farmers holding the older water rights waste water on inefficient agriculture, such as irrigating hay. They also claim that the farm water which is provided by government projects was developed through inappropriate subsidies. Cities are buying water from farmers through water marketing, and farmers are worried that this will put them out of business. Farmers selling water are placed in conflict with those holding on.

The appropriation doctrine varies from state to state, but in general it is based on filing a claim for water (appropriating it); diverting the water from the stream; putting it to a beneficial use; and then having the water right adjudicated in state court. Principal beneficial uses for water were agriculture, mining, urban development, and industrial supply. The doctrine is not considered to be equitable; it is meant to be efficient. During a drought, for example, if there is not enough water for two water rights, rather than apportion the shortage, the doctrine will give all of the water to the senior water right. However, by assigning water according to priority, the doctrine enables water to be "administered" efficiently. The doctrine in its pure form does not provide for leaving water in the stream for fish and natural systems; this would not be a diversion. However, most states are providing modifications to overcome this problem. For example, Colorado has enacted an in-stream flow law that allows one state agency, but not private owners or environmentalists, to file for in-stream flow rights.

Up to now the appropriation doctrine has not satisfied environmentalists, but water developers and associated economic development interests have been generally happy with it. Environmentalists argue that the doctrine provides no answers to the West's environmental problems. These include lack of in-stream flows, deteriorated water quality, endangered species, dried-up wetlands, water for wilderness, and wild and scenic rivers. However, a fierce debate is raging over the appropriation doctrine. The revisionist argument is that it simply is not adequate for today's needs and should be discarded. The protectionist argument is that it is the best system that has been devised and should be retained, albeit modified for today's changing needs.

Interestingly, one of the attractive features of the appropriation doctrine is that it provides for water marketing, the sale of water property rights from one owner to another. Environmentalists generally like this feature because it provides for cities to buy water from "wasteful" farmers and move it to the cities. The environmentalists can then put public pressure on cities to expand supplies through conservation rather than to build new storage reservoirs. This scenario was one of the features of the recent Two Forks project veto in Denver. A problem with water marketing is that it requires high transaction costs. This is a reason for the jokes about so many water lawyers and engineers in the West.

Property rights

The revisionist arguments about the appropriation doctrine suggest that the system should be abandoned for something new that would protect the "public trust"; in other words, the state should hold water rights in the public interest. Of course, what constitutes the public interest is a subject of debate. How water right owners would be compensated for water they would lose under the public trust doctrine has not been explained. This can be a source of much conflict, because a farmer's water rights entitlement may be the main basis of family wealth. Arguments that water right owners do not own the water, just the right to use it, and that the "public" also has rights, seem hollow to water right owners who are bent on protecting what they perceive as their property.

Another property rights issue is subsidies to farmers, specifically subsidies that result in providing low-cost water to them from government projects. If they raise subsidized crops, they are said to be "double dippers." This issue claims a good share of attention from environmentalists and is embroiled in more complex debates over national agricultural policy in general.

Regions and interest groups

Much of the conflict in Western water is between regions, including regions within states and between states. Interstate issues include problems such as allocating the Colorado River, managing the Missouri, and conserving the groundwater in the Ogallala Aquifer.

Time magazine's cover article about the Colorado River stresses the interstate conflict over that key river, and these issues may be the most visible in the nation (Gray, 1991). The Colorado provides water for 20 million people in seven states and for 2 million acres of farmland. Seventy percent of San Diego's water comes from the river. Arizona is taking water for Phoenix and Tucson. The Imperial Valley, grower of a major share of the nation's vegetables, depends on the river. A prolonged California drought reached its fifth year in 1991, and cities are starting to desalt water. The 1922 Colorado River Compact is based on overestimated water supplies: It counts on 16.5 million acre-feet, but the river has produced only 9 million acre-feet during the drought. The lower three states, California, Arizona, and Nevada, have gotten more than their share; now the upper four states worry that this overuse may become institutionalized.

Fights within states over water are also a source of conflict. Water supply for Las Vegas is a current source of conflict. There, water is needed for new casinos and urban settlers and the proposal is to take it from farmers elsewhere in the state. In Colorado the "East Slope–West Slope" battle over water has gone on for decades. The water is on the West Slope and the people are on the East Slope. Recently, a private company, American Water Development, Inc. (AWDI), has proposed exporting groundwater from a largely Hispanic farming region, the San Luis Valley, to Front Range urban users. This private business venture, an example of water marketing, has generated tremendous conflict. The claim was thrown out by the Colorado District Court.

An example of interest-group issues is Native American water rights. Tribes are claiming their water rights, which they say have been pushed aside. Tribal interests can be in conflict with environmental concerns, increasing the complexity of Western water. An example of this is the Animas-La Plata project in southwestern Colorado, where Native American claims are to be settled by new water development that has been alleged to threaten endangered fish species.

Turf between agencies

Another source of conflict is between state and federal agencies and between different water management agencies. The veto of the Two Forks dam near Denver has raised questions in the West about the extent of state and local rights to manage water and make decisions. All of the plans and financing for the project were state and local, and approvals for the proposed water supply reservoir had been obtained from several federal agencies, but the project was vetoed in Washington by the Administrator of the Environmental Protection Agency. Other examples of state–federal issues include federal reserved water rights, federal water quality law overlays over state water rights, and enforcement of the Endangered Species Act.

Local skirmishes over water management are common in the West. At stake is the right to control development. In the Denver region, for example, the Denver Water Department had teamed with over 20 suburban water agencies to plan and develop the Two Forks project. Getting to that point had required years of negotiation, a governor's roundtable for conflict mitigation, and much personal political work. One of the arguments for the project was that it was required for metropolitan cooperation, to hold the cooperators together. This kind of regional conflict is typical of water management in general, and is described in my article, "Regionalization in Water Supply Systems: Status and Needs" (Grigg, 1989; see also Chap. 21).

National, Not Only Regional, Problems

Western water is to some extent a subset of national water problems and issues. The national issues found in the West are conflicts between environment and development, pressure to find water supplies for urban areas, surface water quality, and groundwater contamination. The differences are in the scale of the problems, in the aridity in the West, in the reliance in the West on irrigation, and in special regional concerns such as Native American water rights, public lands, and wildlife issues.

National water problems have been reviewed frequently in the past decade, many of the reviews being front page media events asking if there is a "water crisis." Most of those working with water consider it a policy and financial crisis, not a water crisis. Most of the problems, contamination or supply, can be solved with good plans and policy and enough money. However, Western water issues may be more intractable than those in the East due to the tight hold on water property rights in the West.

Although there are many stories and articles, the issues reviewed remain the same. One review, by Harvard Professor Peter Rogers in *Atlantic Monthly* (1983), discussed the following problems: health hazards from drinking water, groundwater contamination, rationing the Colorado River, providing fishable-swimmable waters, depletion of groundwater in the Ogallala Aquifer, poor public planning, hazardous wastes management, excessive consumption in the United States compared to Europe, poor pricing policies and too much subsidy, losses from leaky pipes, and expensive and wasteful projects.

The problems can be placed into four categories of issues: contamination, supply, environment, and cost. Contamination includes concerns about safe drinking water, stream pollution, nonpoint-source runoff, contaminated groundwater and coastal water issues, including bays, estuaries, and ocean pollution. Supply includes municipalities not running out of water, providing security against drought, and conflict over transfer from river basins. Environmental issues cover a broad band of emerging concerns—in-stream flows, wetlands, endangered species. Cost requires changing the management paradigm, to gain public acceptance of higher water and sewer bills and for conservation.

These problems differ across the regions of the country. In the Northeast, high population density, old industrial sites, and smokestack industries have induced water supply problems, deteriorating infrastructure, and groundwater contamination. One of the most visible problems has been Massachusetts's $6 billion Boston Harbor cleanup, a problem that became an issue in the 1988 Presidential campaign. In the Southeast, Florida leads in visibility for water problems, with the focus on South Florida's population increase and pressure on the fragile Everglades environment and underground supplies. Atlanta's water supply problems, caused by growth of a supercity on watershed divides, are placing stress on regional supplies and river basins, and may lead to a problem much like the interstate water wars of the West. Several states in the Southeast have spotty water issues such as groundwater overdrafts and drought problems. Irrigation has increased greatly in South Florida, Georgia, and the coastal areas of the Carolinas, and has placed pressure on groundwater supplies in the delta region of Mississippi and Arkansas.

These Eastern water problems differ in focus and intensity from those of the West, where the needed solutions are different due to the water laws. In the East the debates are mainly over money and policy. In the West, lifestyle changes are needed as well as policy and money, and these will focus on the approach to agriculture and natural resources management.

Complexities and Conflicts

The complexities of water issues cause much of the conflict. Water resources management is multidimensional, multiobjective, multidisciplinary, and interregional. There are many different ways to look at the problems. The trick is to integrate the differing views into a solution, an impossible task if all interest groups are to be completely satisfied. A feasible solution space usually does not exist after all the different objectives are considered, and trade-offs and compromises are required.

Dimensions of water management include discipline, space, political position (horizontal and vertical), ecology, and function. The discipline dimension arises due to the boundaries between the disciplines involved in water issues. Biologists do not speak the same language as engineers or lawyers, and economists have different agendas than political scientists. The space dimension refers to geography and water accounting units, such as cities, regions, river basins, and states. Water accounting units are normally different than political units. Political dimensions span vertical issues such as state–federal, and horizontal conflicts such as interregional problems or problems between interest groups. Ecological dimensions deal with water systems complexities and ecological issues. Functional dimensions refer to water management purposes such as urban water supply, fish and wildlife, and others. The time dimension deals with the stage of problem solving: policy issues, operational issues, for example.

The complexity caused by the multidimensional nature of water generates conflict due to lack of knowledge, ignorance, and uncertainty. The answer is research, analysis, and education at all levels. Dealing with systems complexity requires mathematical models to even begin to make sense of river basins. Take the Platte River system, for example. On the Platte, some of the water comes from its own tributary basins; some comes from the Colorado River system through interbasin transfer. Considering large-scale problems on the Platte, one must analyze issues all the way from Los Angeles' water supply needs to water scenarios for the whooping cranes in the Big Bend region in Nebraska. Ecological complexity involves the natural aspects of the streams, such as interaction between surface water and groundwater, reuse factors, ecological aspects, vegetation, wildlife, and fisheries. Value complexity deals with the many conflicting value systems between and among groups. This is due to interdisciplinary issues and other factors.

While unraveling complexity in water management is a necessary part of any solution, it is not sufficient. Final resolution of water conflicts must be legal, financial, and political.

Policy Prescriptions

There have been many national and Western water policy studies, and it seems that the recommendations are repetitive. Why are they not implemented, and what might be done? This section presents some of the main policy proposals in categories and then suggests how they apply to Western water issues. However, I make no claim to being exhaustive.

Documents used to compile this summary are the recent Harvard study (Foster and Rogers, 1988), a discussion of the National Water Commission report ("Whither Federal Water Policy," 1989), a conference report by the American Water Resources Association (Born, 1989), Marc Reisner's and Sara Bates's book (1990), and a state government policy study (Wilson, 1978).

The national policy suggestion that seems to appear most frequently deals with uncoordinated federal water policies and programs, and relations between federal water agencies and states. This deals with the political dimension of water management. A commonly proposed solution is a federal government coordinating body, some kind of President's water council with an independent chairperson. This was, of course, a goal of the Water Resources Planning Act of 1965.

Another frequently cited problem is that water regions are not the same as political entities. A proposed solution is to have regional councils for each key water problem region of the country. Again, the River Basin Commissions feature of the Water Resources Planning Act dealt with this need.

Another group of problems deals with knowledge, data, research, and education. One problem is uncoordinated and unavailable information and data. A proposed solution is a national water information program, perhaps to establish a clearinghouse of information and policy analysis. Related to this is the problem that there is not enough knowledge or water research. The proposed solution is the renewal of the national water resources research program, possibly including a national water extension service. This solution would aim at another related problem, that water issues are too complex, and not understood by the public, with the solution being better educational efforts at all levels to raise awareness. Another related problem is that policy analysis is not adequate, and the proposed solution is that key inquiries and investigations should be launched.

On the economic front, the problem is that water pricing is not effective. The proposed solution is to introduce modern pricing and water marketing to all federally produced water, and to undertake other needed pricing reforms, such as using conservation rates for urban water.

The policies mentioned above occur mainly in national studies. Reisner and Bates (1990) suggest Western water policy needs and concentrate on policies for water transfers, solving environmental problems, and improving water use efficiency. They propose that federal and state governments adopt policies to enhance the possibility for water transfers, that subsidies be reduced, that special environmental problems be tackled, that basins of origin be protected, that in-stream flows be enhanced, that the public trust doctrine be introduced, and that water use efficiency be improved.

Policy suggestions have been made in other places about a host of technical, financial, legal, economic, and environmental problems. For the most part I view these as tactical measures rather than grand strategies. A few examples include: groundwater, nonpoint sources, infrastructure funding, championing third-party interests, monitoring rules of decision making, supporting planning and management efforts of the states, developing national standards, providing financial and technical assistance, managing investment in water projects better, having state primacy, assigning the federal role in program development, having more comprehensive water management, having federal actions consistent with state and interstate policies, providing continuity of federal support, having greater flexibility in federal support, refining criteria for federal water program and project evaluation, reforming cost sharing, concentrating on water conservation, expanding federally supported water research, and studying Native American and federal reserve water rights within the framework of state legal systems.

Conclusions

Water management in the West is complex due to the many interconnecting policy issues dealing with economic and environmental interest groups, water law, property rights, regional issues, and turf between agencies. There is not enough water in the West to satisfy all of the competition between groups and regions. Many of the issues are national, but the West's problems are larger in scale and contain more complexities and conflict than those in the East.

While unraveling the complexity in Western water management is necessary, it is not sufficient to bring peace in water policy. Final resolution of water conflicts must be legal, financial, and political.

The policy strategies that have been suggested include coordinating federal and state water policies; providing for regional solutions to water problems; providing the necessary research, data management, education, and policy analysis programs; reforming water pricing; and in the West, providing more flexibility in water transfers, increasing efficiency in water use, and remediating past environmental problems. These reveal that the underlying water management integration issues are: roles, finding political integration; regions, finding geographic integration and equity; knowledge about water, including research, education, information, and policy choices; pricing, including efficiency and distribution issues; and equity issues between interest groups.

The indicated policy question is: What is needed in policy to force this integration in water management? The jury remains out on this question, and it will continue to dominate the policy agenda.

The clash of values between environmental and developmental groups and issues surrounding property rights in water seem sure to prevent the visionary dreams of planners from coming true. That is, water conflicts will not be resolved in reasoned negotiation sessions between interest groups; they will be resolved in elections, court battles, bureaucratic rule making and decision making, and water right purchases.

When *Time* magazine added to the Western water controversy by making it the subject of its cover story on July 22, 1991, it concluded: "there's no more urgent task for the West than ensuring that the Colorado survives." This might be said about all of the Western water controversy: There is no more urgent policy agenda for the West than finding the balance in water policy, resolving conflict by unraveling complexity, and finding legal, financial, and political solutions.

Questions

1. Explain the basic cause of Western water conflicts between economic and environmental interest groups, water laws, property rights, regional issues, and turf between agencies.

2. Are these conflicts unique to the West, or do you expect them in the East as well? What are the differences between Western and Eastern conditions?

3. *Time* magazine stated: "there's no more urgent task for the West than ensuring that the Colorado survives." Do you agree with the urgency of this statement?

4. How did California's State Water Project (see Chap. 12) contribute to Western water conflicts and issues? What were the conflicts that it generated?

5. Is the debate over growth management an underlying factor in Western water conflicts? If so, how?

6. Do you believe that new reservoirs can be built in California? Physically? Politically? If not, how can the state accommodate its growth?

7. The Bay-Delta issue (Chap. 22) is an interface between Western water and coastal water issues. What are the positions, roles and activities in the Bay-Delta issue?

References

Born, Stephen (ed.), *Redefining National Water Policy: New Roles and Directions,* American Water Resources Association, AWRA Special Publication 89-1, Bethesda, MD, 1989.

Foster, Charles H. W., and Peter P. Rogers, *Federal Water Policy: Toward an Agenda for Action,* Harvard University, Energy and Environmental Policy Center, John F. Kennedy School of Government, August 1988.

Gray, Paul, The Colorado: The West's Lifeline Is Now America's Most Endangered River, *Time,* July 22, 1991, pp. 20–26.

Grigg, Neil S., Regionalization in Water Supply Systems: Status and Needs, *Journal of the Water Resources Planning and Management Division, American Society of Civil Engineers,* May 1989.

Rogers, Peter, "The Future of Water," *Atlantic Monthly,* July 1983, pp. 80–92.

Reisner, Marc, *Cadillac Desert: The American West and Its Disappearing Water,* Penguin Books, New York, 1987.

Reisner, Marc, and Sara Bates, *Overtapped Oasis: Reform or Revolution for Western Water,* Island Press, Washington, DC, 1990.

Whither Federal Water Policy, *California Water,* University of California Water Resources Center, No. 3, Summer 1989.

Wilson, Leonard U., *State Water Policy Issues,* Council of State Governments, Lexington, KY, 1978.

25

Water Supply and Sanitation in Developing Countries

Introduction

One reference for this chapter is *Water for the Thousand Millions,* by Arnold Pacey (1977). For Americans, a "thousand million" is a billion, but the title captures the immensity of the task of providing water for the millions who live in poverty in developing nations. This chapter deals with the problem of safe water supply and sanitation to serve these people. Sustainable development seems an empty goal without meeting the needs of these people. Figure 25.1, a group of women in Mali using a common handpump, illustrates the kind of safe water systems that can be supplied on a low-cost basis.

While many of the issues are political, budgetary, and socioeconomic, the fundamental answers go to the heart of water management. As we will see in the chapter's conclusions, two general avenues of solutions are needed, institutional and managerial.

At the U.N. Water Conference at Mar del Plata, Argentina, the Intermediate Technology Development Group reported that just over 1000 million people in the world's rural areas lacked access to a safe water supply (Pacey, 1977). As a result of these findings, and others that were just as dramatic, the United Nations proclaimed the 1980s as the "International Drinking Water Supply and Sanitation Decade." The goal was to supply all of the world's population with safe drinking water and sanitation by 1990. Suffice it to say that this goal has

Figure 25.1   Women using water pump in Mali. (*Source: World Bank.*)

not been met; in fact, wars and other social and economic problems made the problem worse in many areas during the 1980s.

Curt Canemark, the World Bank division chief for the water supply sector, reported in 1989 that the coverage had improved, but that the greatest achievement had been the communication, awareness, and priority setting that had occurred to deal with the problem (Canemark, 1989). A statement was adopted in 1990 in New Delhi (at the Global Consultation on Safe Water and Sanitation for the 1990s) which formalized the need to provide sustainable access to safe water and sanitation, emphasizing the "some for all" rather than the "more for some" approach. At the 1992 U.N. Conference on Environment and Development in Rio de Janeiro, drinking water supply and sanitation was featured by Agenda 21 (the action program) as one of seven major program themes for the freshwater sector (United Nations, 1992).

Anatomy of the Problem

The root cause of the problem is, of course, the conditions that cause disorder, lack of opportunity, injustice, poverty, and the other factors behind misery in developing countries. High population growth, lack of opportunity in rural areas, and rural-to-urban migration are key factors that impede nations' abilities to keep up with the problem. These factors result in uncontrolled growth in urban areas, leading to shantytowns that go by many different names: periurban, squatter areas, urban slums, informal settlements, illegal settlements, urban land invasions, *barrios marginales* (Honduras), *tugurios* (El Salvador), *favelas* (Brazil), *pueblos jovenes* (Peru), *asentamientos populares* (Ecuador), *villas miserias* (Argentina), *bustees* (India), *kampung* (Indonesia), and *bidonvilles* (Morocco) (U.S. Agency for International Development, 1990).

According to the Water and Sanitation for Health (WASH) project, unique challenges of the periurban areas include extremely poor site conditions, high population density and diverse populations, lack of legal land tenure or official recognition of plots, high percentage of households headed by single women, bias against the urban poor, and lack of data. Accordingly, the periurban problem is more complex than rural or formal urban sanitation problems, and the complexity requires a comprehensive, interdisciplinary approach to understanding and solving the problem.

Dan Okun (1991), a professor of environmental engineering with over 50 years of experience in working on problems of water and sanitation, summarized his theories of the causes of the problems in urban areas in a 1991 paper to the National Research Council. The paper was for the Abel Wolman Distinguished Lecture, and how appropriate the topic was, given the enormous contributions of Abel Wolman in this field (see Chap. 19 for citations by Wolman).

Okun reported, to his regret, that in his 50 years of work on the water and sanitation problems in urban areas, they had gotten worse. The reasons were an inadequate supply of water in the cities attributable to limited water resources, and/or poor facilities for treating and distributing the water, compounded by an absence of proper sewerage.

What happens is that intermittent supplies of water create opportunities for infiltration of heavily contaminated water into the distribution systems when the pressure is off. Water-borne infectious agents then can reach taps, even when the water is safe as it leaves the treatment plants.

Infrastructure for both water supply and sewerage can be inadequate in the cities of developing countries, even when the city skylines are impressive. Wastewaters that are discharged into drainage channels can pollute wells and the groundwater table, and really create unsanitary conditions. This is especially true in the fringe areas, where many poor and landless families live.

At Colorado State University, Robertus Triweko (1992) did a thesis on A Paradigm of Water Supply Management in Urban Areas of Developing Countries. Triweko attributed the problems in urban areas to two causes: low level of service and inability to improve and maintain the continuity of service. His recommended paradigm would adopt a "comprehensive perspective" that explained a management system with technological, institutional, and financial subsystems. Results were to be measured by management efficiency and level of services. Triweko's "comprehensive perspective" illustrates the systems nature of this complex problem.

Lessons Learned

Responding to the International Drinking Water Supply and Sanitation Decade, the U.S. Agency for International Development (1990) organized the WASH project. The project's lessons learned were organized in terms of principles and lessons as follows:

Principle 1: Technical assistance is most successful when it helps people learn to do things for themselves in the long run.   Lessons were: local institution building is the key to transferring sustainable skills; technical assistance in water supply and sanitation requires an interdisciplinary approach, not a narrow, specialized one; a participatory approach, facilitation and not dictation, maximizes the chance for sustainable programs and projects; coordination and collaboration are important but often depend more on professional networking and personal relationships than on institutional and contractual relationships; and an active information service can expand the reach of technical assistance as well as its visibility and credibility.

Principle 2: Water supply and sanitation development proceeds most effectively when its various elements are linked at all levels.

Lessons were: water supply projects do not achieve their full impact unless they are linked first to hygiene education and then to sanitation; health benefits are the major, but not the sole, justification for support of water and sanitation projects, such projects also having wide economic benefits; behavioral changes combined with greater access to facilities are the basis for health benefits through improved water supply and sanitation; and a participatory approach to planning helps ensure linkages and cooperation in implementation.

Principle 3: The basic measure for success of both the national system for development and the community management systems it creates is sustainability, the ability to perform effectively and indefinitely after donor assistance has been terminated. Lessons were: successful institutional development projects strive for comprehensiveness and wide participation; training yields the best results when it employs participatory, experiential methods; full consideration of appropriate engineering design and application is essential to system sustainability; making plans for operations and maintenance before facilities are constructed and in place helps to ensure that sustainable technologies are selected; and plans for system finance that ignore the cost of long-term operation and maintenance are inadequate.

Principle 4: Sustainable development is more likely to occur if each of the key participants recognizes and assumes its appropriate role and shoulders its share of the responsibility. Lessons were: the national government role is to assume primary responsibility for sector management, including planning, donor coordination, policy reform, regulation, and institutional and financial aspects of development; the donor role is to provide coordinated support in the context of national plans; the nongovernmental organization (NGO) role is most effective if it is played out in the context of national development plans; the community role is to own and manage the facilities constructed and to be actively involved in decision making in all phases of project development; and private enterprise has a definite role in water supply and sanitation; that role is determined by the overall government strategy for the sector.

The "lessons learned" by WASH began with institutional issues, probably the most critical issue. WASH emphasized the need for broad participation, not just by government. Institutional development includes activities such as establishing autonomy in financial issues, personnel, policies, and planning; providing effective leaders and skilled managers; hiring, training, and retaining personnel; setting up administrative procedures and policies; maintaining a commercial orientation and financially sound practices; making sound technical decisions in all areas; creating a positive organizational culture; and providing quality services to consumers (Interagency Task Force, 1992).

Workers in international development have known for a long time that it is not wise to try to force U.S. standards on developing countries, where the technologies may not be appropriate. Accordingly, the concept of "appropriate technology" has arisen to describe the levels of technology that are correct, affordable, and sustainable for different situations. Albertson (1995) wrote: "Appropriate hard technology relates to engineering techniques, physical structures and machinery, that meet a need defined by the people being served, and utilize as much as possible the materials at hand or ready available." A good example of this technology is the pump shown in Fig. 25.1 or the canal construction in India shown in Fig. 25.2.

Albertson (1995) also explained that "appropriate soft technology" deals with the social structures, human interactive processes, and motivation techniques. It is the structure and process for social participation and action by individuals and groups in analyzing situations, making choices and engaging in choice-implementing behaviors that bring about change."

Figure 25.2   Lining the Rajasthan Canal in India with tiles. (*Source: World Bank.*)

Case Study: Infrastructure for Water and Sanitation in the Western Hemisphere

This case study was prepared for the Inter-American Dialog on Water Management, held in Miami during October 1993 (Grigg et al., 1993). In it, we reviewed the status of water and wastewater investments in the Western Hemisphere, focusing on the United States, Venezuela, and Brazil.

Although the problems in Latin America are formidable, progress has been made since about 1959 when the Inter-American Bank was organized. In fact, the bank's first loan was to expand a water supply and sewerage system in Peru (Inter-American Development Bank, 1992). In spite of progress, problems continue to increase due to population growth, urbanization, and industrialization. Although the percentage of people served has increased, the total number without service has actually increased. Another problem is that because safe drinking water is so important and has received priority, sewage systems have been neglected, at the expense of basic sanitation. An estimated 90 percent of all sewage in the region is still dumped untreated, and adversely affects local populations, especially in urban fringe areas with large, impoverished populations. In Latin America, a significant proportion of all disease is still attributed to polluted drinking water and untreated sewage, and water-related diarrheal diseases continue as the leading cause of infant mortality in many countries.

Venezuela

The case study of Mérida, Venezuela, illustrates the breakup of a national company into regional companies, similar to the case in Great Britain in the 1970s. In 1943, when the National Institute of Sanitation Works (INOS) was created, its main objective was to provide water supply and wastewater collection for the citizens. The administration of INOS became centralized, with a growing bureaucracy, leading to budget problems. Also, union activity became a factor in the operation of the institute. Complaints multiplied, and water supply began to be a topic of great concern.

In 1989, Venezuela began a new economic program that included a reorganization of the water supply sector. The specific objectives were to decentralize service by creating autonomous water companies at regional levels; to reach financial self-sufficiency and to equalize the financial operations of the regional companies; and to strengthen institutional aspects of the planning and management of the systems.

To reach these objectives a new organization was proposed. It was to involve a system of eight regional water and wastewater companies, called Empresas Hidrológicas. These Empresas Hidrológicas were to be coordinated by a national holding company called HIDROVEN. HIDROVEN would not have operating functions, and the regional companies would be free to sign contracts with private companies to perform specific tasks.

In the city of Mérida, located at one end of the Andes Cordillera, the new water supply company (HIDROANDES) is trying to reach the goals set by the restructuring of Venezuela's water supply system. In 1993 the company had about 142,000 customers in the metropolitan zone. All are billed, but only 62 percent of the residential customers actually pay, 85 percent of commercial accounts pay, and 100 percent of industrial users pay. The operation of the system is by HIDROANDES, but maintenance is contracted to a private company with direct inspection by the Empresa Hidrológica.

Mérida reflects the situation of water supply in most of Venezuela: problems with old systems, losses, high unaccounted-for water, administrative problems, insufficient training, budget problems, and others. Mérida has a well-designed treatment plant, but lacks sufficient trained personnel. HIDROANDES will have to either pay competitive salaries to hire skilled technicians or invest to prepare its own personnel. There is a lack of knowledge about the system in Mérida, because the changes that were made in the original designs were not recorded. HIDROANDES is now in the process of recuperating and organizing the information.

The goals in Venezuela have been to provide water supply and sewerage service to every city. In general, the water supply goal has been accomplished, but the service has to be improved. The sewerage goal has not been reached in the whole country, lacking some small towns, but is on the way. With respect to wastewater, interest in the quality of the environment has not yet reached the needed levels. Venezuela is having new experiences with regional authorities and counties, elected directly by the people, with control over the main decisions. The water companies are learning that sometimes decisions that are technically necessary can be rejected for political reasons. Finally, financing is a great problem.

Brazil

Brazil, a vast nation with similarities to both the United States and to Venezuela, also faces a wide diversity of problems related to water and wastewater infrastructure. Brazil is divided into states like the United States, but due to its developing status, depends more on central direction and investment than the United States does. According to the World Bank, Brazil's 1991 population was 151 million, with a growth rate of 2.2 percent, in contrast to the 252 million in the United

States (0.9 percent growth rate) and Venezuela's 20 million and 2.7 percent growth rate (World Bank, 1993).

Brazil's water supply systems cover about 88 percent of the population, up from about 45 percent in 1970. Still, there are some 13 million citizens in urban centers without water supply systems. Some 46 percent of rural residents lack access to water of good quality. Problems of sewage disposal are relatively more severe, with some 73 million Brazilians (65 percent of the urban inhabitants) lacking access to adequate wastewater infrastructure. Only 10 percent of the country's sewage receives adequate treatment, with 90 percent being discharged untreated into the nation's waterways. According to data from the United Nations (1991), Brazil discharges 95 percent of its urban sewage without treatment into water bodies closest to where the sewage is generated, a situation not different from that in developing countries.

Brazil faces different problems in each of the sanitation sectors, water, wastewater, solid wastes, and drainage. Problems include excessive centralization, and little participation of states and municipalities in setting priorities. Imbalance and inequity are major problems. On the one hand, state companies are responsible for services to about 3000 municipalities, but there is a problem preventing the extension of services to the poorest citizens. Inefficiencies are a major worry. These include management inefficiency, unaccounted-for-water losses, and inadequate technologies.

The spillover of problems in the sanitation sector affects Brazil's most basic social problems: quality of life for the general population, low-income populations, and infant health, and it portends future misery unless the problems can be fixed. Cities and rural areas are vulnerable to problems of water-borne disease, infant mortality, dysentery, and general problems of low-income populations.

Institutional problems in Brazil include excessive centralization of decisions and little participation of states and municipalities in defining priorities. The inability of the system to provide services to the poorest populations is a serious indicator. Financing system operation and improvements will be a continuing problem.

With its tremendous size and diversity, Brazil has numerous regional issues to deal with, much as the United States does. Its largest city, São Paulo, illustrates the scale of the problems it faces. São Paulo plans to undertake, through the São Paulo State Basic Sanitation Authority, a $3–4 billion program to solve sanitation problems in the metropolitan area, while in other parts of the nation, rural and urban fringe areas face immense problems of basic sanitation for a rapidly growing population.

São Paulo's forthcoming effort to solve sanitation problems will focus on cleaning up the Tiete River, a heavily polluted waterway that drains the city (São Paulo to Launch Massive River Cleanup, 1992). Some 20 million people live in this river basin, illustrating the massive scale of the environmental issues faced there.

The financing program, said to be the Inter-American Development Bank's largest financing ever at $450 million, will expand the city's sewerage system to serve an additional 1.5 million people, most of them poor. Two new plants will be built, and the proportion of water treated will rise from 19 to 45 percent by 1995. Also included will be training and institutional strengthening benefits. The São Paulo state agency responsible for pollution control will gain the capability to monitor 1250 industries, which are responsible for 90 percent of the area's industrial pollution, and management capabilities to maintain the plants and to improve financial management.

In summary, Brazil faces tremendous challenges in the water and sanitation sector. Its large population and rapid growth rate challenge the public and private sectors to provide the institutional infrastructure and the financing to provide needed infrastructure services. Unless problems of the sector can be solved, the implications for public health and quality of life in both urban and rural areas are extremely significant.

Analysis of cases

The Western Hemisphere is experiencing high levels of migration, and interregional flows of trade, technologies, financing, and expertise. Economic integration may be a key to solving some of the disparities between regions and nations, both inside large nations and from nation to nation. Regardless of future progress in equalizing physical and social environments, the wide disparity of access to safe drinking water and adequate sanitation services is a serious problem needing attention in the hemisphere.

Technologies also vary widely from region to region, not so much because of technological barriers, but because of lack of access to capital. This is a worldwide problem. The support base of the world's water industry includes international consulting firms, contractors, and equipment suppliers who are ready to bring the latest technologies when funding is adequate.

The issue of appropriate technologies is germane to the discussion of equalizing services, because many of the basic technologies needed for water supply and sanitation are not necessarily expensive, but they do require training, expertise, and at least a local manufacturing and management capacity.

Management institutions in the hemisphere vary across the spectrum from purely public to purely private. In the Venezuelan and Brazillian cases, limitations of public authorities are made clear, and the United States is also aware of these limitations, and has given attention to privatization in the water and sanitation sectors. Institutional factors are no doubt the most important in equalizing water and sanitation services in the hemisphere.

Financial capacities constrain national capabilities to invest in each country. External and internal debt structures are such that borrowing will be limited, and the ability of central governments to subsidize regional problems is also quite limited. Improving planning, efficiency, and local attention to problems is a critical issue, as is developing effective institutions to address problems without massive financial infusions.

Perhaps the greatest disparity in levels of service is the gap between those who have service and those who lack it. This is made clear in the Brazil case study, which provided national data on the percentages of citizens who still lack access to safe drinking water and sanitation. This remains a worldwide problem, as evidenced by the data from the International Drinking Water and Sanitation Decade.

With high rates of growth, migration, and urbanization, every nation in the hemisphere faces challenges in basic education, governance, training, and institution building. These problems result in problems with management efficiency in water agencies. Perhaps this problem is most evident in two symptoms: the small water system problem of the United States, mirrored in the rural problems throughout the hemisphere, and the inability of large, state-owned companies to provide access to services throughout Latin America. Improving management efficiency is another serious institutional problem for all nations.

In the final analysis, there is little generic difference in the problems faced by the nations in the hemisphere. As shown by the case studies, they include administrative and budget problems, infrastructure issues, inadequate training of personnel, inadequate mapping and information, treatment plants that may have good technology but need improved operation, high levels of needed investments, political problems such as technical decisions being overruled for political reasons, inadequate charging systems, and general financial problems.

Conclusions

Ismail Serageldin, Vice-President for Environmentally Sustainable Development of the World Bank, states that the first challenge is to address the "old agenda" of providing household services to the 1 billion who lack adequate water and the 1.7 billion who lack access to adequate sanitation. The second challenge is to address the "new agenda" of achieving environmentally sustainable development, especially in the light of the fact that water is so seriously degraded in developing countries and the financial capacity is so limited. He sees two central elements in moving toward solutions: an institutional element and an instrumental element (Serageldin, 1994).

The institutional element ranges from measures at the household level, or putting affected people in charge of the decisions that affect them, to choose the services they receive and the payment levels. At a higher level, it means having the stakeholders in river basins choose their environmental quality and what they will pay for. These measures stress democratization and that decision making should be moved to the lowest possible level. According to Serageldin, this will require a much more sharply defined role for government, and much broader participation of the private sector and nongovernmental organizations (see Chaps. 4 and 8 for more on coordination measures).

For the instrument element, Serageldin sees more extensive use of marketlike instruments at all levels. This means, at the lowest level, greater reliance on user charges for raising revenues and enhancing accountability and efficiency. At the service level, it means greater reliance on the private sector; and at the higher levels, it means greater uses of abstraction charges, pollution charges, and water markets.

A great deal is known about the problem of water supply and sanitation in developing countries. WASH's four principles sum up some 20 "lessons" learned after years of experience. The principles line up well with other principles of water management, as articulated in this book.

Although providing safe water and sanitation services is a different type of challenge for water managers, they must participate. After all, it is a logical program element fitting into the water industry, and the valid principles as outlined in this book apply. The main difference is in the level of service provided to customers, given the wide variation in ability to pay.

Questions

1. In your opinion, what is the root cause of the international problem in water supply and sanitation?

2. In the United States, the term "environmental justice" has arisen to describe the fact that the poor bear a disproportionate burden of pollution. How does this relate to the international problem in water supply and sanitation?

3. In developing countries, a high percentage of households are headed by single women. If they lack access to safe water, what does this say about prospects for the next generation?

4. Elaborate on the principles developed by the WASH project and explain how they relate to water management principles in the United States:

   a. Technical assistance is most successful when it helps people learn to do things for themselves in the long run.

   b. Water supply and sanitation development proceeds most effectively when its various elements are linked at all levels.

   c. The basic measure for success of both the national system for development and the community management systems it creates is sustainability, the ability to perform effectively and indefinitely after donor assistance has been terminated.

   d. Sustainable development is more likely to occur if each of the key participants recognizes and assumes its appropriate role and shoulders its share of the responsibility.

5. Can democratization, that is, moving decision making to the lowest possible level, help resolve the crisis in water supply and sanitation? What are the barriers to it?

References

Albertson, Maurice L., Appropriate Technology for Sustainable Development: Criteria for Civil Engineering Infrastructure, 22nd Annual Conference of Water Resources Planning and Management Division, ASCE, Boston, May 1995.

Canemark, Curt, The Decade and After: Lessons from the 80s for the 90s and Beyond, *World Water 89,* London, November 14, 1989.

Grigg, Neil S., Tomas A. Bandes, Angela Henao, Sara Morales, and Rubem Porto, Water and Wastewater Infrastructure for the Hemisphere, Inter-American Dialog for Water Management, Miami, October 1993.

Interagency Task Force, Protection of the Quality and Supply of Freshwater Resources, Country Report, USA, International Conference on Water and the Environment, January 1992.

Inter-American Development Bank, *Water and Sanitation,* June 1992.

Okun, Daniel A., Meeting the Need for Water and Sanitation for Urban Populations, The Abel Wolman Distinguished Lecture, National Research Council, Washington, DC, May 1991.

Pacey, Arnold (ed.), *Water for the Thousand Millions,* Pergamon Press, Oxford, 1977.

São Paulo to Launch Massive River Cleanup, *The IDB,* December 1992.

Serageldin, Ismail, Keynote address to ministerial conference on implementing Agenda 21 for drinking water and environmental sanitation, March 1994.

Triweko, Robertus, A Paradigm of Water Supply Management in Urban Areas of Developing Countries, Colorado State University, Fall 1992.

United Nations, Agenda 21, Chapter 18, Protection of the Quality and Supply of Freshwater Resources: Application of Integrated Approaches to the Development, Management and Use of Water Resources, 1992.

United Nations, *Global Consultation on Safe Water and Sanitation for the 1990's,* New Delhi, 1991.

U.S. Agency for International Development, Lessons Learned from the WASH Project, USAID, Water and Sanitation for Health Project, Washington, DC, 1990.

World Bank, *World Bank Atlas, 25th Anniversary Edition,* Washington, DC, 1993.

A

Definitions and Concepts

Comprehensive water management means management that embraces all factors that bear on water decision making.

To *cooperate* means to work together with others. Cooperation in water management is any form of working together to manage water.

To *coordinate* means to harmonize the different elements of something.

The *disciplinary viewpoint* includes considerations of different branches of knowledge: technology, law, finance, economics, politics, sociology, life science, mathematics, and others.

Environmental elements are those facilities and processes for water management provided by nature. They involve natural features that produce, store, or convey water. They include the atmosphere, watersheds, stream channels, aquifers and groundwater systems, lakes, estuaries, seas, and the ocean.

The *functional viewpoint* deals with the purposes to which water is applied, as in water supply for cities, wastewater management, irrigation, and other purposes.

The *geographic viewpoint* refers to scale and accounting units: global, river basin, country, water body, locale, region, etc.

The *hydroecological viewpoint* refers to hydrologic and ecological or biological issues.

To *integrate* means to bring together the parts of something. Integration in water resources management is the formulation of plans and actions that take into account different sets of viewpoints by discipline, geographic location, political position, ecological framework, and functional frame of reference.

Nonstructural measures for water management are management programs that do not involve constructed facilities.

The *political viewpoint* is important because much of water resources management is carried out by government agencies. There are both horizontal and vertical political issues. Horizontal issues are between different government agencies at the same level, usually local government, and vertical issues refer to relationships between layers of government agencies, as in state–federal issues.

Purposes of water resources management are expressed as service categories: water supply, wastewater and water quality management, storm- and floodwater control, hydropower, transportation and water for the environment, fish/wildlife, and recreation. The purposes serve four categories of water users: people, industries, farms, and the general environment.

Regional (distinct district of a state, country, or city).

Regionalization in water management (the act or process or regionalizing water management; meaning to manage water with regional cooperation or integration).

Structural measures for water management are constructed facilities used to control water flow and quality.

Water resources management is the application of structural and nonstructural measures to control natural and man-made water resources systems for beneficial human and environmental purposes.

A *water resources system* is a combination of water control facilities and/or environmental elements that work together to achieve water management purposes.

B

Players in the Water Industry

Political Level

Boards of water management agencies

Governors

Elected officials

City and county governments

Legislative committees

State water congresses

Councils of government

Trade and professional associations

Water Providers and Wastewater Agencies

Municipal water agencies

Municipal wastewater agencies

Water and sanitation districts

Data Collection and Assessment Agencies

State water and geological survey agencies

National Oceanic and Atmospheric Administration (NOAA), National Weather Service

U.S. Geological Survey

U.S. Fish and Wildlife Service (National Biological Survey)

State climatologists

Comprehensive Water Management Agencies

Water management districts

Regional authorities

U.S. Army Corps of Engineers

U.S. Bureau of Reclamation

Hydropower, Navigation, and Flood Control Interests

Federal Energy Regulatory Commission (FERC)

Power distributors

Navigation companies

National Water Resources Association

Government energy agencies

Power utilities

U.S. Department of Energy

Tennessee Valley Authority (TVA)

Port authorities

Industrial and Urban Development Interests

U.S. Department of Commerce

Economic development agencies

Manufacturing industries

U.S. Department of Housing and Urban Development (HUD)

Financing programs such as Appalachian Regional Commission (ARC)

Associations of realtors

Chambers of commerce

Tourism boards

Agriculture and Resource Development Sectors

Irrigators

Livestock interests

Miners

Forest/loggers

Farmers

Government agriculture departments

U.S. Soil Conservation Service

University experiment stations and extension offices

Cattlemen's associations

State boards of agriculture

Farm bureaus

Irrigators' associations

Ditch companies

Associations of soil conservation districts

Grain and feed associations

Cattle feeders' associations

U.S. Department of Agriculture (USDA)

Farmer's Home Administration (FmHA)

Government soil and water agencies

State agriculture departments

State forest services

State mining boards and agencies

State departments of natural resources

State divisions of water resources

Environmental, Recreation, and Public Interest Organizations

National Park Service

U.S. Fish and Wildlife Service

State fish and wildlife departments

State parks and recreation departments

Natural resources or conservation departments

Environmental groups

Recreationists

Wildlife enthusiasts

Wilderness advocates

Government coastal management agencies

League of Women Voters

Regulatory Agencies

State environmental protection agencies

State health agencies

U.S. Environmental Protection Agency

U.S. Department of Health and Human Services

Land use commissions

State engineer's offices

Court and Legal Systems

State water courts

Federal court system, including U.S. Supreme Court

State courts

State attorneys general

U.S. Department of Justice

Attorneys

American Bar Association

Native American Rights Fund

Universities, Research Institutes, Publishers

Higher education

Water resources research institutes

Government and policy law centers

Publishers

Scientific and Trade Associations

American Water Resources Association

American Society of Civil Engineers

National Society of Professional Engineers

Consulting Engineers Council

American Water Works Association

Water Environment Federation

Rural water associations

Water users' associations

National Academy of Sciences

Association of State and Interstate Water Pollution Control Administrators (ASIWPCA)

Association of State Dam Safety Officials (ASDSO)

Association of State Drinking Water Administrators (ASDWA)

Scientific organizations

International Water Resources Association (IWRA)

American Public Works Association (APWA)

National Water Resources Association (NWRA)

National Water Congress

State water congresses

Contractors

Utility contractors

Heavy contractors

Water well contractors

Suppliers

Consultants

C

Conversion Factors for Water Computations

Volume

1	acre-foot (AF)	=	43,560	cubic feet (ft^3)
1	acre-foot	=	325,853	gallons (gal)
1	acre-foot	=	1,233.49	cubic meters (m^3)
1	thousand AF (TAF)	=	1.2335	million m^3 (MCM)
1	million AF (MAF)	=	1,233.5	million m^3 (MCM)
1	ft^3	=	7.4806	gallons
1	ft^3	=	28.3170	liters
1	gal	=	3.7854	liters
1	million gal (MG)	=	3.0689	AF
1	m^3	=	35.3145	ft^3
1	m^3	=	1,000	liters
1	m^3	=	264.17	gallons
1	million m^3	=	810.71	AF
1	billion m^3	=	810710	AF
1	billion m^3	=	1	milliard
1	billion m^3	=	1	km^3
1	cfs-yr	=	723.97	AF
1	cfs-day	=	1.9835	AF
1	mi^3	=	3.3792	MAF
1	mi^3	=	4.1682	BCM
1	thousand gal	=	1.337	CCF

Rate

1	cfs	=	448.836	gal/min (gpm)
1	cfs	=	28.317	liters/sec
1	cfs	=	723.970	AF/yr
1	cfs	=	1.9835	AF/day
1	cfs	=	2446.6	CM/day
1	Mgal/day (mgd)	=	1.5472	cfs
1	m^3/sec	=	35.3145	cfs
1	m^3/sec	=	22.8248	mgd

Area

1	acre	=	43,560	ft^2
1	acre	=	0.40469	hectare (Ha)
1	square mi	=	640	acres
1	square mi	=	259.00	hectares (Ha)
1	hectare (Ha)	=	10,000	m^2

Length

1	foot	=	0.3048	meter
1	meter	=	39.3700	inches
1	meter	=	3.2808	feet
1	inch	=	25.400	mm
1	mile	=	1,609.35	meters
1	kilometer	=	0.62137	mile

Chapter 10: Water Supply and Environment: Two Forks Case

Chapter 11: Flood Plain Management
  Colorado Flooding: Mountains and Plains
  Urban Drainage and Flood Control District
  Black Warrior River Flood
  Great Mississippi River Flood of 1993
  Bangladesh

Chapter 12: Water Infrastructure Planning and Management
  Ft. Collins Water Treatment Master Plan
  Colorado Big Thompson Project
  California Water Plan

Chapter 13: Reservoir Operations and Management
  Multiple-Purpose Reservoir
  Drought
  Flood
  Appropriation Doctrine
  Complex Systems
  Enviro Conflicts

Chapter 14: Water Quality Management and Nonpoint Source Control
  State Government

Index

ABOUT THE AUTHOR

Neil S. Grigg is professor and chairman of the civil engineering department at Colorado State University, where he has also served as director of the Colorado Water Resources Research Institute and director of the International School for Water Resources. He was previously assistant secretary for natural resources and director of environmental management for the State of North Carolina, and director of the University of North Carolina Water Resources Research Institute. Professionally, he is the co-founder and former vice president of Sellards & Grigg, Inc., a Denver-based civil engineering consulting firm. He has written several books, as well as numerous journal articles, on a variety of water resources planning and management topics.